AF602138

HANDBOOK OF GEOPHYSICAL EXPLORATION

SEISMIC EXPLORATION

VOLUME 36

INFORMATION-BASED INVERSION AND PROCESSING
WITH APPLICATIONS

HANDBOOK OF GEOPHYSICAL EXPLORATION

SEISMIC EXPLORATION

Editors: Klaus Helbig and Sven Treitel

Volume

1. Basic Theory in Reflection Seismology
2. Seismic Instrumentation, 2nd Edition
3. Seismic Field Techniques
4A. Seismic Inversion and Deconvolution: Classical Methods
4B. Seismic Inversion and Deconvolution: Dual-Sensor Technology
5. Seismic Migration (Theory and Practice)
6. Seismic Velocity Analysis
7. Seismic Noise Attenuation
8. Structural Interpretation
9. Seismic Stratigraphy
10. Production Seismology
11. 3-D Seismic Exploration
12. Seismic Resolution
13. Refraction Seismics
14. Vertical Seismic Profiling: Principles 3rd Updated and Revised Edition
15A. Seismic Shear Waves: Theory
15B. Seismic Shear Waves: Applications
16A. Seismic Coal Exploration: Surface Methods
16B. Seismic Coal Exploration: In-Seam Seismics
17. Mathematical Aspects of Seismology
18. Physical Properties of Rocks
19. Shallow High-Resolution Reflection Seismics
20. Pattern Recognition and Image Processing
21. Supercomputers in Seismic Exploration
22. Foundations of Anisotropy for Exploration Seismics
23. Seismic Tomography
24. Borehole Acoustics
25. High Frequency Crosswell Seismic Profiling
26. Applications of Anisotropy in Vertical Seismic Profiling
27. Seismic Multiple Elimination Techniques
28. Wavelet Transforms and Their Applications to Seismic Data Acquisition, Compression, Processing and Interpretation
29. Seismic Signatures and Analysis of Reflection Data in Anisotropic Media
30. Computational Neural Networks for Geophysical Data Processing
31. Wave Fields in Real Media: Wave Propagation in Anitsotropic, Anelastic and Porous Media
32. Nuclear Magnetic Resonance Petrophysical and Logging Applications
33. Seismic Amplitude Inversion in Reflection Tomography
34. Seismic Waves and Rays in Elastic Wave Media
35. Seismic While Drilling: Fundamentals of Drill-Bit Seismic for Exploration
36. Information-based Inversion and Processing with Applications

SEISMIC EXPLORATION

Volume 36

INFORMATION-BASED INVERSION AND PROCESSING WITH APPLICATIONS

by

Tadeusz J. ULRYCH
Department of Earth and Ocean Sciences
The University of British Columbia
Vancouver, BC, Canada

Mauricio D. SACCHI
Department of Physics
Institute for Geophysical Research
University of Alberta
Edmonton, Alberta, Canada

ELSEVIER

Amsterdam – Boston – Heidelberg – London – New York – Oxford
Paris – San Diego – San Francisco – Singapore – Sydney – Tokyo

ELSEVIER B.V.
Radarweg 29
P.O. Box 211, 1000 AE Amsterdam
The Netherlands

ELSEVIER Inc.
525 B Street, Suite 1900
San Diego, CA 92101-4495
USA

ELSEVIER Ltd
The Boulevard, Langford Lane
Kidlington, Oxford OX5 1GB
UK

ELSEVIER Ltd
84 Theobalds Road
London WC1X 8RR
UK

© 2005 Elsevier Ltd. All rights reserved.

This work is protected under copyright by Elsevier Ltd., and the following terms and conditions apply to its use:

Photocopying
Single photocopies of single chapters may be made for personal use as allowed by national copyright laws. Permission of the Publisher and payment of a fee is required for all other photocopying, including multiple or systematic copying, copying for advertising or promotional purposes, resale, and all forms of document delivery. Special rates are available for educational institutions that wish to make photocopies for non-profit educational classroom use.

Permissions may be sought directly from Elsevier's Rights Department in Oxford, UK: phone (+44) 1865 843830, fax (+44) 1865 853333, e-mail: permissions@elsevier.com. Requests may also be completed on-line via the Elsevier homepage (http://www.elsevier.com/locate/permissions).

In the USA, users may clear permissions and make payments through the Copyright Clearance Center, Inc., 222 Rosewood Drive, Danvers, MA 01923, USA; phone: (+1) (978) 7508400, fax: (+1) (978) 7504744, and in the UK through the Copyright Licensing Agency Rapid Clearance Service (CLARCS), 90 Tottenham Court Road, London W1P 0LP, UK; phone: (+44) 20 7631 5555; fax: (+44) 20 7631 5500. Other countries may have a local reprographic rights agency for payments.

Derivative Works
Tables of contents may be reproduced for internal circulation, but permission of the Publisher is required for external resale or distribution of such material. Permission of the Publisher is required for all other derivative works, including compilations and translations.

Electronic Storage or Usage
Permission of the Publisher is required to store or use electronically any material contained in this work, including any chapter or part of a chapter.

Except as outlined above, no part of this work may be reproduced, stored in a retrieval system or transmitted in any form or by any means, electronic, mechanical, photocopying, recording or otherwise, without prior written permission of the Publisher.
Address permissions requests to: Elsevier's Rights Department, at the fax and e-mail addresses noted above.

Notice
No responsibility is assumed by the Publisher for any injury and/or damage to persons or property as a matter of products liability, negligence or otherwise, or from any use or operation of any methods, products, instructions or ideas contained in the material herein. Because of rapid advances in the medical sciences, in particular, independent verification of diagnoses and drug dosages should be made.

First edition: 2005

ISBN-13: 978-0-08-044721-6
ISBN-10: 0-08-044721-X

ISSN: 0950-1401 (Series)

Transferred to Digital Printing 2007

Working together to grow
libraries in developing countries

www.elsevier.com | www.bookaid.org | www.sabre.org

ELSEVIER BOOK AID International Sabre Foundation

Contents

List of Figures

Dedication

We dedicate this book to our wives, *Jacqueline Ulrych and Luciana Sacchi* and our children *Jason, Lyza and Viviane* from the Ulrych clan, and *Federico and Bianca* from the Sacchi clan, and we thank them for bearing with us.

Acknowledgements

At the outset, we would particularly like to acknowledge Rongfeng Zhang, Sven Treitel and Klaus Helbig for helping us so generously in making the presentation of this material more clear. We also acknowledge our current and former graduate students without whose invaluable input this book would not have been feasible and without whose stimulus our research careers would feel somewhat empty.

So many people have enriched our scientific lives and so many have made this book a possibility that, undoubtedly, we will fail miserably in the task of properly acknowledging all. However, with humble apologies for omissions, we do gratefully acknowledge these, our collegial friends, to whom we owe our unbounded thanks for conversations, support, humor, knowledge and generosity: Michael Bostock, Alberto Comínguez, Vittorio deTomasi, Jacob Fokkema, Doug Foster, Masami Hato, Kris Innanen, Sam Kaplan, Henning Kuehl, Evgeny Landa, Scott Leaney, XinGong Li, Larry Lines, Bin Liu, Gary Margrave, Ken Matson, Toshi Matsuoka, Bill Nickerson, Milton Porsani, Enders Robinson, Don Russell, Brian Russell, Jeffrey Scargle, Ulrich Theune, Daniel Trad, Sven Treitel, Maurice van de Rijzen, Danilo Velis, Kees Wapenaar, Arthur Weglein, Alan Woodbury, Rongfeng Zhang, Paul Zwartjes.

Preface

Perhaps the first thing that we should try to accomplish is to answer a question that will undoubtedly arise, "why did we write this book"? There are, after all, so many excellent books describing time series analysis (Box and Jenkins, 1976; Priestley, 1981; Anderson, 1971), Fourier transforms (Bracewell, 1986; Papoulis, 1962), spectral analysis (Kay, 1988; Marple, 1987), inversion (Menke, 1989; Tarantola, 1987), inversion and processing (Claerbout, 1992), geophysical processing (Claerbout, 1976; Robinson and Treitel, 2002; Kanasewich, 1981) to mention but a few. Yes indeed, these are marvellous texts, but they simply do not do all that we wish to do. We have been teaching senior undergraduate and graduate courses for a century, it seems, and we like to do so in a particular manner. This book describes this manner. It deals with the basics that we require, issues of probability theory, estimation, Fourier transforms, time series models, eigenvector representation and the like. We go on to aspects of information theory and the use of entropy. We tackle inversion in our particular fashion, regularization, sparseness, inference via Bayes and maximum entropy and various topics in data processing and decomposition (2D and 3D eigenimages, parabolic and hyperbolic Radon transforms and a particular wavelet transform for good measure).

We hope that our treatment appeals to a large segment of the scientific population. We must make some issues quite clear, however. First of all, this is not a book for pure mathematicians. If you are accustomed to dealing with functions that are defined as belonging to a Banach space of continuous functions vanishing at infinity then, perhaps, this book is not for you. We use engineering language. Our functions and vectors, all of them, belong to the well understood Hilbert space. As Gilbert Strang, (Strang, 1988), points out, this is the most famous infinite dimensional space. Simply put, the appropriate Hilbert space contains functions or vectors which have "finite length". Exactly what this means will be seen later but, intuitively, this characteristic appears eminently reasonable.

A second point follows from the above discussion. Functions and vectors are connected by means of Parseval's formula (Strang, 1988). It is for this reason mainly, and for others of convenience, that we are cavalier about variables being either continuous or discrete. Virtually everything we do here will be independent of this distinction. Of course, we realize that the world is not always so simple. For example, entropy of a continuous random variable is different to that of a discrete random variable. When such issues arise, however, we deal with them. It so happens, as we will see later, that the simplification of dealing with the continuous version of entropy over the discrete version is huge, and the above mentioned difference may be easily reconciled.

Finally, there are some issues that probably stand out because of their absence. As a specific example, consider the topic of Wiener filtering. Certainly, one of the difficult tasks that authors face is to decide what material, and in what detail, should be included in the text. Since the subject of Wiener filters is so expertly explored in the literature, we have decided to refer the reader to some favourite writings and to merely introduce the basic framework here. Favourite texts, to mention but two are, Robinson and Treitel (2002) for a discrete treatment, specifically with application to seismic problems, and Lathi (1968) for a superb view of the continuous theory. In fact, of course, we do often make use of the Wiener formalism. Autoregressive (AR) processes are one example, prediction error operators and spiking deconvolution are another.

We have attempted to organize this book as logically as possible. Since we believe, however, that in all good lectures some disorganization, repetition, juxtaposition, is essential, the reader will find a measure of all these here. Please bear with us. We believe the bearing will be worth it. We set out the topics covered by each chapter in the itinerary that follows.

- **Chapter 1.** Basic concepts which include, some probability theory (expectation operator and ensemble statistics), elementary principles of parameter estimation, Fourier and z transform essentials, and issues of orthogonality.

- **Chapter 2.** Linear time series models and estimation, which includes, the Wold decomposition theorem, MA, AR and ARMA models, the Levinson recursion, the minimum phase property of the PEO, the Pisarenko estimator and a special ARMA model and, finally, the relationship of the linear models to seismic data modelling and analysis.

- **Chapter 3.** Information theory and relevant issues contains a synopsis of those topics that we use through out the book. Entropy, conditional entropy, Burg's maximum entropy spectral estimator, the beautiful theorem of Edwin Jaynes and mutual information. This chapter also contains a skeletal derivation of a most useful information criterion, AIC, and its relationship to issues such as spectral estimation and whiteness.

- **Chapter 4.** Inverse problems are dealt with, first from a deterministic point of view, then from a probabilistic one. The first part contains, essentially, some thoughts about l_2 inversion and the manner of regularizing the ill conditioned problem. The probabilistic view is illustrated by means of relative minimum entropy and Bayesian inference.

- **Chapter 5.** This chapter studies the enhancemnet of seismic records via FX prediction filters, eigenimages, and Radon transforms.

- **Chapter 6.** This chapter deals with the topic of deconvolution and the inversion of acoustic impedance. The subject is treated in two parts. The first discusses band-limited extrapolation assuming a known wavelet and also considers the issue of wavelet estimation. The second part deals with sparse deconvolution using various 'entropy' type norms.

- **Chapter 7.** Here is a potpourri of topics, three in number, that fascinate us and that are excellent examples of the application of the material presented in the preceding chapters.

Please Read Initially

We use certain simplifications throughout the text for the sake of clarity and to avoid clutter. These are:

- We always use $\int$ for $\int_{-\infty}^{+\infty}$. We believe that the ∞ can be understood implicitly.
- Sums are often written as $\sum_n$ where it is understood what the limits of n are.
- We write log to imply ln, the natural log, since the ln is sometimes lost in a complex equation. When necessary, we stipulate the base of the logarithm.
- We use, f, the temporal frequency and, ω, the angular frequency, interchangeably where convenient. For example, the continuous Fourier transform is symmetric in f, whereas it contains $1/2\pi$ in ω. On the other hand, the use of ω often eliminates the tedium of writing 2π. When using ω, we refer to it as frequency, not angular frequency, for convenience.
- We quite often jump from continuous to discrete series. Continuous series are designated by $x(t)$, for example, whereas discrete versions are denoted by x_k, x_n or x_t, as convenient.
- We use abbreviations where appropriate. Thus,

Singular value decomposition	$\Leftrightarrow$	SVD
Principle component analysis	$\Leftrightarrow$	PCA
Wavelet transform	$\Leftrightarrow$	WT

  and so on.

Chapter 1

Some Basic Concepts

1.1 Introduction

The purpose here is merely to present, without detailed derivation which may be found in any number of excellent texts, some basic concepts which we use throughout this book. One of the central tasks of concern to us entails the separation of signal from noise, each component requiring specific definition which depends on the problem at hand. As an example, one with which we are particularly familiar, we consider the case of a seismic signal which may often be considered as a superposition of a stochastic (random) and a deterministic process. For example, in a marine environment, the recorded seismogram contains, in general, a surface related multiple which, for the zero-offset case, can be viewed as a periodic, predictable and therefore deterministic signal. The stochastic component is represented by the convolution of the seismic wavelet with the random reflectivity of the earth. We will be mostly concerned here with the modelling of the stochastic component and for this purpose, we review some fundamental concepts. In general, we deal with discrete processes and we represent signals by vectors. Sometimes however, it is most convenient to present certain concepts in terms of functions continuous in time. The transition from a continuous to a discrete representation assumes that the required sampling has been properly implemented without aliasing, a concept important in its own right that is discussed in some detail in this chapter.

1.2 Probability Distributions, Stationarity & Ensemble Statistics

The concept of a random variable is central to much of our discussion. So of course is the understanding of such variables in terms of the underlying probability distributions and associated moments. In these sections we briefly review probability concepts and related issues assuming that the reader has a basic understanding from previous readings.

1.2.1 Essentials of Probability Distributions

This section is a very brief review of probability concepts. For a deeper view, we recommend the classic text by Papoulis (1984), a very readable, practical and novel text by Lupton (1993) and two very favorite texts of ours, Hamilton (1964) and Lathi (1968). Much of our book deals with stochastic processes. These are defined in terms of random variables which may be understood as follows. Consider a random experiment which has M outcomes, $x_1, x_2, \ldots, x_M$, corresponding to $\xi = 1, 2, \ldots, M$. The random variable, $\mathbf{x} = \mathbf{x}_\xi$, which describes the experiment, can take on any of the M outcome values. $\mathbf{x}$ is associated with a sample space, $\mathcal{S}$, and a probability density function or pdf $p_{\mathbf{x}}(x)$, which we write as $p(x)$ for convenience.

The pdf defines the probabilities associated with $\mathbf{x}$. Thus, the probability that $\mathbf{x}$ will assume a value, x, in the range $[a, b]$ is

$$\text{Prob}[a \leq x \leq b] = \int_a^b p(x)dx$$

Naturally, $p(x)$ obeys the unimodular constraint

$$\int p(x)dx = 1$$

At this stage we define one of the most useful of operators, the expectation operator, $E[\cdot]$

$$E[g(x)] = \int g(x)p(x)dx \tag{1.1}$$

which we use throughout this book. Some important elementary relations concerning the expectation operator are given in Appendix A.

The two statistics that are most pertinent to the description of $\mathbf{x}$ are the mean and the variance. These are defined as

$$E[x] = \mu_1' = \mu = \int xp(x)dx \tag{1.2}$$

and

$$\sigma^2 = \text{var}(\mathbf{x}) = \int (x - \mu)^2 p(x)dx \tag{1.3}$$

A related statistic is the second moment defined by

$$E[x^2] = \mu_2' = \bar{\mathbf{x}}^2 = \int x^2 p(x)dx$$

related to the variance by

$$\bar{\mathbf{x}}^2 = \sigma^2 + \mu^2$$

which shows that the variance is the second moment about the mean.

Higher order moments are defined by

$$E[x^n] = \mu'_n = \int x^n p(x)dx$$

A very useful concept when dealing with pdf's, is that of the moment generating function, $M(q)$, which is defined as

$$M(q) = E[e^{qx}] = \int e^{qx} p(x)dx$$

and using a Taylor's series expansion

$$= \int \left(1 + qx + \frac{(qx)^2}{2!} + \cdots\right) p(x)dx$$
$$= 1 + \mu'_1 q + \mu'_2 \frac{q^2}{2!} + \cdots$$

which shows that the coefficients of the Taylor expansion of the moment generating function are the moments of the distribution. Of particular interest here is the relationship of the moments of a pdf to its cumulants, which are defined in terms of a transformation on $M(q)$. In particular, consider

$$c(x) = \log M(x) = \kappa_1 + \kappa_2 \frac{q^2}{2!} + \cdots$$

where κ_n is the nth cumulant of $p(x)$. Cumulants have the very useful property, used in Chapter 6, that

$$\kappa_1 = \mu'_1 = \mu$$
$$\kappa_2 = \mu_2$$
$$\kappa_3 = \mu_3$$

and

$$\kappa_4 = \mu_4 - 3\mu_2^2 \tag{1.4}$$

where

$$\mu_n = E[(x-\mu)^n]$$

and represents the nth central moment of $p(x)$. Finally, we define

$$\begin{aligned} \textit{skewness} &= \kappa_3 \\ \textit{kurtosis} &= \kappa_4 = \mu_4 - 3\mu_2^2 \end{aligned} \tag{1.5}$$

and a *normalized* version

$$\bar{\kappa}_4 = \frac{\bar{\kappa}_4}{{\kappa_2}^2} = \frac{\mu_4}{\mu_2^2} - 3 \tag{1.6}$$

An important fact which is easy to show using the above definitions, is that, for a Gaussian distribution

$$\bar{\kappa}_4 = \kappa_4 = 0 \tag{1.7}$$

We now consider two random variables, $\mathbf{x}$ and $\mathbf{y}$. The dependence of these two variables on each other is contained in the joint pdf, $p(x, y)$. Another important distribution that also defines a measure of dependence is the conditional pdf, $p(x|\mathbf{y} = y)$ (written as $p(x|y)$ for short) which defines the probability of $\mathbf{x}$ given that $\mathbf{y} = y$. It is clear intuitively that $p(x, y)$ and $p(x|y)$ are related according to

$$p(x, y) = p(x|y)p(y) \tag{1.8}$$

Since $p(y, x) = p(x, y)$, and obeys a relationship equivalent to that expressed in Eq. (1.8), we have

$$p(x|y)p(y) = p(y|x)p(x)$$

which yields

$$p(x|y) = \frac{p(y|x)p(x)}{p(y)} \tag{1.9}$$

the famous Bayes' theorem that we will have occasion to use very often. An important concept, that of independence, states that, if $\mathbf{x}$ and $\mathbf{y}$ are independent, then

$$p(x, y) = p(x)p(y)$$

or, in words, $\mathbf{x}$ is totally unaffected by $\mathbf{y}$.

This book deals with random vectors. Thus, $\mathbf{x} = [x_1, x_2, \ldots, x_N]^T$ is a vector of N random variables described by the joint density

$$p(\mathbf{x}) = p(x_1, x_2, \ldots, x_N)$$

with mean, for example, given by the vector

$$\boldsymbol{\mu} = \int \mathbf{x}p(\mathbf{x})d\mathbf{x}$$

where $d\mathbf{x} = dx_1 dx_2 \ldots dx_N$. We use the same notation for a random variable and a random vector in order to minimize clutter. Which is which will be clear from the context.

Finally, a word about second order statistics of the random vector, $\mathbf{x}$. We define the autocovariance matrix, $\mathbf{C_{xx}}$, to be the matrix with c_{ij}, the ijth element, given by

$$c_{ij} = \int x_i x_j p(\mathbf{x}) dx_i dx_j$$

where $\mathbf{x}$ is zero mean. Otherwise, we refer to the autocorrelation matrix. In similar fashion, the ijth element of the crosscovariance matrix , $\mathbf{C_{xy}}$, between two random vectors, $\mathbf{x}$ and $\mathbf{y}$, is given by

$$c_{ij} = \int x_i y_j p(\mathbf{x}, \mathbf{y}) dx_i dy_j$$

1.2.2 Ensembles, Expectations etc.

A random process, $\mathbf{x}(t, \xi)$, which will be written as $\mathbf{x}(t) = \mathbf{x}_\xi(t)$ for convenience, is a generalization of the concept of a random variable, presented above, to a random process. A realization of the random process is designated by $x(t) = x_\xi(t)$.

A concept of central importance in the analysis of all stochastic processes is the *ensemble*. An ensemble is the collection of the individual realizations, $x(t)$, M in number, of a random process. A random process is, in consequence, the ensemble, the sample space, $\mathcal{S}$, and the associated pdf. An ensemble of random signals is illustrated in Fig. 1.1.

We now make the following observations.

(i) $\mathbf{x}_\xi(t)$ is a random process and will generally be written as $\mathbf{x}(t)$.

(ii) For a particular value of ξ, say ξ_i, $\mathbf{x}_{\xi_i}(t)$ is a realization of the random process and is written as $x(t)$.

(iii) $\mathbf{x}_\xi(t = t_1)$ is a random variable, $\mathbf{x}$.

(iv) $\mathbf{x}_{\xi_i}(t = t_1)$ is a single number corresponding to the amplitude of the random variable evaluated at $t = t_1$

An example is always useful at this stage. We consider a random process of the form

$$\mathbf{x}(t) = \mathbf{a}\cos(\omega_0 t + \boldsymbol{\theta})$$

where $\mathbf{a}$ and $\boldsymbol{\theta}$ are random variables with pdf's given by

$$p(a) = \frac{1}{\sigma\sqrt{2\pi}} e^{-(a-\mu)/2\sigma^2}$$

$$p(\theta) = \begin{cases} \frac{1}{2\pi} & 0 \leq \theta \leq 2\pi \\ 0 & \text{otherwise} \end{cases}$$

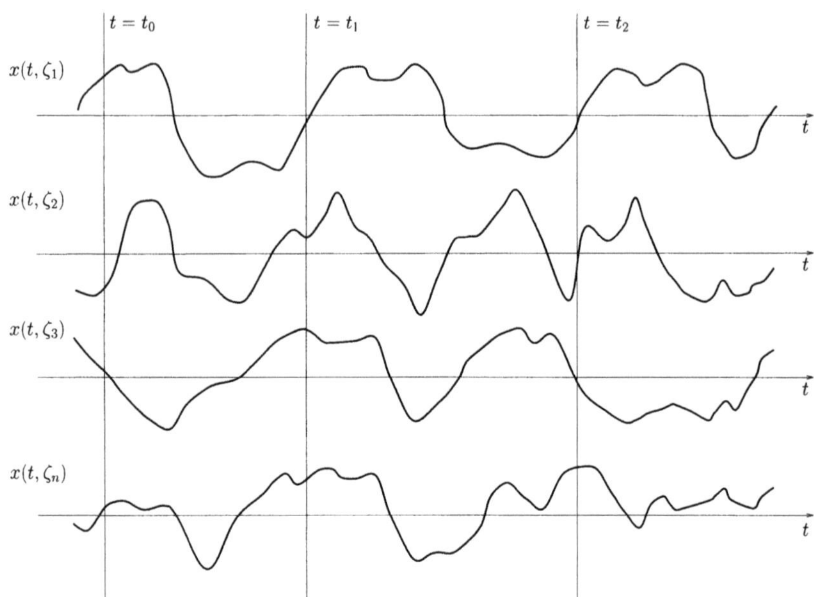

Figure 1.1: An ensemble of random signals.

In this particular case, illustrated in Fig. 1.2, each realization of the stochastic process is, in fact, a deterministic function (in the sense that a future point may be predicted without error from its past).

Statistics associated with the ensemble are dependent on particular characteristics of the pdf which describes the stochastic process. We write the pdf $p_{\mathbf{x}}(x, t)$ as $p(x, t)$ for convenience and remark that, in general, this pdf is a function of both amplitude *and* time. The process is classified in terms of the moments of $p(x, t)$ and, in particular, how t enters into the equation, i.e., can $p(x, t)$ be written as $p(x)$? The process is classified as stationary or nonstationary.

A nonstationary process is one for which the pdf must be expressed as $p(x, t)$, i.e., moments of the pdf, for example, the mean and variance, are time dependent. Although nonstationarity is extremely important, the topics which we cover at present assume stationary processes, a class which may be divided into two parts, ergodic and nonergodic processes. *Strict* stationarity implies that *all* statistics determined in terms of the ensemble are time independent. It is usual to relax this strict definition and to consider processes which are *weakly* or *wide sense* stationary. Such processes have the following properties defined in terms of ensemble averages or *expectations*

$$E[\mathbf{x}(t)] = \int xp(x)dx \tag{1.10}$$

$$E[\mathbf{x}^2(t)] = \int x^2 p(x)dx \tag{1.11}$$

where $E[\cdot]$ in the above equations is the expectation operator. Eq. (1.10) implies that the mean of the process is time independent and may be computed from $p(x)$

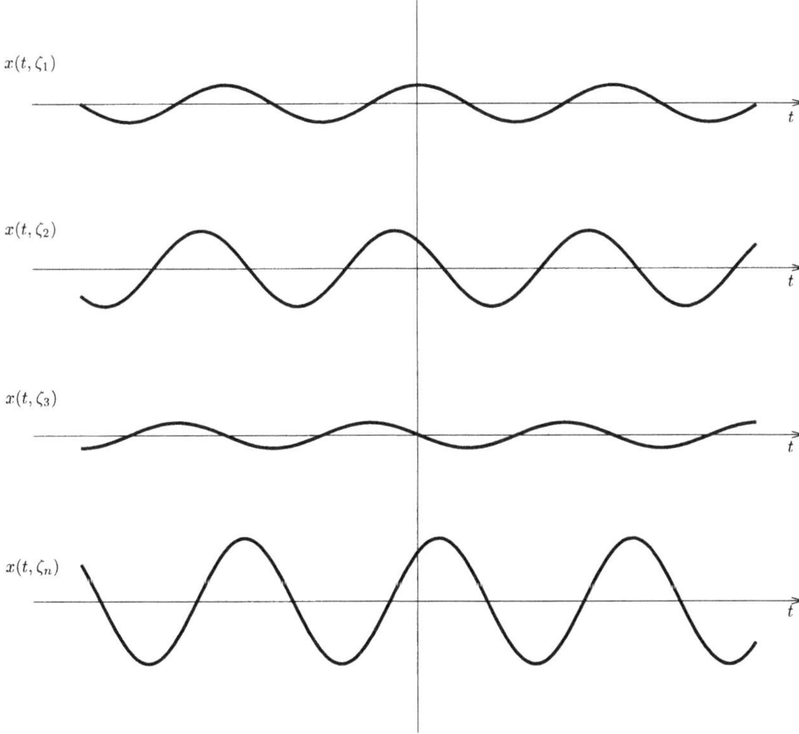

Figure 1.2: A harmonic ensemble.

at any particular value of t. Eq. (1.11) expresses the same property for the second moment of the process or, assuming zero mean, for the variance σ^2. We remark that the expectation operator, $E[\cdot]$, is linear and, in its use, there is an implicit assumption of the knowledge of the underlying pdf. Since, in practice, we seldom if ever have this information at our disposal, we must at some point estimate the expectation in some manner. We will do so, but for now we simply use it with audacity.

For weakly stationary processes we may also write

$$p(x_1, x_2; t_1, t_2) = p(x_1, x_2; t_1 - t_2)$$

i.e., the joint pdf is independent of the actual time, but dependent on the time difference. Now, defining $t_2 - t_1 = \tau$, we may write the autocovariance of the process, $r(t_1, t_2)$, as

$$\begin{aligned} r(t_1, t_2) &= r(t_2 - t_1) = r(\tau) \\ r(\tau) &= \int x_1 x_2 p(x_1, x_2; \tau) dx_1 dx_2 \end{aligned}$$

So far, we have estimated all statistics over an ensemble, a fictitious view of an infinite set of all data. We now imagine a further fictitious hypothesis, interestingly enough, a hypothesis which is central in much of the processing of seismic data. We now enlarge on the *ergodic* hypothesis.

1.2.3 The Ergodic Hypothesis

The ergodic hypothesis assumes that, using any realization of the ensemble, we can estimate the statistics of the entire ensemble using that realization. Consider a particular statistic expressed by

$$E[\phi(\mathbf{x}(t))] = \int \phi[(x)]p(x)dx$$

Then, for an ergodic process

$$E[\phi(\mathbf{x}(t))] = \lim_{T\to\infty} \frac{1}{T} \int_{-T/2}^{T/2} \phi[x(t)]dt$$

In other words, for an ergodic process, an ensemble average is equivalent to a time average. This equivalency requires some very special properties described in terms of higher moments of the associated pdf. Although these conditions are always violated in practice, ergodicity is, in general very often assumed or implied (we will see why a little later) which means that we can utilize the beauty and simplicity of the $E[\cdot]$ linear operator in much of our analysis.

Practically speaking, ergodicity implies that any member of the ensemble takes on all possible values in time with the same relative frequency. Our example in Fig. 1.2 is of value here. Each member of the ensemble does in fact take on all possible values in time and, therefore, given the assumed pdf's, the process is ergodic. Interestingly enough, it can be easily shown that if the pdf for θ is not uniform over 2π, the process is not stationary. We now quote a result of great generality that we make use of in the ensuing discussion. This result is known as Chebychev's inequality .

1.2.3.1 The Chebychev Inequality

Consider a random variable $\mathbf{x}$ with some pdf, $p(x)$, with mean μ and variance σ^2. We know that σ^2 is a measure of the concentration of $p(x)$ around μ. The spread around this value is given by (Papoulis, 1977)

$$\text{Prob}(|\mathbf{x} - \mu| \geq \Delta) \leq \frac{\sigma^2}{\Delta^2} \tag{1.12}$$

Eq. (1.12) tells us that values of $\mathbf{x}$ tend to cluster about their mean value. The probability that a particular value of $\mathbf{x}$ is far from μ is, in fact, small and depends on the associated variance. This inequality allows us to compute, for example, the 95% confidence interval for the mean based on a single observation, x, for any pdf. In this case, $\sigma^2/\Delta^2 = 0.05$, so that $\Delta = 4.47\sigma$ and we obtain the interval for the worst case scenario as

$$x - 4.47\sigma \leq \mu \leq x + 4.47\sigma$$

Eq. (1.12) also formally tells us that, when the variance is zero

$$E[\mathbf{x} - E[\mathbf{x}]] = 0$$

which implies that

$$\mathbf{x} = E[\mathbf{x}]$$

In other words, $\mathbf{x}$ is a constant for all possible outcomes.

1.2.3.2 Time Averages and Ergodicity

We now use Chebychev's inequality in a brief look at the ergodic assumption. Essentially, the question is, under what circumstances can time averages be used to obtain the same results as those given by ensemble averages? For each member, $x_i(t)$, of the ensemble $\{\mathbf{x}(t)\}$, we compute the mean, $\boldsymbol{\mu}$, and the autocovariance, $\mathbf{r}(\tau)$, by means of the relations

$$\boldsymbol{\mu} = \lim_{T\to\infty} \frac{1}{2T} \int_{-T}^{+T} x_i(t)dt$$

$$\mathbf{r}(\tau) = \lim_{T\to\infty} \frac{1}{2T} \int_{-T}^{+T} x_i(t+\tau)x_i(t)dt$$

Note that both $\boldsymbol{\mu}$ and $\mathbf{r}(\tau)$ are random variables. The conditions that are required for equality are

$$\boldsymbol{\mu} = E[\mathbf{x}(t)] = \mu \tag{1.13}$$

and

$$\mathbf{r}(\tau) = E[\mathbf{x}(t+\tau)\mathbf{x}(t)] = r(\tau) \tag{1.14}$$

Eq. (1.13), following Chebychev's inequality, holds if $E[(\boldsymbol{\mu} - \mu)^2] = 0$. Assuming that $\mu = E[\boldsymbol{\mu}]$, which is not difficult to show, we require, therefore, that

$$E[(\boldsymbol{\mu} - E[\boldsymbol{\mu}])^2] = \sigma^2_{\boldsymbol{\mu}} = 0 \tag{1.15}$$

In a similar vein, for Eq. (1.14) to hold, we must show that

$$E[(\mathbf{r}(\tau) - E[\mathbf{r}(\tau)])^2] = \sigma^2_{\mathbf{r}(\tau)} = 0 \tag{1.16}$$

Eqs. (1.15) and (1.16) require knowledge of the 4th moment statistics of the process which are generally not available. The conclusion is, basically, that the ergodic hypothesis is adopted for convenience of computing time averages, and is very seldom fully justifiable in practice.

1.3 Properties of Estimators.

One of the chief concerns in time series modelling, spectral estimation and inversion, is the determination of *good estimators*. What we now explore is the question of what we mean by *good*. For example, when is an autocovariance estimator a good estimator? A related and simple question is, how do we estimate the variance of a time series, given that we were required to estimate the mean value in the first place? These are truly fundamental questions that we now address. The principles that are outlined below are basic in all that follows.

Goodness is, essentially, contained in two qualities: bias and variance. These two qualities are not unrelated. In fact they are related through a universal concept, the principle of uncertainty. This principle takes on diverse forms. In the form of interest in the present application, it tells us that, if we want to estimate a quantity with little bias it will cost us more variance.This is the *bias/variance* tradeoff. Another way of understanding this important concept is by posing the question "is it better to localize parameters correctly but with a degree of uncertainty, or localize parameters incorrectly with certainty". We referred, above, to bias and variance rather loosely. We now define these concepts rigorously.

1.3.1 Bias of an Estimator

Define $\hat{\boldsymbol{\theta}}$ as the estimated value of a random variable $\boldsymbol{\theta}$ with pdf $p(\theta)$ and mean θ, which we call the "true value". $\hat{\boldsymbol{\theta}}$ is itself a random variable. The bias, B, is defined as

$$B = E[\hat{\boldsymbol{\theta}}] - \theta \tag{1.17}$$

$\hat{\boldsymbol{\theta}}$ is unbiased if $B = 0$. Essentially, this implies that the pdf that is conditional on $\hat{\boldsymbol{\theta}}$ is centered around the true value, θ.

1.3.1.1 An Example

Consider the estimate

$$\hat{\boldsymbol{\mu}} = \frac{1}{N}\sum_i x_i \quad i = 1, 2, \ldots, N \tag{1.18}$$

Taking expectations

$$E[\hat{\boldsymbol{\mu}}] = \frac{1}{N}\sum_i E[x_i] = \frac{1}{N}\sum_i \mu = \mu.$$

Hence $\hat{\boldsymbol{\mu}}$ as defined in Eq. (1.18) is an unbiased estimator of μ.
If

$$E[\hat{\boldsymbol{\theta}}] \;\neq\; \theta$$

but

$$\lim_{N\to\infty} E[\hat{\boldsymbol{\theta}}] \;=\; \theta$$

then $\hat{\theta}$ is unbiased in the limit.

1.3.2 Variance of an Estimator

The variance of $\hat{\boldsymbol{\theta}}$ is given by

$$\begin{aligned}\text{var}[\hat{\boldsymbol{\theta}}] &= E[(\hat{\boldsymbol{\theta}} - E[\hat{\boldsymbol{\theta}}])]^2 \\ &= E[\hat{\boldsymbol{\theta}}^2] - (E[\hat{\boldsymbol{\theta}}])^2\end{aligned}$$

If $\lim_{N\to\infty} \text{var}[\hat{\boldsymbol{\theta}}] = 0$, $\hat{\boldsymbol{\theta}}$ is called a consistent estimator.

1.3.2.1 An Example

Consider the variance of $\hat{\boldsymbol{\mu}}$ in Eq. (1.18)

Since

$$\begin{aligned}E[(\hat{\boldsymbol{\mu}})] &= \mu \\ \text{var}[\hat{\boldsymbol{\mu}}] &= E[(\hat{\boldsymbol{\mu}} - \mu)^2] \\ &= E[(\frac{1}{N}\sum_i x_i - \frac{1}{N}\sum_i \mu)^2] \\ &= E[\frac{1}{N^2}\sum_i (x_i - \mu)^2]\end{aligned}$$

we write this as

$$\frac{1}{N^2}\sum_i\sum_j E[(x_i - \mu)(x_j - \mu)] = \frac{1}{N^2}\sum_i \text{var}(x_i) + \sum_{i\neq j} \text{cov}(x_i, x_j)$$

Since, for independent samples

$$\text{Cov}(x_i, x_j) = 0$$

we obtain

$$\text{var}[\hat{\boldsymbol{\mu}}] = \frac{\sigma_x^2}{N^2}$$

Clearly, $\hat{\boldsymbol{\mu}}$ is a consistent and unbiased estimator.

1.3.3 Mean Square Error of an Estimator

Trade-offs exist in all walks of life. In parameter estimation, the most famous is the trade-off between bias and variance expressed by the relation

$$\text{mean square error} = \text{variance} + \text{bias}^2$$

It is interesting to see how this expression arises. Using the definitions of bias and variance above, we anticipate the expression for the mean square error (MSE) by computing

$$\begin{aligned}\mathrm{var}[\hat{\boldsymbol{\theta}}] + B^2 &= E[(\hat{\boldsymbol{\theta}} - E[\hat{\boldsymbol{\theta}}])^2] + (E[\hat{\boldsymbol{\theta}}] - \theta)^2 \\ &= E[\hat{\boldsymbol{\theta}}^2] - (E[\hat{\boldsymbol{\theta}}])^2 + (E[\hat{\boldsymbol{\theta}}])^2 - 2E[\hat{\boldsymbol{\theta}}]\theta + \theta^2 \\ &= E[\hat{\theta}^2] - 2E[\hat{\theta}]\theta + \theta^2 \\ &= E[(\hat{\boldsymbol{\theta}} - \theta)^2] \\ &= \mathrm{MSE} \end{aligned} \tag{1.19}$$

One oft-cherished compromise between variance and bias, a compromise that is often utilized, is to minimize the MSE as defined by Eq. (1.19).

1.4 Orthogonality

When we began writing this book, we soon became aware of a central theme or concept, principle if you wish, which appears everywhere. From the earliest discussion of the expansion of functions in terms of other functions, transformation from one domain to another, time series modelling, matrix inversion, projection operators etc., we meet *orthogonality* and its concomitant companion *uncorrelatedness*. This will all be very simple, in all probability well known to all, and can certainly be skipped.

1.4.1 Orthogonal Functions and Vectors

We use functions and vectors interchangeably. The reasons for this will become clear. We begin with vectors. Consider two vectors denoted by $\mathbf{x}_1$ and $\mathbf{x}_2$. We wish to find the projection or component of $\mathbf{x}_1$ along $\mathbf{x}_2$ which we denote by $c_{12}\mathbf{x}_2$. Geometrically, the projection of one vector onto another is simply obtained by drawing a perpendicular from the end of $\mathbf{x}_1$ onto the vector $\mathbf{x}_2$ as shown in Fig. 1.3.

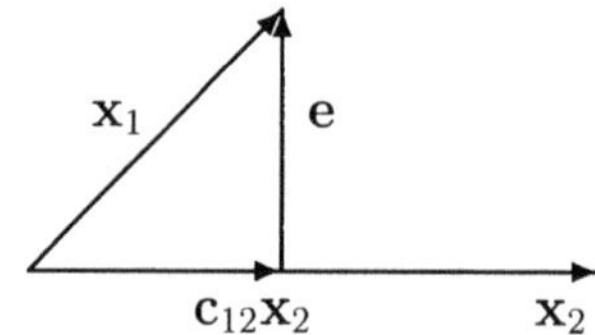

Figure 1.3: The 'smallest' error.

We write the projection in vector form as

$$\mathbf{x}_1 = c_{12}\mathbf{x}_2 + \mathbf{e} \tag{1.20}$$

where $\mathbf{e}$ is the resulting residual or error vector. There is something very special indeed about the way we constructed $\mathbf{e}$. It was constructed so that the error vector is orthogonal to the basis vector, i.e., $\mathbf{e}\perp\mathbf{x}_2$. What this implies, of course, is that of all possible error vectors which could be constructed from the tip of $\mathbf{x}_1$ to $\mathbf{x}_2$, $\mathbf{e}$ is smallest in magnitude. In other words, from a physical point of view, we are looking for the minimum error in the projection. Mathematically, we will find that c_{12} which yields the minimum error vector. Using the principle of least squares, we minimize the error energy, $\mathbf{e}^T\mathbf{e}$, with respect to c_{12}.

$$\begin{aligned}\mathbf{e}^T\mathbf{e} &= (\mathbf{x}_1 - c_{12}\mathbf{x}_2)^T(\mathbf{x}_1 - c_{12}\mathbf{x}_2) \\ &= -2c_{12}\mathbf{x}_1^T\mathbf{x}_2 + c_{12}^2\mathbf{x}_2^T\mathbf{x}_2 + \mathbf{x}_1^T\mathbf{x}_1\end{aligned} \tag{1.21}$$

Minimizing Eq. (1.21) with respect to c_{12} by setting $d\mathbf{e}^T\mathbf{e}/dc_{12} = 0$ we obtain

$$c_{12} = \frac{\mathbf{x}_1^T\mathbf{x}_2}{\mathbf{x}_2^T\mathbf{x}_2} \tag{1.22}$$

Eq. (1.22) expresses the required coefficient in terms of the two vector dot products. The orthogonality of $\mathbf{e}$ and $\mathbf{x}_2$ follows immediately.

$$\begin{aligned}\mathbf{x}_2^T\mathbf{e} &= \mathbf{x}_2^T(\mathbf{x}_1 - c_{12}\mathbf{x}_2) \\ &= \mathbf{x}_2^T(\mathbf{x}_1 - (\frac{\mathbf{x}_1^T\mathbf{x}_2}{\mathbf{x}_2^T\mathbf{x}_2})\mathbf{x}_2) \\ &= \mathbf{x}_2^T\mathbf{x}_1 - \mathbf{x}_1^T\mathbf{x}_2 \\ &= 0\end{aligned}$$

The analogy between vectors and functions allows us to express the orthogonality of two vectors in equivalent terms for two functions. Thus, whereas $\mathbf{x}_1$ and $\mathbf{x}_2$ are orthogonal if $\mathbf{x}_2^T\mathbf{x}_1 = 0$, $x_1(t)$ and $x_2(t)$ are orthogonal on (a, b) if $\int_a^b x_1(t)x_2(t)dt = 0$. Since we will be, mostly, using vectors rather than functions, we concentrate on the vectorial representation.

1.5 Orthogonal Vector Space

We are all familiar with the spatially 3-dimensional world in which we live. The description of a vector $\mathbf{x}$ in this world is conveniently accomplished by projecting it onto three mutually orthogonal coordinates. We call this coordinates system an orthogonal *basis*. The reason why we normally choose an orthogonal basis will become dramatically clear in a moment. In general, our vector need not be limited to 3 dimensions. In fact we can envisage vectors which live in a N dimensional space and we may also wish to describe such a vector in terms of a N dimensional orthogonal basis. Such a description follows the path outlined above.

Consider a general vector $\mathbf{x}$ in a N dimensional space and basis vectors, $\mathbf{v}_1, \mathbf{v}_2, \ldots, \mathbf{v}_N$ along N mutually perpendicular directions. If $c_1, c_2, \ldots, c_N$ are the components of $\mathbf{x}$ along these coordinate axes, then

$$\mathbf{x} = c_1\mathbf{v}_1 + c_2\mathbf{v}_2 + \ldots + c_N\mathbf{v}_N \tag{1.23}$$

For any general vector $\mathbf{x}$ to be represented by Eq. (1.23), the set of basis vectors $\{\mathbf{v}_k\}, k = 1, 2, \ldots, N$ must be *complete*. This means that the maximum number of independent vectors in this N dimensional space is N. A set of vectors is said to be linearly independent if none of the vectors in this set can be represented in terms of a linear combination of the remaining vectors. i.e, for our set $\{\mathbf{v}_k\}$ to be independent means that it is impossible to find constants $a_1, a_2, \ldots, a_N$, not all zero, such that

$$\sum_{k=1}^{N} a_k v_k = 0$$

Since our space has a maximum of N independent vectors, it is clear that any other general vector $\mathbf{x}$ in this space may be expressed as a linear combination of the basis set as expressed by Eq. (1.23). It should also be clear that the particular set represented by the choice of our basis vectors $\{\mathbf{v}\}$ is by no means unique. More on this later.

Let us now determine the required coefficients c_k. In so doing, the beauty and simplicity (often related) of the orthogonality constraint will become apparent. Since the basis set is assumed orthogonal

$$\mathbf{v}_m^T \mathbf{v}_n = \begin{cases} 0 & m \neq n \\ k_m & m = n \end{cases} \tag{1.24}$$

Multiplying Eq. (1.23) by $\mathbf{v}_i^T$ yields $k_i c_i$. Thus, we can obtain all required coefficients as

$$c_i = \frac{\mathbf{v}_i^T \mathbf{x}}{k_i} \qquad i = 1, 2, \ldots, N$$

It is easy to see what would have occurred if the basis were not orthogonal. The multiplication of $\mathbf{x}$ by $\mathbf{v}_i^T$ would have produced cross products for all coefficients and the task of solving for the coefficients would be hopelessly more complicated. Incidentally, it is often possible and desirable to normalize the basis so that the constant $k_i = 1$. In such a case, the basis is called orthonormal.

1.5.1 Gram-Schmidt Orthogonalization

We have mentioned that the orthogonal N dimensional basis set is not unique. We can, in fact, construct an infinite number of such sets. Let us see how to do this. For simplicity we follow a procedure which we will adopt throughout the book. We begin with two vectors and then extend to N vectors. Assume then, that we are given two independent vectors $\mathbf{x}_1$ and $\mathbf{x}_2$ (the two vectors would be dependent if they lay on the same line), and we want to construct two perpendicular vectors $\mathbf{v}_1$ and $\mathbf{v}_2$. We are free to choose $\mathbf{v}_1$ in any way we wish. This is where the non uniqueness of the basis comes in. Let us, for convenience, choose $\mathbf{v}_1 = \mathbf{x}_1$. We now wish to express $\mathbf{v}_2$ in terms of $\mathbf{x}_2$ and $\mathbf{x}_1$. Following our previous discussion concerning the orthogonality associated with the error vector, we see that $\mathbf{x}_2$ is the sum of $\mathbf{v}_2$ (the equivalent error vector) and the projection of $\mathbf{x}_2$ onto $\mathbf{x}_1$. Thus, following Eq. (1.22)

$$\begin{aligned} \mathbf{x}_2 &= \mathbf{v}_2 + c_{12}\mathbf{x}_1 \qquad (1.25) \\ &= \mathbf{v}_2 + c_{12}\mathbf{v}_1 \\ &= \mathbf{v}_2 + \left(\frac{\mathbf{v}_1^T \mathbf{x}_2}{\mathbf{v}_1^T \mathbf{v}_1}\right)\mathbf{v}_1 \end{aligned}$$

Hence

$$\mathbf{v}_2 = \mathbf{x}_2 - (\frac{\mathbf{v}_1^T\mathbf{x}_2}{\mathbf{v}_1^T\mathbf{v}_1})\mathbf{v}_1$$

We now have the two orthogonal basis vectors $\mathbf{v}_1$ and $\mathbf{v}_2$. To obtain $\mathbf{v}_2$, what we have done is to subtract from $\mathbf{x}_2$ its component along the settled direction (see Strang, 1988, for more details). If a third vector is added, we proceed in an exactly similar manner. We obtain $\mathbf{v}_3$ by subtracting from $\mathbf{x}_3$ the components of this vector along the two, previously settled, directions. The procedure for determining N independent and orthonormal basis vectors, $\mathbf{q}_i$, $i = 1, 2, \ldots, N$ from N independent vectors $\mathbf{x}_i$, $i = 1, 2, \ldots, N$ can now be detailed as follows

$$\mathbf{v}_i = \mathbf{x}_i - \frac{\mathbf{v}_1^T\mathbf{x}_i}{\mathbf{v}_1^T\mathbf{v}_1}\mathbf{v}_1 - \ldots - \frac{\mathbf{v}_{i-1}^T\mathbf{x}_i}{\mathbf{v}_{i-1}^T\mathbf{v}_{i-1}}\mathbf{v}_{i-1} \tag{1.26}$$

$$\mathbf{q}_i = \frac{\mathbf{v}_i}{\sqrt{\mathbf{v}_i^T\mathbf{v}_i}} = \frac{\mathbf{v}_i}{||\mathbf{v}_i||} \tag{1.27}$$

Let us go on a bit. We talked about the importance of orthogonalization from various points of view, one of them being matrix inversion. This is an excellent place to illustrate this. Let us rewrite Eq. (1.25), for the first three vectors, in matrix form, where the columns of the matrix $\mathbf{V}$ are the vectors $\mathbf{v}_i$ and the columns of the matrix $\mathbf{X}$ are the vectors $\mathbf{x}_i$. We have, for $i = 1, 2$ and 3

$$\mathbf{X} = \mathbf{VS} \tag{1.28}$$

where the matrices are defined by

$$\begin{bmatrix} \mathbf{x}_1 & \mathbf{x}_2 & \mathbf{x}_3 \end{bmatrix} = \begin{bmatrix} \mathbf{v}_1 & \mathbf{v}_2 & \mathbf{v}_3 \end{bmatrix} \begin{bmatrix} 1 & * & * \\ 0 & 1 & * \\ 0 & 0 & 1 \end{bmatrix} \tag{1.29}$$

The $*$'s in the $\mathbf{S}$ matrix are elements composed of combinations of the coefficients c_{ij}, $i = 1, 2$ and $j = 1, 2, 3$. The actual values are not important here. What is of importance is that the columns of $\mathbf{V}$ are orthogonal and $\mathbf{S}$ is upper diagonal. Normalizing $\mathbf{V}$ by the lengths of the vectors, we obtain the famous QR factorization of $\mathbf{X}$, namely

$$\mathbf{X} = \mathbf{QR} \tag{1.30}$$

Here, the columns of $\mathbf{Q}$ are orthonormal and like $\mathbf{S}$, $\mathbf{R}$ is also upper diagonal.

We are constantly faced with the computation of inverse matrices. Consider, for instance, the linear equation $\mathbf{Am} = \mathbf{d}$. The solution we seek, assuming for simplicity that $\mathbf{A}$ is square and invertible, is $\mathbf{m}$=$\mathbf{A}^{-1}\mathbf{d}$. Using the QR factorization, we write

$$\begin{aligned} \mathbf{m} &= (\mathbf{QR})^{-1}\mathbf{d} \\ &= \mathbf{R}^{-1}\mathbf{Q}^T\mathbf{d} \end{aligned}$$

The solution now requires only the product $\mathbf{Q}^T\mathbf{d}$ and a back substitution in the triangular system $\mathbf{Rm} = \mathbf{Q}^T\mathbf{d}$. As pointed out by Strang (1988), the QR factorization does not save computer time which depends on the operation count. Its strength lies in improved numerical stability which is of fundamental importance in inverse problems.

1.5.1.1 Remarks

Orthogonality of two vectors implies a lack of dependence. This lack of dependence, in turn, implies a lack of correlation. In as much as we saw how important it is to choose an orthogonal basis for representation, so it is equally important to be able to design systems which remove existent correlations. Examples abound, the most widely known being, undoubtedly, Wiener filtering. In order to develop fully the idea of correlation cancelling, we turn to concepts developed in the previous chapter concerning random variables, probability functions,expectations and the like.

1.5.2 Orthogonality and Correlation

When two vectors are orthogonal they are also, clearly, uncorrelated. Inasmuch as correlation plays a central role in all that we do in this book, correlation cancelling (Orfanidis 1985), the equivalent of orthogonalization, plays an equally central role. Consider two random vectors $\mathbf{x}$ and $\mathbf{y}$ that are, for convenience, zero mean and of equal length. We assume that $\mathbf{x}$ and $\mathbf{y}$ are correlated such that $\mathbf{C_{xy}} = E[\mathbf{xy}^T] \neq 0$. We seek a linear transformation to optimally remove the correlations and define the error vector by

$$\mathbf{e} = \mathbf{x} - \mathbf{Hy}$$

where $\mathbf{H}$ is the transformation matrix. We are going to estimate $\mathbf{H}$ in two ways. The first approach is to minimize the mean squared error, $\mathcal{E}$, with respect to $\mathbf{H}$. To do this we form

$$\begin{aligned} \mathcal{E} = E[\mathbf{e}^T\mathbf{e}] &= E[(\mathbf{x}-\mathbf{Hy})^T(\mathbf{x}-\mathbf{Hy})] \\ &= E[\mathbf{x}^T\mathbf{x}] - 2\mathbf{H}\,E\,[\mathbf{x}^T\mathbf{y}] + E[\mathbf{y}^T\mathbf{H}^T\mathbf{Hy}] \end{aligned} \tag{1.31}$$

Differentiating $\mathcal{E}$ with respect to each row vector, $\mathbf{h}_i$, of $\mathbf{H}$, setting the result to zero, and gathering terms into a matrix, produces

$$\mathbf{H}_{opt} = \mathbf{C_{xy}}\mathbf{C_{yy}}^{-1} \tag{1.32}$$

The second approach follows from the a priori requirement that

$$\mathbf{C_{ey}} = 0$$

(Note that we use $\mathbf{C_{ey}}$ rather than $\mathbf{R_{ey}}$. This is merely to affirm that we are dealing with zero mean processes, an assumption that is convenient and ubiquitous). By definition, we have

$$\mathbf{C_{yy}} = E[\mathbf{yy}^T]$$

and

$$\mathbf{C_{ey}} = E[\mathbf{ey}^T] = E[(\mathbf{x} - \mathbf{Hy})\mathbf{y}^T] \quad (1.33)$$
$$= \mathbf{C_{xy}} - \mathbf{HC_{yy}} \quad (1.34)$$

Since our constraint is that $\mathbf{C_{ey}} = 0$, we obtain

$$\mathbf{H}_{opt} = \mathbf{C_{xy}}\mathbf{C_{yy}}^{-1} \quad (1.35)$$

Eqs. (1.32) and (1.35) are identical as they should be. Minimization of the mean squared error is equivalent to the orthogonal projection of $\mathbf{x}$ on $\mathbf{y}$, giving rise to orthogonality equations. This is a reoccurring theme throughout the book. In later chapters we will see that this is also called the *innovation* approach giving rise to *normal* equations.

1.5.3 Orthogonality and Eigenvectors

This section, on various issues associated with orthogonality, would simply not be complete without some thoughts about eigenvectors and eigenfunctions. We find these entities throughout the scientific domain, in many marvellous guises. We cannot do justice, here, to the myriad of characteristics that are associated with the eigen domain, so we will satisfy ourselves with our choice of some.

We are not, primarily, concerned with the physical meaning of eigenfunctions as, for example, the functions describing the vibration of a drumhead. We are principally interested in the meaning of eigenvectors as they relate to the analysis and processing of our data. With this in mind, we begin with a special case of eigenfunctions, that is associated with a square $(N \times N)$ matrix, $\mathbf{A}$. Consider the linear equation

$$\mathbf{Ax} = \mathbf{y}$$

where $\mathbf{x}$ and $\mathbf{y}$ represent $(N \times 1)$ vectors. The vector $\mathbf{y} \in \mathbf{R}^N$ will, in general, point in a direction that is different to that of $\mathbf{x}$. For a typical matrix $\mathbf{A}$, however, there will be N different vectors $\mathbf{x}$ such that the corresponding $\mathbf{y}$ will have the *same direction* as $\mathbf{x}$ and, likely, a different length. We can express this notion by

$$\mathbf{Ax} = \lambda\mathbf{x} \quad (1.36)$$

where λ is a constant. To clarify the above assertion, we rewrite Eq. (1.36) as

$$(\mathbf{A} - \lambda\mathbf{I})\mathbf{x} = 0 \quad (1.37)$$

where $\mathbf{I}$ is the identity matrix. Remembering our linear algebra, Eq. (1.37) has a solution other than zero only if

$$|\mathbf{A} - \lambda\mathbf{I}| = 0 \quad (1.38)$$

The determinant in Eq. (1.38) is a polynomial in λ of order N and, in general, has N distinct roots, real or complex, which give rise to $\lambda_k,\ k = 1, 2, \ldots, N$. Each λ_k, called an eigenvalue, is associated with a corresponding solution vector $\mathbf{x}_k$, called an eigenvector, which is determined to within a scale factor. The reason for the great importance of this eigenvalue-eigenvector decomposition (from now on we call it the eigenvector or eigen decomposition), is the fact that these N (in general) eigenvectors may be shown to be linearly independent, hence orthogonal, and serve as a basis for the decomposition of an arbitrary vector, $\mathbf{w} \in \mathbf{R}^N$. i.e., we have found a natural coordinate system for the representation of any vector in $\mathbf{R}^N$.

Let us now look at a different interpretation of eigenvectors. Eq. (1.36) shows that an eigenvector (eigenfunction) is a vector (function) that emerges again after some transformation. Consider the complex exponential that we met at the very beginning of the book, the one describing the famous Euler identity

$$e^{i\omega t} = \cos(\omega t) + i\sin(\omega t)$$

and, for convenience, let

$$x(t) = e^{i\omega t}$$

Now consider the process of translation, i.e., we form $x(t+\tau)$. We have

$$x(t+\tau) = e^{i\omega(t+\tau)} = e^{i\omega t}e^{i\omega\tau} = \lambda(\omega)x(t)$$

where

$$\lambda(\omega) = e^{i\omega\tau}$$

is the eigenvalue, and

$$x(t) = e^{i\omega t}$$

is the eigenfunction of the translation.

We proceed by considering a linear, time-invariant system (Section 1.7.6), the impulse response of which we denote by $h(t)$. Such a system is defined by the process of convolution. If $x(t)$ is the input and $y(t)$ is the output, we have

$$y(t) = \int h(u)x(t-u)du$$

and the substitution

$$x(t) = e^{i\omega t}$$

produces

$$y(t) = e^{i\omega t}\int h(u)e^{-i\omega u}du$$

$$y(t) = H(\omega)e^{i\omega t}$$

We see, therefore, that the linear system which is excited with an input $e^{i\omega t}$ produces an output that is the input multiplied by a factor $H(\omega)$. The linear system can, therefore, be described by eigenfunctions $e^{i\omega t}$ associated with eigenvalues $H(\omega)$, which turn out to be the values of the transfer function at frequency ω. Further, it is true and important that $e^{i\omega t}$ are also the eigenfunctions associated with differentiation, integration and differencing (Hamming, 1983). This is clear from

(1) differentiation

$$\frac{d}{dt}e^{i\omega t} = i\omega e^{i\omega t} = \lambda(\omega)e^{i\omega t}$$

(2) integration

$$\int e^{i\omega t}dt = \frac{e^{i\omega t}}{i\omega} = \lambda(\omega)e^{i\omega t}$$

(3) differencing

$$\Delta e^{i\omega t} = e^{i\omega(t+1)} - e^{i\omega t} = (e^{i\omega} - 1)e^{i\omega t} = \lambda(\omega)e^{i\omega t}$$

Lastly, an important property of eigenvectors, is in relation to the quadratic form Q of a symmetric matrix $\mathbf{A}$ defined by

$$Q = \mathbf{x}^T\mathbf{A}\mathbf{x} \tag{1.39}$$

where $\mathbf{x}$ is any non-zero vector.

The quadratic form in Eq. (1.39), a representation of which is a weighted error, is found in many problems, particularly optimization formulations and so we deal with the optimization of Q. Naturally, we have to impose a constraint on $\mathbf{x}$ so that we do not obtain zero or infinity. The problem is now formulated as:

Find the vector $\mathbf{x}$ that maximizes (it so turns out) Q subject to $\mathbf{x}^T\mathbf{x} = 1$. Introducing the Lagrange multiplier, λ (Appendix B), a cost function J can be defined as

$$J = \mathbf{x}^T\mathbf{A}\mathbf{x} - \lambda(\mathbf{x}^T\mathbf{x} - 1) \tag{1.40}$$

Differentiating and equating to zero

$$2\mathbf{A}\mathbf{x} = 2\lambda\mathbf{x} \tag{1.41}$$

Once again we see eigenvectors appearing. The vector that maximizes the quadratic form associated with the symmetric matrix $\mathbf{A}$, is the dominant (clearly) eigenvector corresponding to this matrix. We can now appreciate the ubiquitous role that eigenvector decomposition plays in our field. We will meet this decomposition in all its glory in Chapter 5, where we describe the singular value decomposition, a very general eigenvector decomposition.

1.6 Fourier Analysis

1.6.1 Introduction

Fourier analysis plays such a central role in all that we deal with, that it is incumbent upon us to review this topic at some length. Readers familiar with this subject are encouraged to skip this section. We will first return to the topic of orthogonal expansions and present some continuous thoughts. We will also deal with Fourier series and with the Fourier transform. We begin with a continuous formulation and treat the discrete case in the next section.

1.6.2 Orthogonal Functions

Fourier series is all about the use of orthogonal functions so, without further ado, we proceed to the expansion of a function (in general a time dependent signal) in terms of a superposition of orthogonal functions.

A set of functions $\phi_j(t)$, $j = 1, 2, 3, \ldots$ is said to be orthogonal in the interval $[t_1, t_2]$ if the following condition is satisfied:

$$\int_{t_1}^{t_2} \phi_i(t)\phi_j(t)dt = k_i\delta_{ij} \tag{1.42}$$

where δ_{ij} is the Kronecker operator

$$\begin{aligned} \delta_{ij} &= 0 \quad \text{if} \quad i \neq j \\ \delta_{ij} &= 1 \quad \text{if} \quad i = j \end{aligned}$$

In signal processing, we usually want to represent a signal as a superposition of simple functions (sines, cosines, boxcar functions). The convenience of this procedure will become clear in due course. In general, one can say that the representation should be in terms of functions possessing either some attractive mathematical properties or some physical meaning.

Let us assume that a function $f(t)$ is to be approximated by a superposition of N orthogonal functions

$$f(t) \approx \sum_i c_i\phi_i(t) \quad i = 1, 2, \ldots, N$$

The N coefficients c_i can be obtained by minimizing the mean square error, $\mathcal{E}$, defined as

$$\mathcal{E} = \frac{1}{t_2 - t_1}\int_{t_1}^{t_2} (f(t) - \sum_i \phi_i(t))^2 dt$$

an equation that can be expanded to obtain

$$\mathcal{E} = \frac{1}{t_2 - t_1} \int_{t_1}^{t_2} (f(t)^2 + \sum_i c_i^2 \phi_i(t)^2 - 2 \sum_i c_i \phi_i(t) f(t)) dt$$

The parsimonious beauty of orthogonality is now evident. There are no cross-product terms of the form $\phi_i(t)\phi_j(t)$ since, according to Eq. (1.42), they cancel. The last equation can be written as

$$\mathcal{E} = \frac{1}{t_2 - t_1} \int_{t_1}^{t_2} f(t)^2 dt + \sum_i c_i^2 k_i - 2 \sum_i c_i \gamma_i \tag{1.43}$$

where

$$\gamma_i = \int_{t_1}^{t_2} \phi_i(t) f(t) dt$$

The term outside the integral in Eq. (1.43) can be rewritten as

$$\sum_{i=1}^{N} (c_i^2 k_i - 2 c_i \gamma_i) = \sum_i (c_i \sqrt{k_i} - \frac{\gamma_i}{\sqrt{k_i}})^2 - \sum_i \frac{\gamma_i^2}{k_i}$$

which allows us to write $\mathcal{E}$ as

$$\mathcal{E} = \frac{1}{t_2 - t_1} \int_{t_1}^{t_2} f(t)^2 dt + \sum_i (c_i \sqrt{k_i} - \frac{\gamma_i}{\sqrt{k_i}})^2 - \sum_i \frac{\gamma_i^2}{k_i}$$

It is clear that $\mathcal{E}$ is minimum when the second term on the RHS in the above equation is zero, hence

$$c_i \sqrt{k_i} = \frac{\gamma_i}{\sqrt{k_i}}$$

or, in other words, the coefficient c_i is given by

$$c_i = \frac{\gamma_i}{k_i} = \frac{\int_{t_1}^{t_2} f(t) \phi_i(t) dt}{\int_{t_1}^{t_2} \phi(t)^2 dt}$$

We have managed to obtain the expression for the N coefficients of the expansion of $f(t)$. If the c_i are chosen according to the last equation, $\mathcal{E}$ becomes

$$\mathcal{E} = \frac{1}{t_2 - t_1} \int_{t_1}^{t_2} f(t)^2 dt - \sum_i c_i^2 k_i$$

It can be shown that as $N \to \infty$ the mean square error vanishes ($\mathcal{E} \to 0$), in which case, the last expression is known as "Parseval's Theorem"

$$\int_{t_1}^{t_2} f(t)^2 dt = \sum_{i=1}^{\infty} c_i^2 k_i$$

1.6.3 Fourier Series

Consider the set of functions given by

$$e^{in\omega_0 t}, \qquad n = 0, \pm 1, \pm 2, \ldots, \pm\infty$$

a set that may be easily shown to be orthogonal on $t \in [t_0, t_0 + \frac{2\pi}{\omega_0}]$. Expanding a signal in terms of this set, we obtain a Fourier series

$$f(t) = \sum_n F_n e^{in\omega_0 t}$$

where the coefficients of the expansion are given by [1]

$$F_n = \frac{2\pi}{\omega_0} \int_{t_0}^{t_0+2\pi/\omega_0} f(t) e^{-in\omega_0 t} dt$$

F_n is the complex spectrum of Fourier coefficients. The periodic signal $f(t)$ has been decomposed into a superposition of complex sinusoids of frequency $n\omega_0$ and amplitude F_n. It is important to remember that a continuous and periodic signal has a discrete spectrum of frequencies given by

$$\omega_n = n\omega_0$$

To analyze non-periodic signals we need to introduce the Fourier transform, FT for short, in which case, the signal is represented in terms of a continuous spectrum of frequencies.

1.6.4 The Fourier Transform

So far we have found an expression that allows us to represent a periodic signal of period T in terms of a superposition of basis functions (complex exponentials). We have seen that the Fourier series can be used to represent periodic or non-periodic signals. We have to realize, however, that the Fourier series does not properly represent a non-periodic signal outside the interval of $[t_0, t_0 + T]$. In fact, outside $[t_0, t_0 + T]$, the Fourier series provides a periodic extension of $f(t)$. We have also proved that a

[1] We have already obtained this result for an arbitrary set $\phi_i(t)$.

periodic signal has a discrete spectrum given by the coefficients of the expansion in terms of the Fourier series, which we have called F_n, $n = 0, \pm 1, \pm 2, \ldots, \pm\infty$.

In this section we provide a representation for a non-periodic signal $f(t)$ in $t \in [-\infty, +\infty]$ by means of a continuous spectrum of frequencies. Composing a periodic signal with period $[-T/2, T/2]$ in terms of a Fourier series, we have

$$f(t) = \sum_n F_n e^{in\omega_0 t}, \qquad \omega_0 = \frac{2\pi}{T} \tag{1.44}$$

with coefficients given by

$$F_n = \frac{1}{T}\int_{-T/2}^{T/2} f(t) e^{-in\omega_0 t} dt \tag{1.45}$$

Substituting Eq. (1.45) into Eq. (1.44) obtains

$$f(t) = \sum_n \frac{1}{T}\int_{-T/2}^{T/2} f(t) e^{-in\omega_0 t} dt e^{in\omega_0 t}$$

Now, letting $T \to \infty$ (thus extending the period to infinity), $\omega_0 = 2\pi/T \to d\omega$, where $d\omega$ is a differential frequency. In this case, the discrete variable $n\omega_0$ transforms into a continuous one, ω, and consequently, the summation becomes an integral

$$f(t) = \int \frac{1}{2\pi}\left(\int f(t) e^{-i\omega t} dt\right) e^{i\omega t} d\omega \tag{1.46}$$

The integral in brackets is called the Fourier transform of $f(t)$

$$F(\omega) = \int f(t) e^{-i\omega t} dt \tag{1.47}$$

Substituting Eq. (1.47) into Eq. (1.46) gives the equation that represents the signal in terms of $F(\omega)$

$$f(t) = \frac{1}{2\pi}\int F(\omega) e^{i\omega t} d\omega \tag{1.48}$$

The pair of equations, Eqs. (1.47) and (1.48), are used to compute the FT and it's inverse, respectively. Eq. (1.48) is also referred to as the inverse FT[2].

It is important to stress that the signal in $[-\infty, +\infty]$ has now a continuous spectrum of frequencies. The FT is in general a complex function that can be written as follows

$$F(\omega) = |F(\omega)| e^{i\Phi(\omega)}$$

where $|F(\omega)|$ is the amplitude spectrum and $\Phi(\omega)$ is the phase spectrum. We will return to the important issues of amplitude and phase when dealing with the processing of seismic signals.

---

[2]In fact, one can think of Eq. (1.47) as a forward transform, or a transform *to go* to a new domain (the frequency domain), whereas Eq. (1.48) is an inverse transform, or a transform *to come back* to the original domain (time).

1.6.5 Properties of the Fourier Transform

We are not going to prove the properties outlined below, most can be easily proved using the definition of the FT. We use the following notation to indicate that $F(\omega)$ is the FT of $f(t)$

$$f(t) \Leftrightarrow F(\omega)$$

Symmetry

$$F(t) \Leftrightarrow 2\pi f(-\omega)$$

Linearity
If

$$f_1(t) \Leftrightarrow F_1(\omega)$$
$$f_2(t) \Leftrightarrow F_2(\omega)$$

then

$$f_1(t) + f_2(t) \Leftrightarrow F_1(\omega) + F_2(\omega)$$

Scale

$$f(at) \Leftrightarrow \frac{1}{|a|} F(\frac{\omega}{a})$$

Convolution
If

$$f_1(t) \Leftrightarrow F_1(\omega)$$
$$f_2(t) \Leftrightarrow F_2(\omega)$$

then

$$\int f_1(u) f_2(t-u) du \Leftrightarrow F_1(\omega) F_2(\omega)$$

or in words: *time convolution* $\Leftrightarrow$ *frequency multiplication*[3].

Convolution in frequency
In a similar vein, but considering signals multiplied in time, we have

$$f_1(t) f_2(t) \Leftrightarrow \frac{1}{2\pi} \int F_1(v) F_2(\omega - v) dv$$

or, in words: *time multiplication* $\Leftrightarrow$ *frequency convolution*[4].

[3]This is a very important property that we use extensively. Many physical systems can be described by, or at last approximated by, linear, time invariant systems and, consequently, obey a convolution relationship.

[4]This property is extensively used in the computation of the FT of a signal recorded in a finite temporal window (considered in Section 1.6.7).

Time delay

$$f(t-\tau) \Leftrightarrow F(\omega)e^{-i\omega\tau}$$

Modulation
This property makes your AM radio work

$$f(t)e^{i\omega_0 t} \Leftrightarrow F(\omega-\omega_0)$$

Time derivative

$$\frac{df(t)}{dt} \Leftrightarrow i\omega F(\omega)$$

It is clear that, taking the derivative of $f(t)$ is equivalent to amplifying the high frequencies. This property can be extended to higher derivatives

$$\frac{d^n f(t)}{dt^n} \Leftrightarrow (i\omega)^n F(\omega)$$

1.6.6 The FT of Some Functions

Some functions are so often used in our work, that it is useful to present the respective transforms here.

A Boxcar
The boxcar, or gate, function is defined as

$$f(t) = \begin{cases} 1 & |t| < T/2 \\ 0 & \text{otherwise} \end{cases}$$

Substituting $f(t)$ into the definition of the FT (Eq. (1.47)) obtains the integral

$$\begin{aligned} F(\omega) &= \int_{-T/2}^{T/2} e^{-j\omega t} dt \\ &= \frac{1}{i\omega}(e^{i\omega T/2} - e^{-i\omega T/2}) \\ &= T\text{sinc}(\frac{\omega T}{2}) \end{aligned}$$

where, in the last equation, $\text{sinc}(x) = \sin(x)/x$. The importance of the sinc function will become evident when we deal with the spectrum of signals that have been truncated in time. Figs. 1.4 and 1.5 display the FTs of two boxcar functions of width $T = 10$ and 20 seconds respectively.

Delta function

$$f(t) = \delta(t)$$

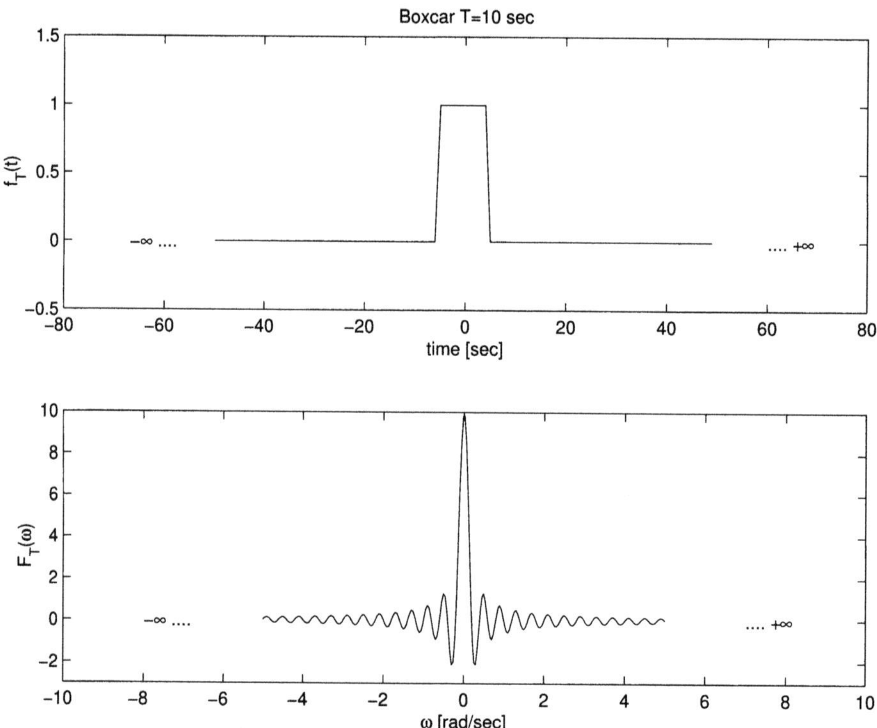

Figure 1.4: The FT of a boxcar function of $T = 10\,\text{sec}$.

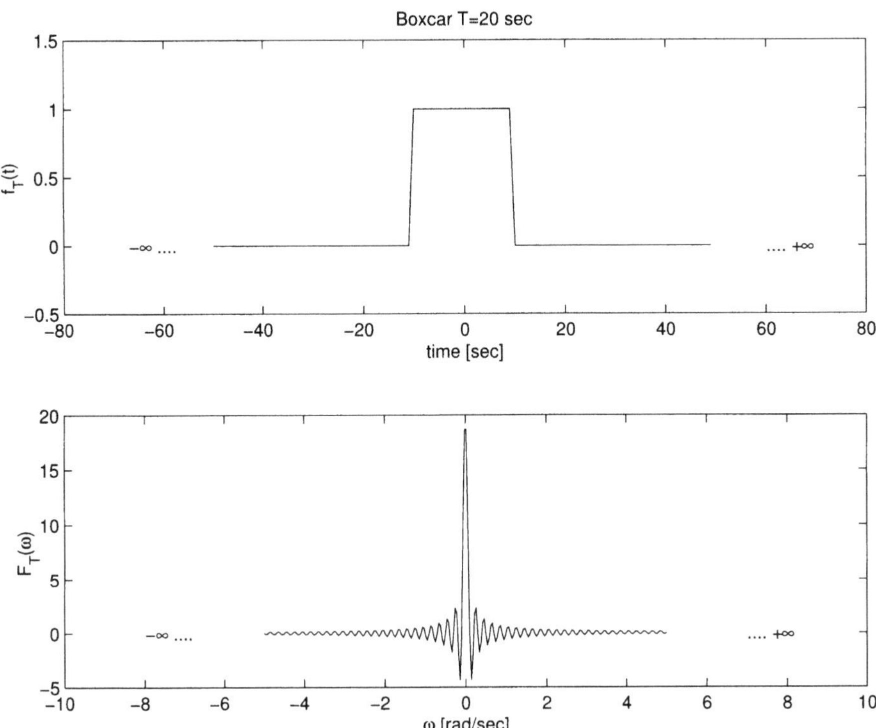

Figure 1.5: The FT of a boxcar function of $T = 20\,\text{sec}$.

The δ function is defined according to

$$\int g(u)\delta(u)du = g(0)$$

The FT of the delta function is

$$F(\omega) = \int \delta(t)e^{-i\omega t}dt = 1$$

and, consequently

$$\delta(t) \Leftrightarrow 1$$

We now apply the "time delay" property

$$\delta(t-\tau) \Leftrightarrow e^{-i\omega\tau}$$

It is clear that the δ function has a constant amplitude spectrum at all frequencies. It is also the seismic wavelet that one would, ideally, like to have in seismic exploration.

A complex harmonic

We can combine the FT of the delta function with the symmetry property to obtain the FT of a complex harmonic. Since, as we have shown above

$$\delta(t-\tau) \Leftrightarrow e^{-i\omega\tau}$$

applying the symmetry property

$$F(t) \Leftrightarrow 2\pi f(-\omega)$$

obtains

$$e^{i\omega_0 t} \Leftrightarrow 2\pi\delta(\omega-\omega_0)$$

In other words, the FT of a complex harmonic of frequency ω_0 is a delta at the corresponding frequency $\omega = \omega_0$.

1.6.7 Truncation in Time

The question we consider here is, given $f(t)$, $t \in [-\infty, +\infty]$, with $f(t) \Leftrightarrow F(\omega)$, how do we obtain the FT of the signal when it is recorded in a finite interval $[-T/2, +T/2]$? Letting $f_T(t)$ be the observed signal in $[-T/2, +T/2]$ and $f(t)$ the original signal in $[-\infty, +\infty]$, it is easy to see that

$$f_T(t) = f(t)b_T(t)$$

where $b_T(t)$ is a boxcar function considered in Section 1.6.6. Using the frequency convolution theorem (Section 1.6.5) we can write

$$F_T(\omega) = F(\omega) * B_T(\omega) = \frac{1}{2\pi}\int F(v)B_T(\omega - v)dv$$

where $B_T(\omega) = T\text{sinc}(\omega T/2)$. This is a very interesting result. Our observation window is (not surprisingly) affecting the FT of the signal. What is interesting is the manner in which the effect occurs. We would like to compute $F(\omega)$ but, since the signal has been recorded in a finite time interval, we have only access to $F_T(\omega)$, a distorted version of $F(\omega)$

$$F_T(\omega) = \frac{T}{2\pi}\int F(u)\text{sinc}((\omega - u)T/2)du$$

It is clear from the above that the best we can do is to obtain $F(\omega)$ convolved with a sinc function. If $f(t) = e^{i\omega_0 t}$, it is easy to see that the truncated version of the complex harmonic has the following FT

$$\begin{aligned} F_T(\omega) &= \frac{T}{2\pi}\int 2\pi\delta(\omega - \omega_0)\text{sinc}((\omega - u)T/2)du \\ &= T\text{sinc}((\omega - \omega_0)T/2) \end{aligned}$$

This is a sinc function with a peak at $\omega = \omega_0$.

Fig. 1.6 illustrates the superposition of 2 complex harmonics of the form

$$f(t) = e^{i\omega_1 t} + e^{i\omega_2 t} \qquad t \in [-10, 10] \text{ sec.}$$

The FT of such a signal (if measured in an infinite interval) is given by two delta functions at frequencies ω_1 and ω_2. Since we are observing the signal in a finite interval, however, the *ideal* FT of $f(t)$ is convolved with the FT of the boxcar function. In this example $\omega_1 = 0.5\,\text{rad/sec}$ and $\omega_2 = 1.0\,\text{rad/sec}$.

1.6.8 Symmetries

Before continuing with the FT and its applications, a few words about the symmetries of the FT are needed. We begin by restating the definition of the FT, Eq. (1.47), for convenience

$$F(\omega) = \int f(t)e^{-i\omega t}dt$$

If $f(t)$ is real, we can write

$$F(\omega) = G(\omega) + iB(\omega)$$

where

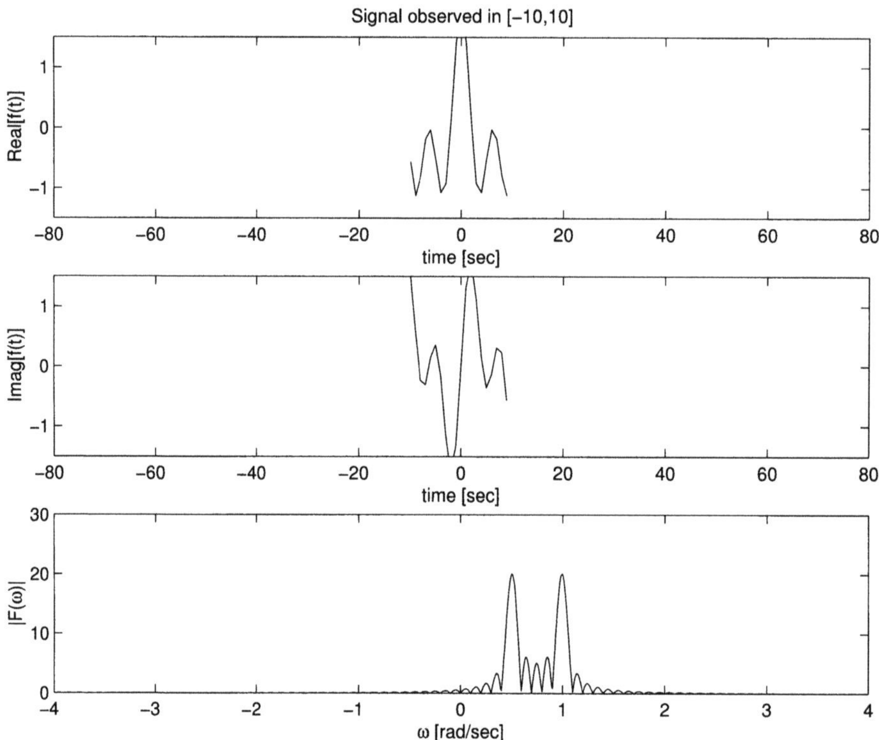

Figure 1.6: The FT of the superposition of two complex harmonics observed in a window of length $T = 20$ sec. Top panel: real part of the signal. Center panel: Imaginary part of the signal. Bottom panel: Amplitude spectrum, $|F(\omega)|$.

$$G(\omega) = \int f(t)\cos(\omega t)dt$$

and

$$B(\omega) = -\int f(t)\sin(\omega t)dt$$

Since cos is an even function and sin an odd function

$$G(\omega) = G(-\omega) \tag{1.49}$$

and

$$B(\omega) = -B(-\omega) \tag{1.50}$$

If $F(\omega)$ is known for $\omega \geq 0$, $F(\omega)$ may be computed for $\omega < 0$ by applying the above identities. In fact, one can always write

$$F(-\omega) = G(-\omega) + iB(-\omega)$$

and by combining the last equation with Eqs. (1.49) and (1.50) we obtain

$$F(-\omega) = G(\omega) - iB(\omega)$$

This equation can be used to compute the negative semi-axis of the FT. This property is often referred to as the Hermitian symmetry of the FT. One can also write

$$F(-\omega) = F(\omega)^*$$

where the $*$ is used to denote complex conjugate. This property, valid only for real time series, explains why one semi-axis only (in general the positive one) is plotted when displaying the Fourier spectrum of a real signal.

The symmetry properties of the real and imaginary parts of the FT can also be used to obtain the symmetry properties of the amplitude and phase spectra of the Fourier transform. Since

$$F(\omega) = |F(\omega)|e^{i\Phi(\omega)} = A(\omega)e^{i\Phi(\omega)} \tag{1.51}$$

it is easily proved that the amplitude spectrum, $A(\omega)$, is an even function

$$A(\omega) = A(-\omega)$$

and that the phase spectrum, $\Phi(\omega)$, is an odd function

$$\Phi(\omega) = -\Phi(-\omega)$$

1.6.9 Living in a Discrete World

So far we have described the FT of continuous (or analog) signals. These signals are, however, always recorded as a series of discrete values and it is vital to understand the connection between these discrete signals and their continuous counterparts.

We represent the analog signal by $f(t)$ and the discretized signal by $f_s(t)$. One obtains $f_s(t)$ by sampling $f(t)$ every Δt seconds. Thus, if $k = 0, \pm 1, \pm 2, \ldots \pm \infty$

$$f_s(t) = f(t) \sum_k \delta(t - k\Delta t) . \qquad (1.52)$$

Using the frequency convolution property, the FT of the sampled signal is obtained as

$$F_s(\omega) = \frac{1}{2\pi} F(\omega) * \omega_0 \sum_k \delta(\omega - k\omega_0) , \qquad \omega_0 = \frac{2\pi}{\Delta t}$$

and, following a few algebraic steps, the desired result is[5]

$$F_s(\omega) = \frac{1}{\Delta t} \sum_k F(\omega - k\omega_0)$$

One can observe that the FT of the sampled signal is a periodic function with period ω_0.

If one wants to compute $F_s(\omega)$ in such a way that $F(\omega)$ can be completely recovered, the signal $f(t)$ must be band-limited. That is, a signal where the spectral components outside the interval $[-\omega_{max}, +\omega_{max}]$ are zero. If the condition

$$\omega_0 \geq 2\omega_{max} \qquad (1.53)$$

is satisfied, there is no overlap of spectral contributions, and therefore $F_s(\omega)$, $\omega \in [-w_{max}, +w_{max}]$ is equivalent, within a scale factor $1/\Delta t$, to the FT of the analog signal $F(\omega)$. The last condition can be re-written as

$$\frac{2\pi}{\Delta t} \geq 2 \times 2\pi f_{max}$$

which reduces to

$$\Delta t \leq \frac{1}{2 f_{max}} \qquad (1.54)$$

where $f_{max} = \omega_{max}/2\pi$. Eq. (1.54) represents the Sampling or Nyquist theorem and specifies the maximum allowable sampling interval for recovery of the FT of the original signal. If this condition is violated, a phenomenon called aliasing results.

---

[5]the reader should derive this result. It is an excellent exercise in manipulating FT's.

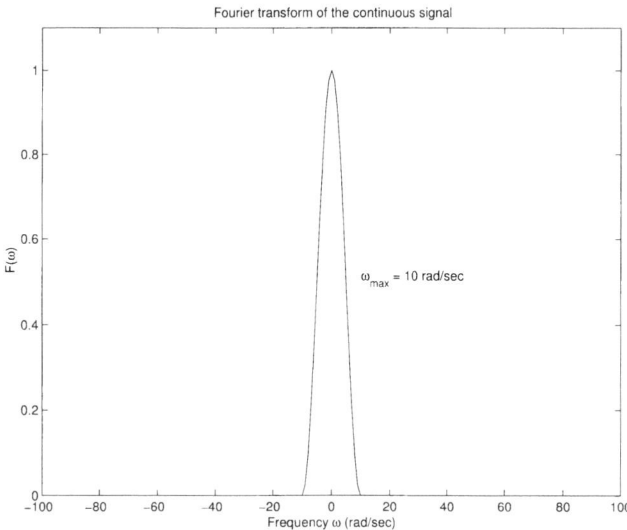

Figure 1.7: The $FT_{\Re}$ of a continuous signal.

1.6.10 Aliasing and the Poisson Sum Formula

The Shannon sampling theorem, symbolized by Eq. (1.54) (Shannon, 1948), together with its various extensions and applications, reviewed in Jerri (1977), has been of fundamental importance both in information and communications theory. A related theorem, intimately connected to the phenomenon of aliasing, and expressed by the Poisson sum formula, (Papoulis, 1977), is also of considerable interest.

Aliasing is so well covered in many excellent texts, e.g., Bracewell (1986), that we do not intend here to go into any detail. We do, however, wish to explore the Poisson sum formula in its relationship to aliasing. First, we present a pictorial view of aliasing, illustrated in Figs. 1.7-1.10. For purposes of illustration, Fig. 1.7 shows the real part of the FT, designated as $FT_{\Re}$, of some particular continuous signal. We note that, to properly recover the FT of this signal, we require a sampling interval specified in accord with Eq. (1.53). The result of such sampling is illustrated in Figs. 1.8 and 1.9. These two figures demonstrate that $FT_{\Re}$ (and, in fact, the FT) of the original (continuous) signal is well represented by that of the discretized signal in the interval $[-\omega_{max}, +\omega_{max}]$. Fig. 1.10 shows an example where the analog data have been under-sampled and, consequently, the FT of the continuous signal can no longer be recovered from that of the discretized signal.

As we have discussed and illustrated above, aliasing is caused by the sampling of a continuous function of time, $x(t)$, at intervals of time, $\Delta t > \frac{1}{2f_{max}}$. It is the result of the addition to $X(f)$ (the Fourier transform of $x(t)$), from zero to the Nyquist frequency, of all the contributions of $X(f)$ which are replicated at intervals of $1/\Delta t$ in the frequency domain (Bracewell, 1986). The fact that the fully aliased spectrum is equal to the Discrete Fourier transform, DFT, of the sampled data is expressed by the Poisson sum formula which has been used in the past to derive a variety of

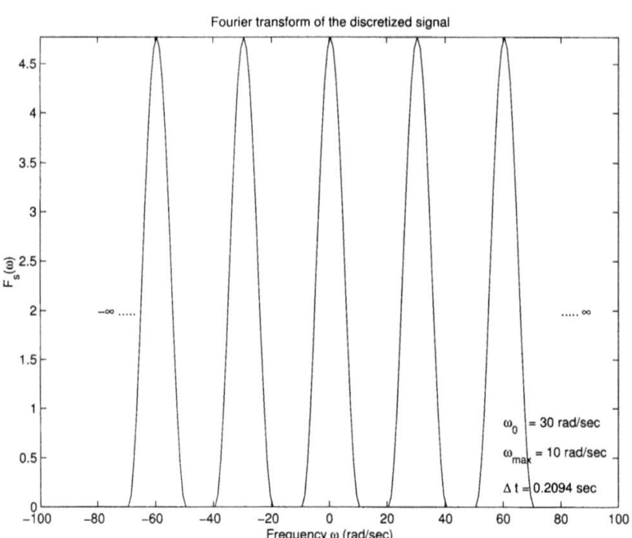

Figure 1.8: The FT$_\Re$ of the continuous signal after discretization. $\omega_{max} = 10$ and $\omega_0 = 30$

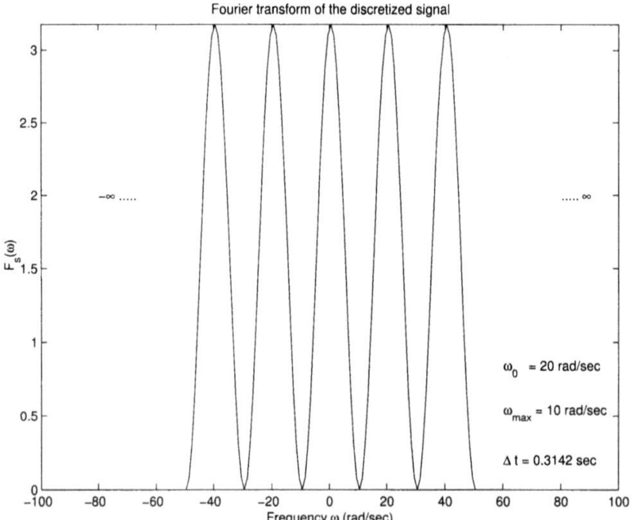

Figure 1.9: The FT$_\Re$ of the continuous signal after discretization. $\omega_{max} = 10$ and $\omega_0 = 20$. The —FT— of the continuous signal is perfectly represented in the interval $[-\omega_{max}, \omega_{max}]$.

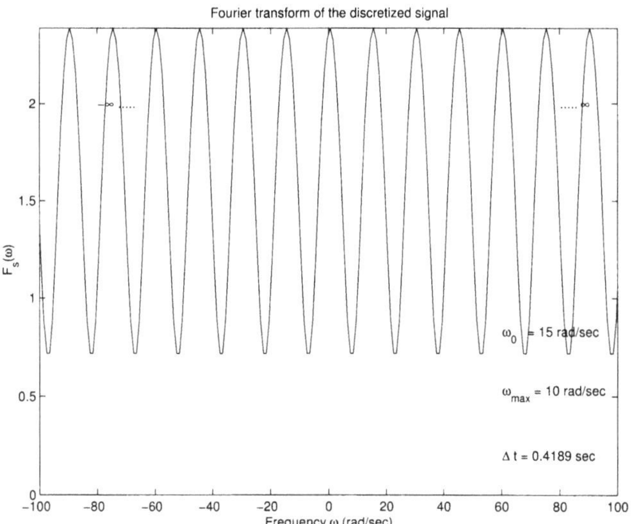

Figure 1.10: The $FT_{\Re}$ of the continuous signal after discretization. $\omega_{max} = 10$ and $\omega_0 = 15$, the Nyquist theorem is not satisfied and the signal is aliased. The FT of the continuous signal cannot be recovered from the FT of the sampled signal.

interesting results (Papoulis, 1984). This formula may be interpreted as stating that, what we need a very long time to compute as an infinite sum, is computed in $Nlog_2N$ operations using the Fast Fourier Transform, FFT, algorithm. We will present here, for amusement if nothing else, an application of this formula which we believe to be both interesting and unusual (Ulrych and Fokkema, 1995).

1.6.10.1 Some Theoretical Details

The Poisson sum formula is generally written as, (Papoulis, 1977)

$$\sum_k X(f + \frac{k}{\Delta t}) = \Delta t \sum_n x(n\Delta t) e^{-i2\pi f n \Delta t} \tag{1.55}$$

where $x(n\Delta t)$ is $x(t)$ sampled at intervals Δt and where the summation indices, $k, n = 0, \pm 1, \pm 2, \ldots, \pm\infty$. Since we will be concerned with functions of time which are causal and may be discontinuous at the origin, Eq. (1.55) must be slightly modified. Specifically, since a causal time function may be denoted by $x(t) = x(t)H(t)$, where $H(t)$ is the Heaviside step function and is defined to be equal to 1/2 at $t = 0$, Eq. (1.55) becomes

$$\sum_k X(f + \frac{k}{\Delta t}) = \Delta t \sum_n x(n\Delta t) e^{-i2\pi f n \Delta t} - \frac{1}{2}\Delta t\, x(0) \tag{1.56}$$

We emphasize at this point that the interpretation of Eq. (1.56) is that the infinite sum which arises from the superposition of the displaced transforms is equal to the

DFT of x(t) which we will represent by X_f and which, naturally, is evaluated at discrete frequencies.

1.6.10.2 Limits of Infinite Series

Ulrych and Fokkema (1995) describe an application of the Poisson sum formula, that of the evaluation of the limits of certain infinite series. We consider here only two particular examples concerned with power series.

Consider a causal function impulsive at the origin, $x(t) = e^{-\alpha t}$, with transform $X(f) = 1/(\alpha + i2\pi f)$. Sampling $x(t)$ at intervals Δt and performing the DFT we obtain

$$X_f = \Delta t \sum_n e^{-\alpha n \Delta t} e^{-i2\pi f n \Delta t} - \frac{\Delta t}{2} \tag{1.57}$$

If we let $r = e^{-(\alpha+i2\pi f)\Delta t}$ and since $\sum_{n=0}^{\infty} r^n = 1/(1-r)$ we may write Eq. (1.57) as

$$X_f = \frac{\Delta t}{1-r} - \frac{\Delta t}{2} = \frac{\Delta t}{2}\left(\frac{1+r}{1-r}\right)$$

Equating the aliased sum to X_f given by the above expression we obtain

$$\sum_k \frac{1}{(\alpha + i2\pi(f + \frac{k}{\Delta t}))} = \frac{\Delta t}{2}\left(\frac{1+r}{1-r}\right) \tag{1.58}$$

Let us, for convenience, assume $\Delta t = 2\pi$. Letting $\beta = 2\pi f$ we obtain from Eq. (1.58) for $|\beta| \leq 1/2$

$$\sum_k \frac{\alpha - i(\beta + k)}{\alpha^2 + (\beta + k)^2} = \pi \left(\frac{1 + e^{-\alpha 2\pi} e^{-i2\pi\beta}}{1 - e^{-\alpha 2\pi} e^{-i2\pi\beta}}\right) \tag{1.59}$$

Splitting Eq. (1.59) into real and imaginary parts we derive the two following expressions

$$\sum_k \frac{\alpha}{\alpha^2 + (\beta + k)^2} = \pi \left(\frac{1 - e^{-\alpha 4\pi}}{1 + e^{-\alpha 4\pi} - 2e^{-\alpha 2\pi}\cos 2\pi\beta}\right) \tag{1.60}$$

and

$$\sum_k \frac{\beta + k}{\alpha^2 + (\beta + k)^2} = \pi \left(\frac{2e^{-\alpha 2\pi}\sin 2\pi\beta}{1 + e^{-\alpha 4\pi} - 2e^{-\alpha 2\pi}\cos 2\pi\beta}\right) \tag{1.61}$$

As two particular examples, we derive the following limits:

setting $\alpha = 1/(2\pi)$ and $\beta = 0$ in Eq. (1.60)

$$\frac{2\sum_k \frac{1}{1+(2\pi k)^2}+1}{2\sum_k \frac{1}{1+(2\pi k)^2}-1} = e$$

and, setting $\alpha = 0$ and $\beta = 1/4$ in Eq. (1.61)

$$\sum_k \frac{1}{\frac{1}{4}+k} = \pi \tag{1.62}$$

a charming result.

Evaluating the series in Eq. (1.62) for $k = -\infty$ to $+\infty$ we obtain

$$\pi = 4\left(1 - \frac{1}{3} + \frac{1}{5} - \frac{1}{7} + \cdots\right)$$

which is the Leibnitz formula (Thomas and Finney, 1979).

Setting $\alpha = 0$ in Eq. (1.61) is justified in Ulrych and Fokkema (1995). Thus, the famous Leibnitz formula follows from the phenomenon of aliasing as expressed by the Poisson sum formula.

1.6.10.3 Remarks

The relationship of aliasing to the DFT as expressed by the Poisson sum formula is very rich in possibilities, one of which we have described here. As pointed out by Ulrych and Fokkema (1995), the DFT can play an important role in the evaluation of infinite sums. Thus, assuming that we wish to compute the value of $\sum_k Q(u+k)$ to some given accuracy, where $Q(u)$ may be expressed as the Fourier transform of some continuous function of time, $q(t)$, then the limit may be evaluated in $N \log_2 N$ operations using the FFT algorithm to an accuracy allowed for by the particular computer. For example, suppose we wish to determine $\sum_k \tan^{-1} 1/(2\pi(\beta+k))^2$ for $|\beta| \leq 1/2$ to an accuracy of one part in 10^7. Since $\tan^{-1} 1/(2\pi f)^2$ is the Fourier transform of $x(t) = e^{-|t|} \sin t/t$ the infinite sum may be evaluated by discretizing $x(t)$ at unit time intervals and evaluating the FFT at the particular value of β. Since an accuracy of one part in 10^7 requires the summation index k to run between -10^6 and $+10^6$ the ratio of the times required for the computation using the sum itself and the FFT is 10^3. Of course another advantage of the FFT approach, where it is in fact an option, is that the FFT at the same time evaluates the sum at all other values of β depending only on the number of points used in the FFT.

1.7 The z Transform

All real data, with which we are concerned here, come in discrete form. The discrete form of Eqs. (1.47) and (1.48), particularly when evaluated using the FFT, play a central role in the analysis of such data and many excellent texts serve as reference, e.g., Bracewell (1986) . A transform which is extremely useful in the manipulation of discrete data and which allows many derivations to be arrived at very simply, is the z transform. Here we give a simple description of its properties and relationship to the FT.

Consider a continuous time series, $x(t)$, which is sampled periodically every Δt seconds by an ideal sampler producing a series of impulses. The strength of each impulse is equal to the value of the input at the time of sampling. The sampled time series, x_t, may be expressed analytically in terms of a series of delayed impulses as

$$\begin{aligned} x_t &= x(0)\delta(t) + x(\Delta t)\delta(t - \Delta t) + x(2\Delta t)\delta(t - 2\Delta t) + \cdots \\ &= \sum_n x(n\Delta t)\delta(t - n\Delta t), \quad n = 0, 1, 2, \ldots, \infty \end{aligned} \tag{1.63}$$

Taking Laplace transforms and using the shift theorem, we obtain

$$\mathcal{L}[x_t] = X(s) = \sum_n x(n\Delta t)e^{-n\Delta ts}$$

The z transform, $X(z)$, is now defined as $X(s)$ with $e^{-\Delta ts}$ replaced by a new variable z. By definition then

$$X(z) = \mathcal{Z}[x_t] = X(s)|_{s=-(1/\Delta t)\log z} = \sum_n x(n\Delta t)z^n \tag{1.64}$$

A word of caution here. Electrical engineers, people in speech and image processing and others, replace $e^{-\Delta ts}$ by $-z$ to obtain

$$X(z) = \sum_n x(n\Delta t)z^{-n} \tag{1.65}$$

What this does, is to rearrange the location of the zeros of $X(z)$ with respect to $|z| = 1$. When you see expressions like 'requires all zeroes to lie *outside* $|z| = 1$' in this book, translate to read 'requires all zeroes to lie *inside* $|z| = 1$' in literature with definition given by Eq. (1.65).

1.7.1 Relationship Between z and Fourier Transforms

The z transform provides a convenient framework for establishing relationships for discrete data between the recorded and transform domains. The connection which exists between the z and the Fourier transforms makes the numerical evaluation of these relationships simple and economical. We establish this connection here.

We have

$$z = e^{-\Delta ts} \tag{1.66}$$

with s in general being the complex variable

$$s = \sigma + i\omega$$

where σ and ω carry their own algebraic sign. We have then

$$z = e^{-(\sigma+i\omega)\Delta t} = e^{-\sigma\Delta t}e^{-i\omega\Delta t}$$

To simplify the discussion, we take $\Delta t = 1$ from now on without loss of generality, and we evaluate for the magnitude and phase angle of z

$$|z| = e^{-\sigma} \quad \text{and} \quad \angle z = -\omega \tag{1.67}$$

From Eq. (1.67) we note the following

$$\text{for } \sigma > 0,\ |z| > 1$$

$$\text{for } \sigma = 0,\ |z| = 1$$

$$\text{for } \sigma < 0,\ |z| < 1$$

Therefore, Eq. (1.66) transforms the entire left-half of the s plane into the space outside of the unit circle in the z plane and the entire right-half plane is mapped into the interior of the unit circle. The imaginary axis transforms into the circumference of $|z| = 1$. The mapping is illustrated in Fig. 1.11.

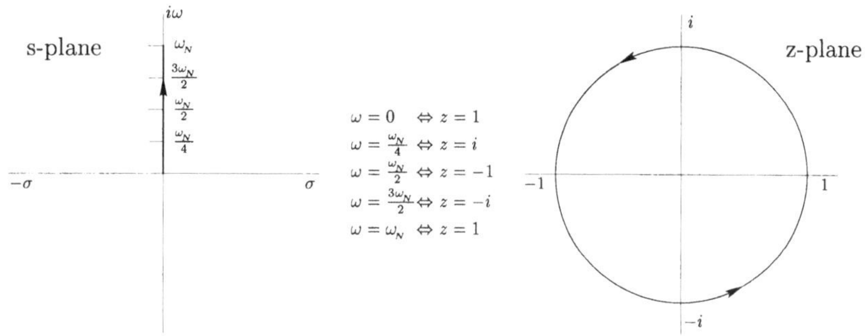

Figure 1.11: Relationship between the Laplace, Fourier and z transforms.

Examining the z transform, ZT, of a signal $x_t = [x_0, x_1, \ldots, x_N]$ which is given by

$$X(z) = x_0 + x_1 z + x_2 z^2 + \ldots + x_N z^N \tag{1.68}$$

we see that we may obtain the FT from the ZT by substituting $z = e^{-i\omega}$ in the polynomial expression. Further, Eq. (1.68) makes it clear why the variable z plays the role of a unit delay operator. Note that the summation in Eq. (1.64) is from 0 to ∞. This is the one sided z transform and is particularly applicable to *causal* signals which are of concern in the analysis of physically realizable systems. Causal systems have important and special properties which we will describe briefly in Section 1.7.7. It goes without saying, that the relationship which exists between the FT and the ZT implies that the various theorems concerning the FT, described above, translate to the z domain.

1.7.2 Discrete Fourier Transform

We have explored the relationship between the z and Fourier transforms in Section 1.7.1 above. In this section we develop the transition from the ZT to the Discrete Fourier Transform, the DFT, which is used to compute the Fourier transform of discrete data. Substituting $z = e^{-i\omega}$ into the z transform, $X(z)$, defined by Eq. (1.64), maps $X(z)$ into the unit circle. We emphasize that $z = e^{-i\omega}$ is a complex variable of unit magnitude and phase given by ω, which is also the frequency variable in units of radians. The z-transform now becomes

$$X(\omega) = \sum_n x_n e^{-i\omega n}, \quad n = 0, 1, \ldots, N-1$$

Remembering that the FT is

$$X(\omega) = \int x(t) e^{-i\omega t} dt$$

(where the frequency is given in radians/sec when time is measured in seconds), we see that in the DFT, integration is replaced by summation, and the Fourier kernel is replaced by $e^{-i\omega n}$. Since n does not have units, ω has units of radians.

The DFT maps a discrete signal into the frequency domain. Thus far ω is a continuous variable, but let us discretize ω in the same manner that we have discretized the temporal variable t. The limits of ω, our angular frequency, are given by $[0, 2\pi)$, and we discretize the frequency axis as

$$\omega_k = k\, 2\pi/N\,, \quad k = 0, 1, \ldots, N-1$$

The DFT is now defined as

$$X_k = X(\omega_k) = \sum_n x_n e^{-i\omega_k n}, \quad k = 0, 1, \ldots, N-1 \tag{1.69}$$

The DFT is a transformation of a N point signal into N Fourier coefficients, X_k . In convenient matrix form

$$\begin{pmatrix} X_0 \\ X_1 \\ X_2 \\ \vdots \\ X_{N-1} \end{pmatrix} = \begin{pmatrix} 1 & 1 & 1 & \ldots & 1 \\ 1 & e^{-i2\pi/N} & e^{-i2\pi 2/N} & \ldots & e^{-i2\pi(N-1)/N} \\ 1 & e^{-i2\pi 2/N} & e^{-i2\pi 4/N} & \ldots & e^{-i2\pi 2(N-1)/N} \\ \vdots & \vdots & \vdots & \vdots & \vdots \\ 1 & e^{-i2\pi(N-1)/N} & e^{-i2\pi 2(N-1)/N} & \ldots & e^{-i2\pi(N-1)(N-1)/N} \end{pmatrix} \begin{pmatrix} x_0 \\ x_1 \\ x_2 \\ \vdots \\ x_{N-1} \end{pmatrix}$$

and in compact form

$$\mathbf{X} = \mathbf{F}\mathbf{x}$$

It is clear that the DFT can be interpreted as a matrix that maps an N-dimensional vector into another N-dimensional vector. The remaining task entails the invertibility of the DFT. We need a transform to return to the time domain, in other words we require $\mathbf{F}^{-1}$.

1.7.3 Inverse DFT

We consider the following inverse

$$x_n = \sum_l \alpha_l e^{i2\pi ln/N}, \quad l = 0, 1, \ldots, N-1$$

where the coefficients α_l must be determined. This formula is analogous to the one used to invert the FT, except that, because of the discrete nature of the problem, integration has become a summation. The parameters α_l are our unknowns and are determined as follows. First we substitute the last equation into Eq. (1.69),

$$X_k = \sum_n \sum_l \alpha_l e^{i2\pi n(l-k)/N}$$

which can be rewritten as

$$X_k = \sum_l \alpha_l \sum_n e^{i2\pi n(l-k)/N} = \sum_l \alpha_l s_{l-k} \tag{1.70}$$

where the sequence s_{l-k} is given by

$$s_{l-k} = \sum_n e^{i2\pi n(l-k)/N} \tag{1.71}$$

At this point we realize the last equation is a geometric series with a sum given by

$$\sum_n u^n = \begin{cases} N & \text{if} \quad u = 1 \\ \frac{u^N}{1-u} & \text{if} \quad u \neq 1 \end{cases}$$

We can set $u = e^{i2\pi n(l-k)/N}$ in Eq. (1.71) and, therefore

$$s_{l-k} = \begin{cases} N & \text{if} \quad l = k \\ 0 & \text{if} \quad l \neq k \end{cases}$$

Substituting this result into Eq. (1.70), we obtain

$$X_k = N\alpha_k, \quad k = 0, \ldots, N-1$$

and the inversion formula becomes

$$x_n = \frac{1}{N} \sum_l X_l e^{i2\pi ln/N}$$

or, equivalently

$$\mathbf{x} = \frac{1}{N}\mathbf{F}^H\mathbf{X}$$

where $\mathbf{F}^H$ is the Hermitian transpose of $\mathbf{F}$. It is clear that the $N \times N$ matrix, $\mathbf{F}$, is orthogonal

$$\mathbf{F}^H\mathbf{F} = N\mathbf{I}_N$$

Finally, we have the pair of transforms, the DFT and the IDFT (inverse DFT), given by

$$X_k = \sum_n x_n e^{-i2\pi kn/N}, \quad k = 0, 1, \ldots, N-1$$

and

$$x_n = \frac{1}{N}\sum_k X_k e^{i2\pi kn/N}, \quad n = 0, 1, \ldots, N-1 \tag{1.72}$$

Since the DFT is an orthogonal transformation, the inverse is computed using the Hermitian operator. This is important, since the cost of inverting an $N \times N$ matrix is proportional to N^3, whereas the cost of multiplying a matrix by a vector is proportional to N^2. We will further diminish the computational cost of multiplying a matrix by a vector by using the FFT .

1.7.4 Zero Padding

The DFT allows us to transform a N point time series into N frequency coefficients, X_k, where the index k is associated with the discrete frequency ω_k

$$\omega_k = \frac{2\pi k}{N} = \Delta\omega k, \quad k = 0, 1, \ldots, N-1$$

The frequency axis is sampled every $\Delta\omega$ radians. At this point it appears that $\Delta\omega$ is controlled by the number of samples of the time series N. The frequency interval can be decreased, however, by appropriate zero padding. We define a new time series that consists of the original time series followed by $M - N$ zeros,

$$x = [x_0, x_1, x_2, \ldots, x_{N-1}, \underbrace{0, 0, \ldots, 0}_{M-N}]$$

This new time series is M points long with a DFT given by

$$X_k = \sum_{n=0}^{N-1} x_n e^{-i2\pi nk/M} = \sum_{n=0}^{M-1} x_n e^{-i2\pi nk/M}, \quad k = 0, \ldots, M-1$$

The sampling interval of the frequency axis is now

$$\Delta\omega = \frac{2\pi}{M} < \frac{2\pi}{N}$$

In general, zero padding is used to oversample the frequency axis at the time of plotting the DFT. It is also important, as we will see, to pad with zeros at the time of performing discrete convolution using the DFT. Figs. 1.12 and 1.13 illustrate the effect of padding a time series with zeros. Fig. 1.12 shows the original time series and the associated DFT (the real and imaginary parts). Fig. 1.13 shows the result after zero padding the original time series with 20 zeros.

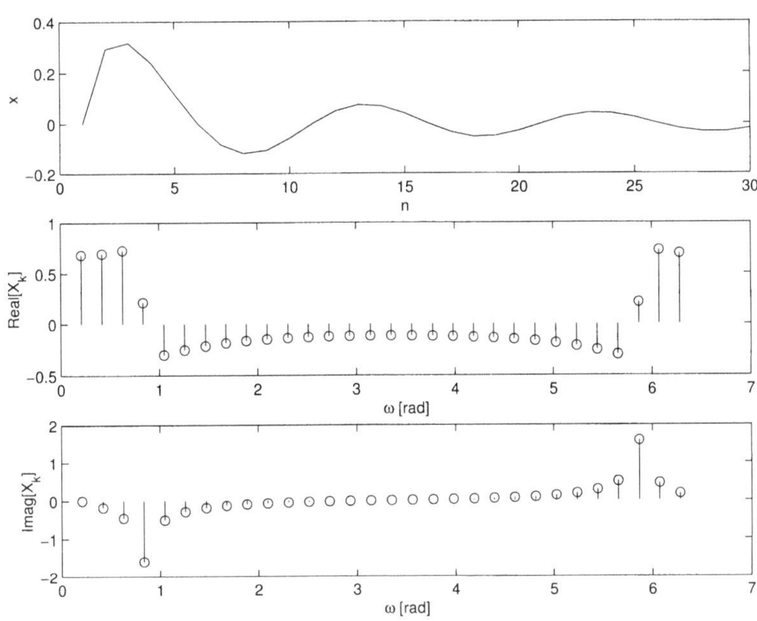

Figure 1.12: A time series and the real and imaginary parts of the DFT. Note that frequency axis is given in radians $(0, 2\pi)$

1.7.5 The Fast Fourier Transform (FFT)

The FFT is not a new transform, it is just a fast algorithm to compute DFTs. The FFT is based on a halving trick, that is, a trick to compute the DFT of a length N time series using the DFT of two sub-series of length N/2. Following Claerbout (1976), we begin with a time series of length $2N$

$$[z_0, z_1, z_2, z_3, \ldots, z_{2N-1}]$$

The DFT of z is

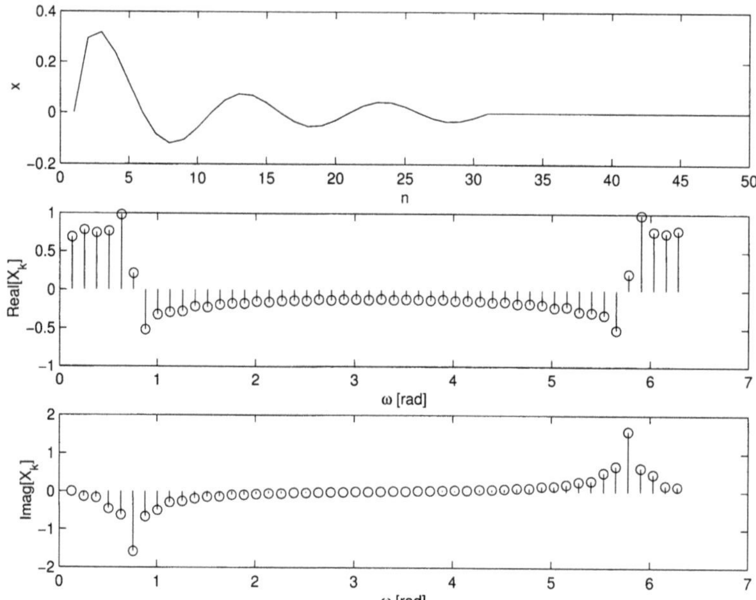

Figure 1.13: A time series and the real and imaginary parts of the DFT. In this case the time series was padded with zeros in order to decrease the frequency interval $\Delta\omega$.

$$Z_k = \sum_{n=0}^{2N-1} z_n e^{-i2\pi nk/(2N)}, \quad k = 0, 1, \ldots, 2N-1$$

We can rewrite the last equation as two equations, one composed of even samples, $x = z_0, z_2, z_4, \ldots$ and the other of odd samples, $y = z_1, z_3, z_5, \ldots$. The first equation will be valid for $k = 0, 1, \ldots N-1$, and the second for $k = N, N+1, \ldots, 2N-1$. We have

$$Z_k = \sum_{n=0}^{N-1} z_{2n} e^{-i2\pi 2nk/(2N)} + \sum_{n=0}^{N-1} z_{2n+1} e^{-i2\pi(2n+1)k/(2N)}$$

and writing the RHS in terms of the DFTs of x and y

$$Z_k = X_k + e^{-i2\pi k/(2N)} Y_k, \quad k = 0, 1, \ldots, N-1 \tag{1.73}$$

The second formula retrieves the second half of the samples of the DFT of z

$$Z_k = \sum_{n=0}^{2N-1} z_n e^{-i2\pi nk/(2N)}, \quad k = N, \ldots, 2N-1$$

and applying the substitution, $j = k - N$

$$Z_{j+N} = \sum_{n=0}^{2N-1} z_n e^{-i2\pi n(j+N)/(2N)}, \quad k = N, \ldots, 2N-1$$

Rewriting the last expression in terms of x and y, obtains

$$Z_{j+N} = X_j - e^{-i2\pi k/(2N)} Y_j, \quad j = 0, 1, \ldots, N-1 \tag{1.74}$$

We now have two expressions to compute the DFT of a series of length $2N$ as a function of two time series of length N. Good FFT algorithms repeat this trick until the final time series are of length 1. The recursions given in (1.73) and (1.74) are applied to recover the DFT of the original time series. It can be proved that the total number of operations of the FFT is proportional to $N \log_2(N)$ (for a time series of length N). This is an important saving with respect to the standard DFT which involves a number of operations proportional to N^2. A small modification to Eqs. (1.73) and (1.74) permits the computation of the IDFT.

1.7.6 Linearity and Time Invariance

The signals of interest to us are generally output signals of some system. For example, a seismogram is the response of the earth, the system in this case, to some type of energy input. We are, in principle, going to be concerned with *linear* systems and we proceed to define this very important class here. Let $y_1(t)$ be the response of a system to an input $x_1(t)$ and $y_2(t)$ be the response to $x_2(t)$. Then, a linear system follows the principle of superposition which, given the above definitions, is

$$\begin{array}{lccc} \text{If} & x_1(t) & \Rightarrow & y_1(t) \\ \text{and} & x_2(t) & \Rightarrow & y_2(t). \\ \text{Then} & x_1(t) + x_2(t) & \Rightarrow & y_1(t) + y_2(t). \end{array}$$

For *time invariant* systems, we have, for some arbitrary time shift τ

$$\begin{array}{lccc} \text{If} & x_1(t) & \Rightarrow & y_1(t) \\ \text{Then} & x_1(t-\tau) & \Rightarrow & y_1(t-\tau). \end{array}$$

Let us now see what all this implies for our system. Because of linearity, the response may be represented in terms of the most general linear functional of the input (Bracewell, 1986)

$$y(t) = \int G(t, t')x(t')dt' \tag{1.75}$$

Because of time invariance we may write Eq. (1.75) as

$$y_1(t-\tau) = \int G(t,t')x_1(t'-\tau)dt' \tag{1.76}$$

which may be written as

$$y_1(t) = \int G(t+\tau, t'+\tau)x_1(t')dt' \tag{1.77}$$

Hence, $G(t+\tau, t'+\tau) = G(t,t')$ is the condition required by time invariance. This implies that the functional $G(t,t')$ is a function only of the time *difference*, $t-t'$, and we write this functional as $h(t-t')$. Consequently, Eq. (1.77) becomes

$$y_1(t) = \int h(t-t')x_1(t')dt' \tag{1.78}$$

which is the integral expression for the convolution of $h(t)$ with $x_1(t)$. Symbolically we write

$$y_1(t) = h(t) * x_1(t)$$

We emphasize this important conclusion

$$\text{Linearity} + \text{Time invariance} \implies \text{Convolution}$$

To complete this section, a word concerning the definition of the FT pair often used throughout the book. In general, we prefer symmetry of the direct and inverse transform and omit the 2π inherent in definitions using the angular frequency (Eqs. (1.47) and (1.48). We use the temporal frequency and write

$$X(f) = \int x(t)e^{-i2\pi ft}dt$$

and

$$(1.79)$$

$$x(t) = \int X(f)e^{i2\pi ft}df \tag{1.80}$$

Symbolically

$$x(t) \Leftrightarrow X(f)$$

and we express the input/output relationships for a linear system by

$$\begin{array}{ccccccc} y(t) & = & h(t) & * & x(t) \\ \Updownarrow & & \Updownarrow & \Updownarrow & \Updownarrow \\ Y(f) & = & H(f) & \times & X(f) \end{array}$$

We refer to $h(t)$ as the impulse response and to $H(f)$ as the transfer function of the linear system, respectively.

1.7.7 Causal Systems

Consider a causal system with impulse response $h(t)$. The transfer function $H(f)$, which is given by $H(f) \Leftrightarrow h(t)$, has special properties. To see this (and, following Bracewell, 1986), split $h(t)$ into even and odd parts

$$\begin{aligned} h(t) &= e(t) + o(t) \\ &= \frac{1}{2}[h(t) + h(-t)] + \frac{1}{2}[h(t) - h(-t)] \end{aligned}$$

Since $o(t) = \mathrm{sgn(t)}e(t)$, where sgn(t) is the signum function, and $\mathrm{sgn}(t) \Leftrightarrow -i/\pi f$, we obtain

$$h(t) = (1 + \mathrm{sgn}(t)e(t)) \Leftrightarrow E(f) + i(\frac{-1}{\pi f}) * E(f) \tag{1.81}$$

where $E(f) \Leftrightarrow e(t)$. Eq. (1.81) shows that the transfer function of a causal system is of the form

$$H(f) = E(f) + iO(f)$$

where $O(f)$ is the Hilbert transform of $E(f)$. That is

$$O(f) = \frac{1}{\pi} \int \frac{E(f')}{f' - f} df' \tag{1.82}$$

and

$$E(f) = -\frac{1}{\pi} \int \frac{O(f')}{f' - f} df'$$

A very important and useful relationship exists between the phase and amplitude spectra, $\Phi(f)$ and $A(f)$ respectively, for causal systems. Writing $H(f)$ as (please refer to Eq. (1.51))

$$H(f) = A(f)e^{i\Phi(f)}$$

and taking natural logarithms, we obtain

$$\log H(f) = \log A(f) + i\Phi(f)$$

We now identify $\log A(f)$ with $E(f)$ and $\Phi(f)$ with $O(f)$. Using Eq. (1.82) obtains

$$\Phi(f) = \frac{1}{\pi} \int \frac{\log A(f')}{f' - f} df' \tag{1.83}$$

Eq. (1.83) tells us that, for causal systems, a phase spectrum may be determined as the Hilbert transform of the natural logarithm of the amplitude spectrum. It turns out, as we will see later, that the phase computed using Eq. (1.83) is a very special phase indeed. It is called a *minimum phase* and plays a central role in such issues as, for example, the design of deconvolution filters. The reason why Eq. (1.83) gives rise to a minimum phase function lies in the location of poles and zeros in the s plane of system functions. Physically realizable systems may not contain poles in the right hand s plane. Since, for every zero in the s plane, $\log A(f)$ has a pole, the equivalent transfer function which corresponds to Eq. (1.83) must have no poles *or zeroes* in the right hand s plane. It is, consequently, a minimum phase transfer function.

1.7.8 Discrete Convolution

Most of the systems that we deal with are linear and the inputs and outputs are discrete. As we have seen, this implies convolution and, specifically, discrete convolution. The continuous integral equation now becomes an infinite summation

$$y_k = \sum_n h_n x_{k-n}$$

In general, we will be concerned with finite length signals and we define

$x_n, \ n = 0, 1, \ldots, Nx - 1$ a signal of length Nx

$y_n, \ n = 0, 1, \ldots, Ny - 1$ a signal of length Ny

$h_n, \ n = 0, 1, \ldots, Nh - 1$ a signal of length Nh

The convolution sum is composed of samples defined in the above intervals and is written as

$$y_k = \sum_{n=N_1}^{N_2} h_n x_{k-n}$$

where N_1 and N_2 denote the lower and upper limits of the sum. These limits depend on the output time sample k as we can see in the following example. Let us assume, for purposes of illustration, that $x = [x_0, x_1, x_2, x_3, x_4]$ and $h = [h_0, h_1, h_2]$. Then, the convolution sum leads to the following system of equations

$$\begin{array}{lclll} y_0 & = & x_0 h_0 & & \\ y_1 & = & x_1 h_0 & +x_0 h_1 & \\ y_2 & = & x_2 h_0 & +x_1 h_1 & +x_0 h_2 \\ y_3 & = & x_3 h_0 & +x_2 h_1 & +x_1 h_2 \\ y_4 & = & x_4 h_0 & +x_3 h_1 & +x_2 h_2 \\ y_5 & = & & x_4 h_1 & +x_3 h_2 \\ y_6 & = & & & x_4 h_2 \end{array} \tag{1.84}$$

The output time series is given by $y = [y_0, y_1, y_2, ..., y_6]$ of length $Ny = Nx + Nh - 1$. The above system of equations can be written in a very convenient matrix form

$$\begin{pmatrix} y_0 \\ y_1 \\ y_2 \\ y_3 \\ y_4 \\ y_5 \\ y_6 \end{pmatrix} = \begin{pmatrix} x_0 & 0 & 0 \\ x_1 & x_0 & 0 \\ x_2 & x_1 & x_0 \\ x_3 & x_2 & x_1 \\ x_4 & x_3 & x_2 \\ 0 & x_4 & x_3 \\ 0 & 0 & x_4 \end{pmatrix} \begin{pmatrix} h_0 \\ h_1 \\ h_2 \end{pmatrix} \tag{1.85}$$

We will make much use of this matrix formulation in the sections to come.

1.7.9 Convolution and the z Transform

The utility of the z transform is well illustrated in application to convolution and deconvolution. Consider the series, $x = [x_0, x_1, x_2, x_3, x_4]$ and $h = [h_0, h_1, h_2]$. The z transforms are

$$\begin{aligned} X(z) &= x_0 + x_1 z + x_2 z^2 + x_3 z^3 + x_4 z^4 \\ H(z) &= h_0 + h_1 z + x_2 z^2 \end{aligned}$$

Computing the product of the above polynomials, we obtain

$$\begin{aligned} X(z)H(z) &= z^0(x_0 h_0) + & (1.86) \\ & z^1(x_1 h_0 + x_0 h_1) + \\ & z^2(x_2 h_0 + x_1 h_1 + x_0 h_2) + \\ & z^3(x_3 h_0 + x_2 h_1 + x_1 h_2) + \\ & z^4(x_4 h_0 + x_3 h_1 + x_2 h_2) + & (1.87) \\ & z^5(x_4 h_1 + x_3 h_2) + \\ & z^6(x_5 h_2) \end{aligned}$$

It is clear, from Eq. (1.84), that the coefficients of this new polynomial are the samples of the time series $y = [y_0, y_1, \ldots, y_6]$ obtained by convolution of x_n and h_n. In other words, $X(z)H(z)$ is the also the z transform of the time series y_n, or

$$Y(z) = X(z)H(z)$$

Therefore, convolution of two time series is equivalent to the multiplication of their z transforms. Of course, given the relationship between the Fourier and z transforms (Section 1.7.1), this is hardly surprising.

1.7.10 Deconvolution

Convolution represents what is often termed, a 'forward' problem. In the discrete and finite case, such problems are always stable. Deconvolution attempts to recover one of the inputs to the forward problem from the output and the other input. It is termed to be an 'inverse' problem. In many instances, and in particular in the instance of deconvolution, the problem is highly unstable. We will have occasion to meet this topic throughout the book, particularly in Chapter 6, and this section sets the stage.

Our discussion, thus far, may be cast as

$$y_k = h_k * x_k \quad \Leftrightarrow \quad Y(z) = H(z)X(z)$$

In the deconvolution process we will attempt to estimate x_k from y_k and h_k[6]. In the z domain, deconvolution is equivalent to polynomial division

[6] We could equally well attempt to estimate h_k from y_k and x_k.

$$X(z) = \frac{Y(z)}{H(z)}$$

Defining the inverse operator as

$$F(z) = \frac{1}{H(z)}$$

we obtain $X(z)$ from

$$X(z) = F(z)Y(z)$$

Since $F(z) = \sum_k f_k z^k$, the coefficients, f_k, define the discrete inverse filter in the time domain that recovers x_k via convolution

$$x_k = f_k * y_k$$

This procedure is important in data processing. In seismology, for example, we can assume that the observed seismogram is composed of two time series; the Earth's impulse response, and the seismic wavelet (also called the source function). Defining

s_k: the seismogram (the measured data)

q_k: the Earth's impulse response (the unknown)

w_k: the Wavelet (assumed known for this example)

we have the ubiquitous linear seismic model

$$s_k = w_k * q_k \Leftrightarrow S(z) = W(z)Q(z) \tag{1.88}$$

In the deconvolution process we attempt to design our inverse filter, f_k, to remove the wavelet. Applying the inverse filter of the wavelet to both sides of Eq. (1.88), obtains

$$f_k * s_k = f_k * w_k * q_k \Leftrightarrow F(z)S(z) = F(z)W(z)Q(z)$$

and it is clear that, if $F(z) = 1/W(z)$, the output sequence is the desired, unknown, impulse response

$$q_k = f_k * s_k$$

In the following sections we will analyze the problem of estimating and inverting the source signature, w_k. In particular, we now combine our discussion of the z transform with that of the concept of minimum phase to consider *dipole filters*, the building blocks of time series.

1.8 Dipole Filters

As we have seen, the z transform of a time series $x_t = [x_0, x_1, \ldots, x_N]$ is simply expressed as

$$X(z) = x_0 + x_1 z + x_2 z^2 + \ldots + x_N z^N \tag{1.89}$$

According to the Fundamental Theorem of Algebra, $X(z)$, which is a polynomial of order N, may be factored into N roots. Understanding the properties of a typical root leads to the understanding of the properties of the polynomial and of the underlying time series. We write a typical root as

$$H(z) = h_0 + h_1 z$$

where the coefficients may be complex. The amplitude and phase spectra of our typical root are easily calculated by substituting $z = e^{-i\omega}$

$$\begin{aligned} H(z) &= h_0 + h_1 e^{-i\omega} \\ &= h_0 + h_1(\cos\omega - i\sin\omega) \end{aligned}$$

Substituting in Eq. (1.51), we have

$$A(\omega) = [h_0^2 + 2h_0 h_1 \cos\omega + h_1^2]^{1/2} \tag{1.90}$$

$$\Phi(\omega) = \tan^{-1}\left[\frac{h_1 \sin\omega}{h_0 + h_1 \cos\omega}\right] \tag{1.91}$$

Let us evaluate Eq. (1.90) for two specific dipoles, $H_1(z) = 1 + 0.5z$ and $H_2(z) = 0.5 + z$. Fig. 1.14 shows the amplitude spectrum of both dipoles. That the amplitude spectra are the same, is easily deduced from Eq. (1.90). The phase spectra, as illustrated in Fig. 1.15, are different, however. In fact, $H_1(z)$ has a 'smaller' phase spectrum, $\Phi_1(\omega)$, than $\Phi_2(\omega)$, the phase spectrum of $H_2(z)$.
The simplicity of the dipole representation allows us to classify general filters in terms of dipole properties. Specifically,

> If $|h_0| > |h_1|$, the root of the dipole filter (or wavelet) lies outside $|z| = 1$ and the dipole filter is said to be *minimum delay* or, equivalently, *minimum phase*.
>
> If $|h_0| < |h_1|$, the root of the dipole filter (or wavelet) lies inside $|z| = 1$ and the dipole filter is said to be *maximum delay* or, equivalently, *maximum phase*.

The phase properties may be explored conveniently by the mapping $W = H(z)$. Typically, for a simple filter, this mapping may appear as illustrated in Fig. 1.16. The unit circle $|z| = 1$ maps into the curve Γ and the inside of $|z| = 1$ maps into the inside of Γ. We can now define minimum phase in the W plane in the following manner:

> The $W = H(z)$ mapping of a minimum phase function h_t with z transform $H(z)$ is such that the curve Γ, which is the mapping of $|z| = 1$, does not touch or enclose the origin, $W = 0$.

The W plane mappings for $H_1(z)$ and $H_2(z)$ are shown in Fig. 1.17.
Clearly, since the curve represented by Γ_1 does not touch or enclose the origin, $H(z)$ is minimum phase. The curve Γ_2 shown in Fig. 1.17b, on the contrary, does enclose the origin, which implies that $H_2(z)$ is not minimum phase.

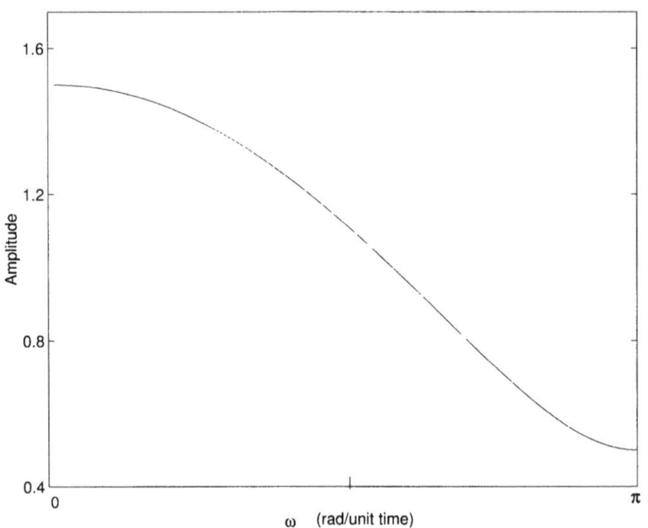

Figure 1.14: Amplitude spectrum of $H_1(z)$ and $H_2(z)$

1.8.1 Invertibility of Dipole Filters

In as much as convolution is ubiquitous to linear systems, so the inverse operation, that of deconvolution, is often required. In the frequency (or z) domain, deconvolution is simply division and so we examine this operation for $H_1(z)$ and $H_2(z)$.

$$\begin{aligned} H_1^{-1}(z) &= (1+\frac{1}{2}z)^{-1} \\ &= 1-\frac{1}{2}z+\frac{1}{4}z^2-\frac{1}{8}z^3+\cdots \\ &= \sum_{n=0}^{\infty}\beta_n z^n \end{aligned}$$

Clearly, the inverse is *stable* in the sense that $\sum_{n=0}^{\infty}|\beta_n| < \infty$. On the other hand, considering $H_2^{-1}(z)$, we have

$$\begin{aligned} H_2^{-1}(z) &= 2(1+2z)^{-1} \\ &= 2(1-2z+4z^2-8z^3+\cdots). \end{aligned}$$

and, in the same sense, we see that the inverse is *unstable*. The vital point is that the causal inverse of a minimum phase dipole is stable (and causal) whereas it is unstable for a maximum phase dipole. $H_2^{-1}(z)$ may be computed in a stable manner by

$$\begin{aligned} H_2^{-1}(z) &= \frac{1}{z(\frac{1}{2}z^{-1}+1)} \\ &= z^{-1}(1+\frac{1}{2}z^{-1})^{-1} \end{aligned}$$

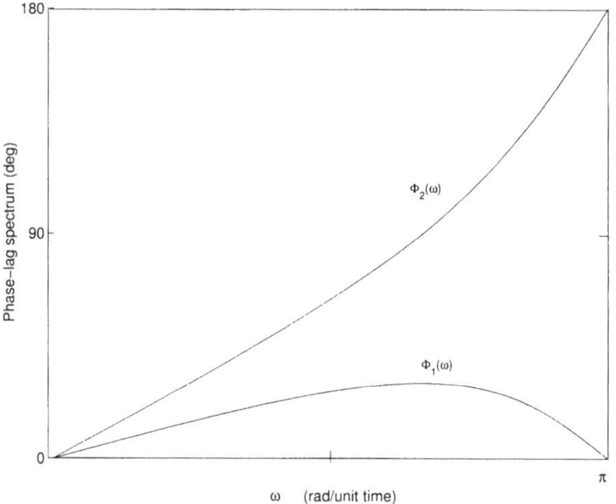

Figure 1.15: Phase spectra of $H_1(z)$ and $H_2(z)$

$$= z^{-1}(1 - \frac{1}{2}z^{-1} + \frac{1}{4}z^{-2} - \frac{1}{8}z^{-3} + \cdots).$$

The inverse is stable but *acausal.* We can now make the following generalizations.

1.8.2 Properties of Polynomial Filters

Consider the filter

$$H(z) = h_0 + h_1 z + h_2 z^2 + \cdots + h_N z^N$$

We can write

$$H(z) = h_n (z + z_1)(z + z_2) \cdots (z + z_N)$$

From the properties of the dipole filters and the Fundamental Theorem, it now follows that:

(a) Any of the dipoles may be reversed without altering the amplitude spectrum (and therefore the autocorrelation) of $H(z)$ [7]. There are 2^N filters with the same autocorrelation.

[7]Note: For real signals, the autocorrelation is computed as

$$R(z) = H(z)\, H(\frac{1}{z})$$

e.g., for $H_1(z) = h_0 + h_1 z$

$$\begin{aligned} R_1(z) &= (h_0 + h_1 z)\,(h_0 + h_1 z^{-1}) \\ &= h_0 h_1 z^{-1} + h_0^2 + h_1^2 + h_0 h_1 z. \\ &\quad \text{(a symmetric function)} \end{aligned}$$

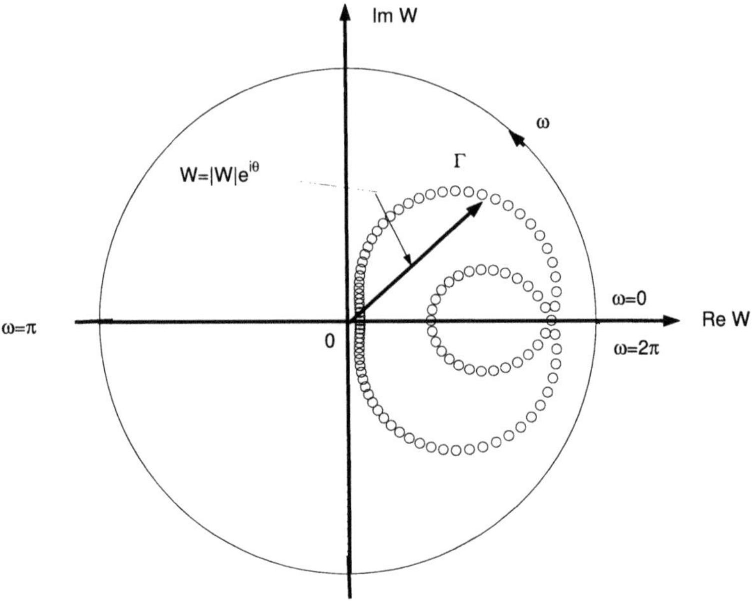

Figure 1.16: The $W = H(z)$ mapping for a simple filter

(b) Only one such filter, the one with all dipoles being minimum phase, is minimum phase.

(c) All other filters are mixed phase except one which is maximum phase.

(d) Only the minimum phase filter has a stable, causal inverse.

1.8.2.1 Some Toy Examples for Clarity

Our discussion thus far has been rather general. Sven Treitel and Enders Robinson have taught us all, however, how important toy examples are. We follow suit by analyzing the deconvolution of very simple signals. By understanding how to work with such signals we will gain the experience required to deal with much more complicated signals and, at the same time, become more familiar with the important concept of minimum phase. We begin by considering the dipole decomposition of the series $x_k = [4, 12, -1, 3]$. The z transform is

$$X(z) = 4 + 12z - 1z^2 + 3z^3 = 4(1 + \frac{1}{2}z)(1 - \frac{1}{2}z)(1 + 3z)$$

As we have seen, multiplication of z transforms is equivalent to the convolution of the respective time series. Therefore, the above expression can also be expressed as a convolution of elementary dipoles

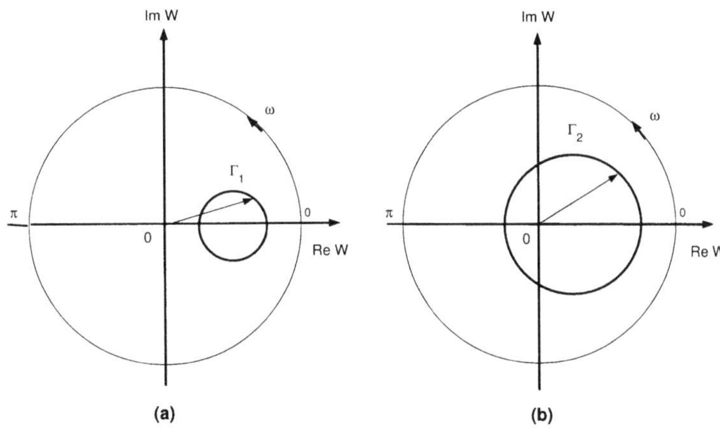

Figure 1.17: The W plane mapping for (a) $H_1(z)$ and (b) $H_2(z)$.

$$[4, 12, -1, 3] = 4[\,(1, \frac{1}{2}) * (1, -\frac{1}{2}) * (1, 3)\,]$$

Each dipole can be treated separately, and the properties of $X(z)$ (and, equivalently, x_k) can then be determined by multiplication. Our typical dipole is

$$H(z) = 1 + az$$

with inverse given by

$$F(z)H(z) = 1 \tag{1.92}$$

Minimum phase dipoles

If $|a| < 1$, $F(z)$ in Eq. (1.92) can be expanded in a convergent geometric series

$$F(z) = 1 - az + (az)^2 - (az)^3 + (az)^4 \ldots$$

where $F(z)$ is the z transform of the time series f_k, $k = 0, 1, \ldots, \infty$, the inverse filter of the dipole. The convolution of the dipole with the filter expressed as a multiplication in the z domain is

$$(1 + az)(1 - az + (az)^2 - (az)^3 + (az)^4 \ldots) = (\underset{\uparrow}{1} + 0z + 0z^2 + 0z^3 + 0z4 + \ldots)$$

which represents a single spike at $n = 0$. If $|a|$ is close to 1, the coefficients of the inverse filter will tend to zero very slowly. On the other hand, if $|a| << 1$, only a few coefficients will be required to properly model the inverse of the dipole. For example, let us compute the inverses of the following dipoles, $h1_k = [1, 0.9]$ and $h2_k = [1, 0.1]$. In the first case we have $a = 0.9$ and the coefficient of the z^4 term of $F(z)$ is 0.6561. In the second case, on the other hand, this coefficient is 0.0001 Clearly, when $a = 0.1$ we can rapidly truncate the expansion without affecting the performance of the filter. To show the last statement we convolve the dipoles with their truncated inverses. In both examples, we truncate the inverse after 5 coefficients.

$$
\begin{aligned}
[1, 0.9] * [1, -0.9, 0.81, -0.729, 0.6561] &= [1, 0.0, 0.0, 0.0, 0.59] \\
[1, 0.1] * [1, -0.1, 0.01, -0.001, 0.0001] &= [1, 0.0, 0.0, 0.0, 0.0]
\end{aligned}
$$

The truncation is negligible when $a = 0.1$ but is certainly not negligible when $a \approx 1$. In this case a long filter is needed to properly invert the dipole. The aforementioned shortcoming can be overcome by adopting a least squares strategy to compute the inverse filter (which is the basis of *spiking deconvolution*).

So far we have considered a minimum phase dipole, the z transform of which has a root, ξ, that lies outside the unit circle, $|z| = 1$

$$
X(z) = 1 + az \Rightarrow X(\xi) = 1 + a\xi = 0 \Rightarrow \xi = -\frac{1}{a}
$$

and $|\xi| > 1$.

Any signal worthy of respect is, of course, more complicated than a simple dipole. We can always, however, factor its z transform in terms of elementary dipoles. If the signal is minimum phase, the decomposition is in terms of minimum phase dipoles

$$
X(z) = x_0 + x_1 z + x_2 z^2 + x_3 z^3 \ldots = k(1 + a_1 z)(1 + a_2 z)(1 + a_3 z) \ldots.
$$

where k is a constant. If $|a_i| < 1$, $\forall i$, the signal is minimum phase and all the zeros lie outside the unit circle

$$
X(\xi) = 0 \Rightarrow \xi_i = -\frac{1}{a_i} \Rightarrow |a_i| < 1 \Rightarrow |\xi_i| > 1
$$

In this case

$$
\begin{aligned}
&X(z)F(z) = 1 \\
&(1 + a_1 z)(1 + a_2 z)(1 + a_3 z) \ldots . F(z) = 1
\end{aligned}
\tag{1.93}
$$

and we can write

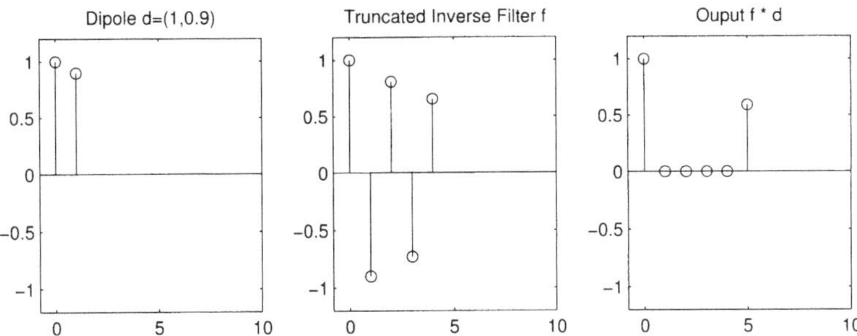

Figure 1.18: Inversion of minimum phase dipoles. The slow convergence of the inverse filter is a consequence of a zero close to the unit circle.

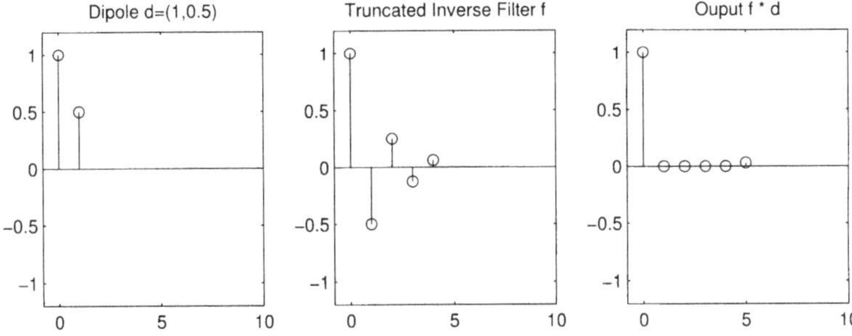

Figure 1.19: Inversion of minimum phase dipoles. The zero is located at z=-2

$$
\begin{aligned}
F(z) &= (1+a_1 z)^{-1}(1+a_2 z)^{-1}(1+a_3 z)^{-1}\ldots \\
&= (1-a_1 z+(a_1 z)^2-(a_1 z)^3\ldots)(1-a_2 z+(a_2 z)^2-(a_2 z)^3\ldots)\times \\
&\quad (1-a_3 z+(a_3 z)^2-(a_3 z)^3\ldots)\ldots
\end{aligned} \quad (1.94)
$$

with the inverse operator given by

$$[f_0, f_1, f_2, f_3, \ldots] = (\,[1, -a_1, a_1^2, -a_1^3] * [1, -a_2, a_2^2, -a_2^3] * [1, -a_3, a_3^2, -a_3^3] * \ldots)$$

Figs. 1.18, 1.19 and 1.20 illustrate the inverses of various minimum phase dipoles and the reconstructed outputs. In the first case the root is close to the unit circle and the inverse filter would require a large number of coefficients to avoid the truncation artifact that is apparent in the output. In Figs. 1.19 and 1.20, we have used dipoles with roots at $|\xi| = 2$ and $|\xi| = 10$, respectively, and the truncation artifacts are minimal.

Maximum phase dipoles

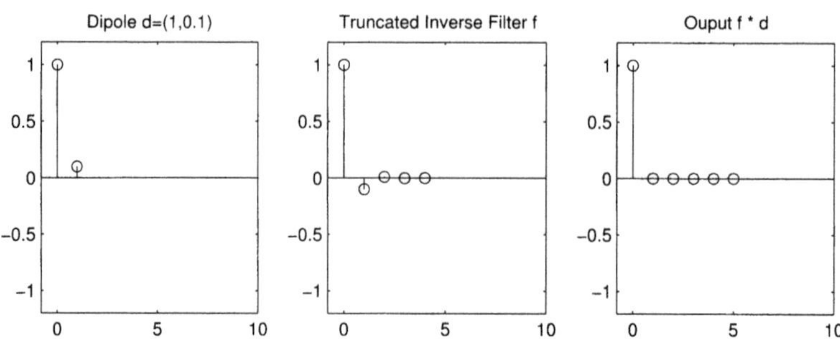

Figure 1.20: Inversion of minimum phase dipoles. Fast convergence of the inverse filter is a consequence of the zero of being far from the unit circle.

Elementary signals of the form $[1, b]$, $|b| > 1$ are called maximum phase dipoles, the zero of which lies inside the unit circle

$$H(z) = 1 + bz \Rightarrow H(\xi) = 1 + b\xi = 0 \Rightarrow \xi = -1/b\,.$$

$|b| < 1$ implies that $|\xi| < 1$.

As we have seen in Section 1.8.1, the causal inverse of such dipoles is unstable and, for stability, a non-causal inverse is required and takes the form

$$F(z) = (bz)^{-1}(1 - (bz)^{-1} + (bz)^{-1} - (bz)^{-3}\dots)$$

The associated operator is given by

$$f_k = [\dots, -b^{-3}, b^{-2}, -b^{-1}, \underset{\uparrow}{0}]$$

where the vertical arrow indicates a value of particular interest. The following example clarifies the discussion. First, given the maximum phase dipole $[1, 2]$, we compute the non-casual inverse sequence (truncated to 6 coefficients)

$$f_k = [-0.0156, 0.0312, -0.0625, 0.125, -0.25, 0.5, \underset{\uparrow}{0}]$$

The convolution of f_k with the maximum phase dipole produces the output sequence

$$\begin{aligned} h_k * f_k &= [-0.0156, 0.0312, -0.0625, 0.125, -0.25, 0.5, \underset{\uparrow}{0}] * [\underset{\uparrow}{1}, 2] \\ &= [-0.0156, 0, 0, 0, 0, 0, \underset{\uparrow}{1}, 0] \end{aligned} \tag{1.95}$$

Fig. 1.21 illustrates our results. The non-causal nature of the inverse filter is clear from this figure.

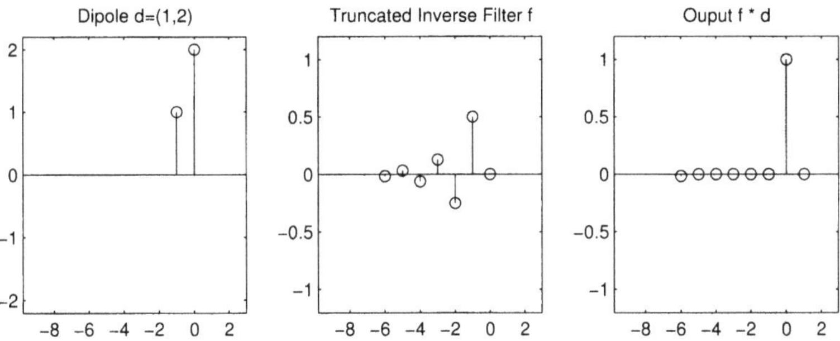

Figure 1.21: A maximum phase dipole, its non-casual truncated inverse, f_k, and the output, $h_k * f_k$.

Dipole autocorrelation function

We briefly described the computation of the autocorrelation function in Section 1.8.2. Here, we delve a little deeper. The autocorrelation function of a sequence with z transform $X(z)$ is defined as

$$R(z) = X(z)X^*(z^{-1})$$

and allows us to analyze some properties of minimum and maximum phase dipoles that are very useful in the design of deconvolution operators. We consider two dipoles, a minimum phase dipole of the form $(1, a)$, $|a| < 1$ and a maximum phase dipole of the form $(a^*, 1)$[8]. In the z domain we have

$$\begin{aligned} H_{min}(z) &= 1 + az \\ H_{max}(z) &= a^* + z \end{aligned}$$

The autocorrelation function for the minimum phase sequence is given by

$$R_{min}(z) = a^* z^{-1} + (1 + |a|^2) + az$$

Similarly, the autocorrelation function for the maximum phase dipole is given by

$$R_{max}(z) = a^* z^{-1} + (1 + |a|^2) + az$$

We have arrived at the very important conclusion to which we alluded previously

$$R_{max}(z) = R_{min}(z) = R(z)$$

[8]Note that for real dipoles, $a^* = a$

In other words, two different sequences can have the same autocorrelation function, which is given by the sequence

$$[a^*, \underset{\uparrow}{(1+|a|^2)}, a]$$

or

$$r_k = \begin{cases} a & \text{if } k = 1 \\ 1+|a|^2 & \text{if } k = 0 \\ a^* & \text{if } k = -1 \\ 0 & \text{otherwise} \end{cases}$$

If the dipoles are real ($a = a^*$), the autocorrelation function is symmetric. Note that $R(z)$, the ZT r_k, is

$$R(z) = r_1 z^{-1} + r_0 + r_1 z^{-1} = a^* z^{-1} + (1 + a^2) + a z^{-1}$$

In general, for more complicated signals (so far we have only considered dipoles), the autocorrelation functions, r_k and $R(z)$, are given by

$$\begin{aligned} r_k &= \sum_n x_n^* x_{n+k} \\ R(z) &= X(z) X^*(z^{-1}) \end{aligned}$$

where k is the time-lag of the autocorrelation function, are a transform pair. The importance of the fact that both minimum and maximum phase dipoles have the same autocorrelation will become very apparent, when we deal with the problem of spectral factorization.

1.8.3 Least Squares Inversion of Minimum Phase Dipoles

We have seen that one of the problems of inverting a dipole via a series expansion is that the inverse filter can result in a long operator. Our objective is to find a filter the output of which, when applied to a dipole, resembles the ideal output one would have obtained by using an infinite number of terms in the series expansion. It turns out that the solution to this problem is a Wiener optimum spiking filter, a subject further briefly explored in Section 1.8.5. We have met the method of least squares that is used in the solution before, (Section 1.6), when dealing with Fourier analysis, and we will explore it in much more detail in Chapter 4.

We begin with the inverse of a minimum phase dipole $[1, a]$ ($|a| < 1$) and we assume that a is real. As we have seen, the ideal inverse filter satisfies

$$H(z)F(z) = 1$$

Our task is to construct a finite length filter with the property that

$$H(z)F_N(z) \approx 1$$

where $F_N(z)$ denotes the z transform of the finite length operator. The above equation, for $N = 3$, becomes

$$[1, a] * [f_0, f_1, f_2] \approx [\underset{\uparrow}{1}, 0, 0, 0]$$

and in matrix form

$$\begin{pmatrix} 1 & 0 & 0 \\ a & 1 & 0 \\ 0 & a & 1 \\ 0 & 0 & a \end{pmatrix} \begin{pmatrix} f_0 \\ f_1 \\ f_2 \end{pmatrix} \approx \begin{pmatrix} 1 \\ 0 \\ 0 \\ 0 \end{pmatrix}$$

or

$$\mathbf{Cf} \approx \mathbf{b}$$

where $\mathbf{C}$ is the matrix whose columns contain the dipole padded with zeros in order to properly represent the convolution. The unknown inverse filter is denoted by $\mathbf{f}$ and the desire output by $\mathbf{d}$. The last system of equations is over-determined and can be solved using LS. It is clear that the solution vector is the one that minimizes the mean squared error given by

$$\mathcal{E} = ||\mathbf{Cf} - \mathbf{b}||^2$$

The solution is obtained by solving the following system of normal equations

$$\mathbf{C}^T\mathbf{Cf} = \mathbf{C}^T\mathbf{b}$$

with the result that

$$\mathbf{f} = \mathbf{R}^{-1}\mathbf{C}^T\mathbf{b} \tag{1.96}$$

and

$$\mathbf{R} = \mathbf{C}^T\mathbf{C} \tag{1.97}$$

The results of the application of a five point inverse filter, given by Eq. (1.96), to two minimum phase dipoles, which differ in the proximity of the zero to $|z| = 1$, is

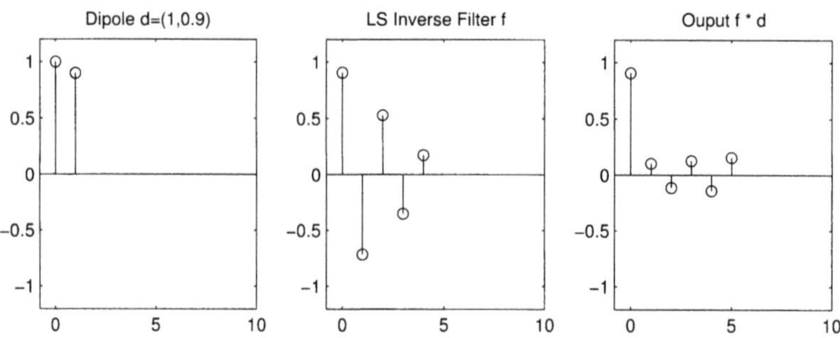

Figure 1.22: Inversion of a minimum phase dipole, $d_k = [1, 0.9]$, using LS.

shown in Figs. 1.22 and 1.23. This toy example demonstrates what should already be intuitively clear from the previous discussions. A minimum phase wavelet can indeed be spiked by means of a LS inverse filter, **f**. The length of **f**, however, depends on the distribution of zeros with respect to $|z| = 1$.

The matrix **R** in Eq. (1.97) has a very special structure indeed, and we meet it repeatedly

$$\mathbf{R} = \begin{pmatrix} 1+a^2 & a & 0 \\ a & 1+a^2 & a \\ 0 & a & 1+a^2 \end{pmatrix}$$

Each row of **R** is composed of elements of the autocorrelation sequence, r_k, and is characterized by

$$\mathbf{R} = \begin{pmatrix} r_0 & r_1 & 0 \\ r_1 & r_0 & r_1 \\ 0 & r_1 & r_0 \end{pmatrix}$$

R is a symmetric, Toeplitz, matrix. One interesting feature of a Toeplitz matrix is that only one row is required to define all the elements of the matrix. This special symmetry is used by a fast algorithm, the Levinson algorithm, which we describe in Section 2.4.3, to invert **R**.

It is interesting to note that the condition number of the Toeplitz matrix[9] increases with N_f, the filter length. This is shown in Fig. 1.24. Similarly, Fig. 1.25 portrays the condition number of the Toeplitz matrix for a dipole of the form $[1, a]$ for different values of the parameter a. It is clear that when $a \to 1$, the zero of the dipole moves towards the unit circle and the system of equation becomes ill-conditioned.

[9]The condition number of the matrix **R** is the ratio $\lambda_{max}/\lambda_{min}$ where λ_{max} and λ_{min} are the largest and smallest eigenvalues of **R**, respectively. A large condition number indicates that the problem is ill-conditioned (numerical problems will arise at the time of inverting the matrix.)

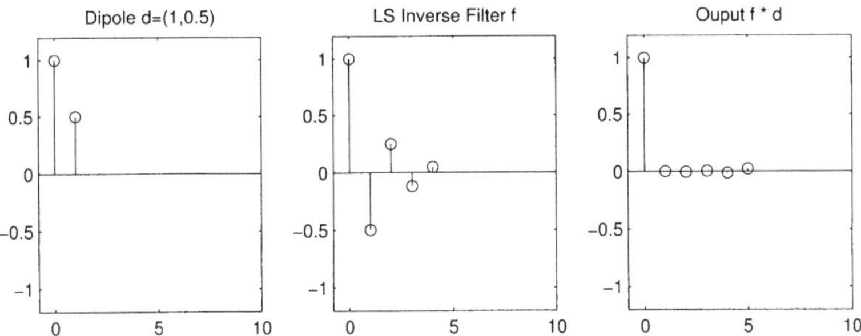

Figure 1.23: Inversion of a minimum phase dipole, $d_k = [1, 0.5]$ using LS.

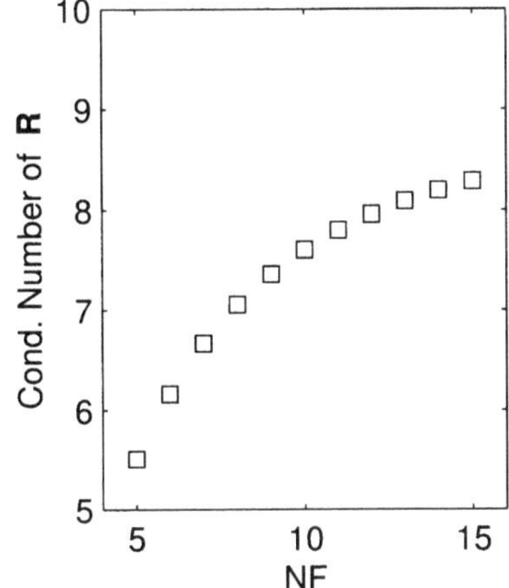

Figure 1.24: Condition number of the Toeplitz matrix versus N_f (filter length). The dipole is the minimum phase sequence $(1, 0.5)$.

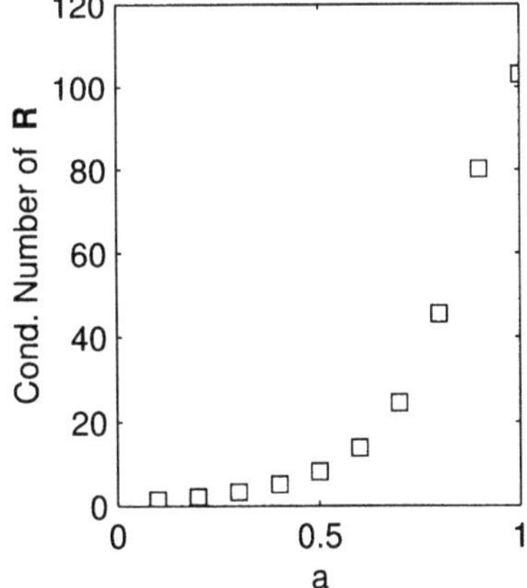

Figure 1.25: Condition number of the Toeplitz matrix versus a for a minimum phase dipole $[1, a]$. The length of operator is fixed to $N_f = 15$.

1.8.4 Inversion of Minimum Phase Sequences

Thus far, we have discussed the problem of inverting elementary dipoles, and we have observed that minimum phase dipoles yield causal and stable inverses. This is also valid for more complicated signals (i.e., a seismic wavelet), where the columns of the convolution matrix are wavelets of length N_w instead of dipoles of length 2. Since a minimum phase wavelet can be factored into minimum phase dipoles[10], we can, therefore, in similar fashion, find the inverse operator that converts the wavelet into a spike. Given the wavelet $w_k,\ k = 1, 2, \ldots, N_w$, we wish to compute a filter, $f_k,\ k = 1, 2, \ldots, N_f$, that satisfies

$$[w_0, w_1, \ldots w_{Nw-1}] * [f_0, f_1, \ldots, f_{Nf-1}] \approx [1, 0, \ldots, 0]$$

In matrix form, assuming $N_w = 7$ and $N_f = 4$ for purpose of illustration

$$\begin{pmatrix} w_0 & 0 & 0 & 0 \\ w_1 & w_0 & 0 & 0 \\ w_2 & w_1 & w_0 & 0 \\ w_3 & w_2 & w_1 & w_0 \\ w_4 & w_3 & w_2 & w_1 \\ w_5 & w_4 & w_3 & w_2 \\ w_6 & w_5 & w_4 & w_3 \\ 0 & w_6 & w_5 & w_4 \\ 0 & 0 & w_6 & w_5 \\ 0 & 0 & 0 & w_6 \end{pmatrix} \begin{pmatrix} f_0 \\ f_1 \\ f_2 \\ f_3 \end{pmatrix} \approx \begin{pmatrix} 1 \\ 0 \\ 0 \\ 0 \\ 0 \\ 0 \\ 0 \\ 0 \\ 0 \\ 0 \end{pmatrix}$$

or

$$\mathbf{Cf} \approx \mathbf{d}$$

The mean square error is

$$\begin{aligned} \mathcal{E} &= ||\mathbf{e}||^2 = ||\mathbf{Cf} - \mathbf{b}||^2 \\ &= \mathbf{e}^T\mathbf{e} = (\mathbf{Cf} - \mathbf{b})^T(\mathbf{Cf} - \mathbf{b}) \end{aligned}$$

and is minimized when

$$\frac{d\mathcal{E}}{d\mathbf{f}} = 0$$

Taking derivatives with respect to **f** and equating to zero, we obtain the well known system of normal equations

$$\mathbf{C}^T\mathbf{Cf} = \mathbf{Cb}$$

or, as perviously defined

[10]In other words, all the zeros of the z transform of the wavelet lie outside the unit circle.

$$\mathbf{Rf} = \mathbf{Cb}$$

If $\mathbf{R}$ is ill-conditioned, i.e., there exist eigenvalues that are zero or close to zero, numerical instabilities occur in the inversion. This can be avoided by using a regularization strategy. Instead of minimizing $\mathcal{E}$, we minimize a penalized objective function

$$J = \mathcal{E} + \mu||\mathbf{f}||^2$$

The solution is now given by a penalized LS estimator where the parameter μ is the regularization parameter (also, called a ridge regression, tradeoff, pre-whitening or hyperparameter). The condition

$$\frac{dJ}{d\mathbf{f}} = 0$$

leads to

$$\mathbf{f} = (\mathbf{R} + \mu\mathbf{I})^{-1}\mathbf{C}^T\mathbf{d}$$

It is clear that μ protects against small eigenvalues which may lead to an unstable filter. Two objectives are entailed in the function J. On the one hand we want to minimize $\mathcal{E}$, on the other, we wish to keep the energy of the filter bounded. When $\mu \to 0$, the error function will be minimum but the filter may have an undesirable oscillatory behavior. When μ is large, the energy of the filter will be small but $\mathcal{E}$ will be large and, in this case, since $(\mathbf{R} + \mu\mathbf{I}) \approx \mu\mathbf{I}$, a matching filter results and is given by

$$\mathbf{f} = \mu^{-1}\mathbf{C}^T\mathbf{d}$$

Fig. 1.26 illustrates the effect of μ in the filter design and in the actual output of the deconvolution. When μ is small, the output sequence is a spike, when μ is increased, the output sequence becomes a band limited signal. This concept is very important when dealing with noisy signals, and we will meet it again in Chapters 4 and 5.

The tradeoff curve is shown in Fig. 1.27. This displays the misfit, $\mathcal{E}$, versus the norm of the filter, $||\mathbf{f}||^2$, for various values of the tradeoff parameter, μ. This curve, also called the Tikhonov or the L curve, is a popular method to analyze the tradeoff between resolution and variance in linear inverse problems.

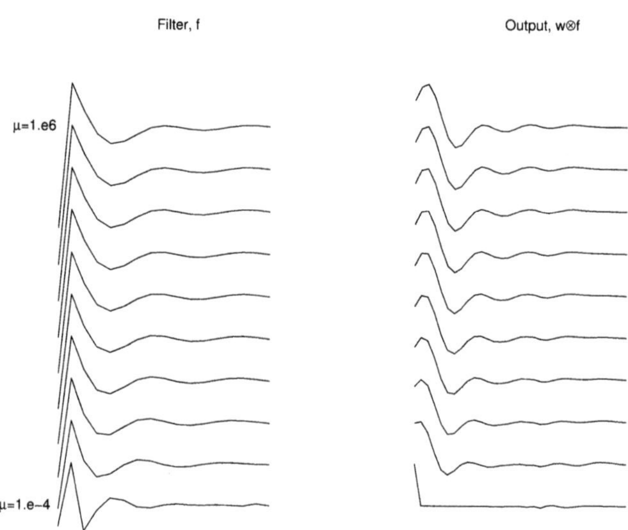

Figure 1.26: A minimum phase wavelet inverted using different values of the tradeoff parameters, (μ).

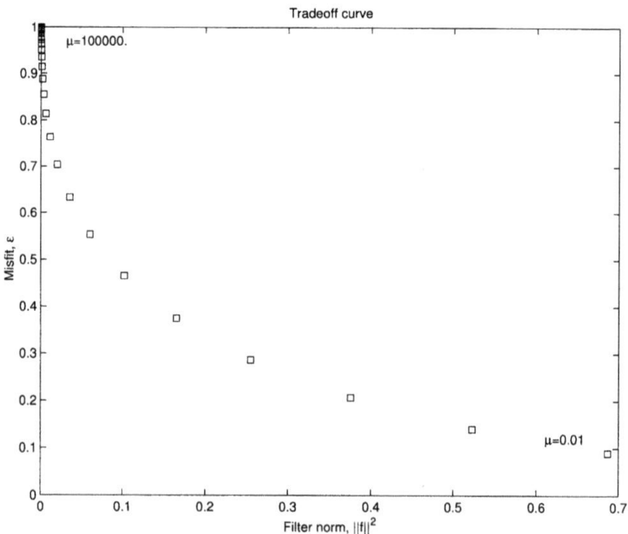

Figure 1.27: Tradeoff curve for the previous example. The vertical axis indicates the misfit and the horizontal axis the norm of the filter.

1.8.5 Inversion of Nonminimum Phase Wavelets: Optimum Lag Spiking Filters

Minimum phase wavelets are inverted using least squares and the resulting filter is often called the Wiener filter or a spiking deconvolution operator. In general, of course, seismic wavelets are not minimum phase but are, in fact, mixed phase. If the wavelet is not minimum phase, the actual output for a filter designed with a desired output, $d_k = [1, 0, 0, \ldots]$, does not resembles the desired output. We do not intend to go into detail here concerning this issue and, instead, refer the reader to the excellent text, Robinson and Treitel (2002) [11]. Suffice it to say that the problem can be alleviated by defining an optimum lag Wiener filter. This is an inverse filter, where the desired output is

$$[0, 0, \ldots 0, 0, 0, 1, 0, 0, \ldots]$$

i.e, a spike that has been delayed by an amount L, called the lag. The optimum lag, L_{opt}, is given by the value L where the actual output optimally resembles the desired output. We need to define some measure of closeness between the actual and desired outputs. This is done by defining a filter performance norm

$$P = 1 - \mathrm{E}$$

where E is the normalized mean square error which can be shown to obey

$$0 \leq \mathrm{E} \leq 1$$

We choose $L = L_{opt}$, a lag for which P attains a maximum.

1.9 Discrete Convolution and Circulant Matrices

Convolution is a central operation in all that we do. A particularly useful form of discrete convolution is its representation in terms of a circulant matrix (Strang, 1988) which leads to a formulation in terms of the eigenstructure of a circulant matrix. This formulation is both useful and illuminating and is particularly relevant to the form of Eq. (1.85).

1.9.1 Discrete and Circular Convolution

Consider two discrete series, x_n and h_n, both of length N (for convenience), which are assumed to be zero for $n < 0$ and $n > N - 1$. Discrete convolution is denoted by

$$y_n = x_n * h_n$$

[11]However, please see Section 2.7 for a discussion of the important topics of spiking deconvolution of nonminimum phase wavelets and predictive deconvolution.

$$= \sum_{k=0}^{N-1} h_{n-k} x_k \quad n = 0, 1, \ldots, 2N-2$$

Now consider series which are periodic with period N, i.e., $\tilde{x}_n = \sum_{l=-\infty}^{+\infty} x_{n+lN}$, or $x_{n+N} = x_n$. Circular convolution is expressed as

$$\tilde{y}_n = \sum_k \tilde{h}_{n-k} \tilde{x}_k \quad n = 0, 1, \ldots, \infty$$

where $\sum_k$ indicates $\sum_{k=o}^{N-1}$. Some remarks are in order.

Remark [1] The Discrete Fourier Transform, DFT, *implies periodicity*, i.e., $y_{n+N} = y_n$. Y_k, the DFT of y_n, is computed from y_n as if the time series was periodic with period N. To see this, we write

$$y_n = \frac{1}{N} \sum_k Y_k W^{-kn}$$

where $W = e^{-i2\pi/N}$. Consequently

$$\begin{aligned} y_{n+N} &= \frac{1}{N} \sum_k Y_k W^{-k(n+N)} \\ &= \frac{1}{N} \sum_k Y_k W^{-kn} W^{-kN} \end{aligned}$$

But, $W^{-kN} = 1$ which yields $y_{n+N} = y_n$.

Remark [2] Let us examine the relationship between discrete and circular convolution in more detail. Computing the convolution sum in the frequency domain and transforming back to time, we write

$$\begin{aligned} y_n &= \frac{1}{N} \sum_k H_k X_k W^{-kn} \\ &= \frac{1}{N} \sum_k H_k [\sum_l x_l W^{kl}] W^{-kn} \\ &= \frac{1}{N} \sum_l x_l \sum_k H_k X_k W^{-k(n-l)} \\ &= \sum_l x_l h_{n-l} \end{aligned} \tag{1.98}$$

An examination of the second term, y_1, which is

$$y_1 = x_0 h_1 + x_1 h_0 + x_2 h_{-1} + x_3 h_{-2} + \ldots + x_{N-1} h_{2-N} \tag{1.99}$$

is instructive. In discrete convolution, all terms with a negative subscript are zero. Because of the periodic property of the DFT, however, $h_1 = h_{N-1}, \ldots, h_{2-N} = h_2$ and y_1 becomes

$$y_1 = \underbrace{x_0 h_1 + x_1 h_0}_{\mathcal{D}} + \underbrace{x_2 h_{N-1} + \ldots + x_{N-1} h_2}_{\mathcal{R}}$$

$\mathcal{D}$ is the discrete convolution term. $\mathcal{R}$ is a term which is due to the periodic extension of the series. In order to perform discrete convolution by means of the DFT, $\mathcal{R}$ must be eliminated. This can, clearly, be accomplished by padding each series with zeros.

1.9.2 Matrix Notation for Circular Convolution

Eq. (1.98) may be written in matrix form as

$$\mathbf{y} = \tilde{\mathbf{H}}\mathbf{x} \tag{1.100}$$

The vector $\mathbf{x}$ is made up of the elements of x_n with a zero extension if discrete convolution is required. $\tilde{\mathbf{H}}$ is a square, circulant matrix, the form of which is best seen by means of an example. Consider the discrete convolution of two series, $x_n, \quad n = 0, 1, \ldots, 3$ and $h_n, \quad n = 0, 1, 2$. We form the matrix equation, writing the circulant matrix out in full.

$$\mathbf{y} = \begin{bmatrix} h_0 & 0 & 0 & 0 & h_2 & h_1 \\ h_1 & h_0 & 0 & 0 & 0 & h_2 \\ h_2 & h_1 & h_0 & 0 & 0 & 0 \\ 0 & h_2 & h_1 & h_0 & 0 & 0 \\ 0 & 0 & h_2 & h_1 & h_0 & 0 \\ 0 & 0 & 0 & h_2 & h_1 & h_0 \end{bmatrix} \begin{bmatrix} x_0 \\ x_1 \\ x_2 \\ x_3 \\ 0 \\ 0 \end{bmatrix}$$

All rows of the matrix are circular shifts of rows directly above and the first row is a circular shift of the last row. We can see that, if the series are not zero padded, the convolution is indeed circular as illustrated in Eq. (1.99). The zero padding has allowed us to perform discrete convolution using the circulant matrix. One may wonder why use the circulant form and not the conventional convolution matrix in the latter case. The answer lies in the properties of the circulant matrix which we explore below.

1.9.3 Diagonalization of the Circulant Matrix

First, we show that the N eigenvectors of a $N \times N$ circulant matrix are linearly independent and the matrix can, therefore, be diagonalized by a similarity transformation. Define a general $N \times N$ circulant matrix, $\tilde{\mathbf{C}}$, by

$$\tilde{\mathbf{C}} = \begin{bmatrix} c_0 & c_1 & \cdots & c_{N-1} \\ c_{N-1} & c_0 & \cdots & c_{N-2} \\ \vdots & \vdots & \ddots & \vdots \\ c_1 & c_2 & \cdots & c_0 \end{bmatrix}$$

Define a coefficient λ_k as

$$\lambda_k = c_0 + c_1 W^k + \ldots + c_{N-1} W^{k(N-1)} \tag{1.101}$$

Since $W^{Nk} = 1$, it is evident from Eq. (1.101) that λ_k satisfies the following set of equations

$$\begin{array}{llllll} \lambda_k & = c_0 & + \; c_1 W^k & + \ldots & + \; c_{N-1} W^{k(N-1)} \\ \lambda_k W^k & = c_{N-1} & + \; c_0 W^k & + \ldots & + \; c_{N-2} W^{k(N-1)} \\ \vdots & = \vdots & \vdots & & \vdots \\ \lambda_k W^{k(N-1)} & = c_1 & + \; c_2 W^k & + \ldots & + \; c_0 W^{k(N-1)} \end{array}$$

which we write as

$$\tilde{\mathbf{C}}\mathbf{w}_k = \lambda_k \mathbf{w}_k \tag{1.102}$$

where

$$\mathbf{w}_k = [1, e^{i2\pi/N}, \ldots, e^{i2\pi k(N-1)/N}]^H \tag{1.103}$$

and ^{H} indicates the Hermitian transpose. Eq. (1.102) represents an eigenproblem where the eigenvalue λ_k may be expressed as

$$\lambda_k = \sum_j c_j e^{i2\pi jk/N} \quad k = 0, 1, \ldots, N-1$$

The eigenvectors given by Eq. (1.103) may be collected into an eigenvector matrix, $\mathbf{W}$

$$\mathbf{W} = [\mathbf{w}_0, \mathbf{w}_1, \ldots, \mathbf{w}_{N-1}]$$

with the kjth element given by

$$w_{kj} = e^{i2\pi jk/N}$$

By inspection of the elements of $\mathbf{W}$ we can easily write an inverse, $\mathbf{W}^{-1}$. We first form the elements of $\mathbf{W}^{-1}$ as

$$w_{jk}^{-1} = \frac{1}{N} e^{-i2\pi jk/N}$$

and using the orthogonality property of the complex exponential, it is clear that

$$\mathbf{W}\mathbf{W}^{-1} = \mathbf{W}\mathbf{W}^H = \mathbf{I}$$

Hence, $\mathbf{W}$ is nonsingular and the eigenvectors are a linearly independent set. We now form $\mathbf{\Lambda}_C$, a diagonal matrix composed of the N eigenvalues of $\tilde{\mathbf{C}}$ and, using Eq. (1.102), we obtain

$$\tilde{\mathbf{C}}\mathbf{W} = \mathbf{W}\mathbf{\Lambda}_C$$

or

$$\tilde{\mathbf{C}} = \mathbf{W}\mathbf{\Lambda}_C\mathbf{W}^H \tag{1.104}$$

This is the diagonalized representation of a circulant. We now apply these results to convolution and deconvolution.

1.9.4 Applications of the Circulant

1.9.4.1 Convolution

We write Eq. (1.100) in terms of the diagonalization of the circulant $\tilde{\mathbf{H}}$ as

$$\mathbf{y} = \mathbf{W}\mathbf{\Lambda}_H\mathbf{W}^H\mathbf{x}$$

and consequently

$$\mathbf{W}^H\mathbf{y} = \mathbf{\Lambda}_H\mathbf{W}^H\mathbf{x} \tag{1.105}$$

Let us look at Eq. (1.105) in some detail and, in particular, the product $\mathbf{W}^H\mathbf{y}$

$$\mathbf{W}^H\mathbf{y} = \begin{bmatrix} 1 & 1 & \cdots & 1 \\ 1 & e^{-i2\pi/N} & \cdots & e^{-i2\pi(N-1)/N} \\ \vdots & \vdots & \vdots & \vdots \\ 1 & e^{-i2\pi(N-1)/N} & \cdots & e^{-i2\pi(N-1)(N-1)/N} \end{bmatrix} \begin{bmatrix} y_0 \\ y_1 \\ \vdots \\ y_{N-1} \end{bmatrix}$$

which may be written as

$$Y_k = [\mathbf{W}^H\mathbf{y}]_k = \frac{1}{N}\sum_j y_j e^{-i2\pi jk/N} \qquad k = 0, 1, \ldots, N-1 \tag{1.106}$$

Y_k is, by definition, the DFT of y_n. Now, we examine $\mathbf{\Lambda}_H$. The diagonal elements, $\Lambda_{kk} = \lambda_k$, are computed from the first row of $\tilde{\mathbf{H}}$ (c.f. Eq. (1.101). Note, however, that the first row of $\tilde{\mathbf{H}}$ is h_{-n} (since the matrix is circulant) and hence

$$\begin{aligned} \Lambda_{kk} &= \sum_j h_{-j} e^{i2\pi jk/N} \qquad k = 0, \ldots N-1 \\ &= \sum_j h_j e^{-i2\pi jk/N} \qquad k = 0, \ldots N-1 \\ &= NH_k \end{aligned}$$

Using Eqs. (1.105) and (1.106) we obtain the well known result

$$Y_k = NH_kX_k$$

the frequency domain equivalent of convolution.

1.9.4.2 Deconvolution

As discussed above, due largely to the bandlimited nature of the seismic pulse, deconvolution is, in general, a highly ill-posed inverse problem. This is particularly true when noise is present. Some form of regularization is essential. The circulant formulation which we have explored lends itself naturally to a particular form of regularization which is often applied in the time domain. We write the forward problem as

$$x_n = b_n * q_n \tag{1.107}$$

where x_n is the observed trace, b_n is the known system impulse response which we wish to deconvolve, and q_n is the required and unknown system input. Writing Eq. (1.107) in the form of the matrix Eq. (1.100), yields

$$\tilde{\mathbf{B}}\mathbf{q} = \mathbf{x}$$

or

$$\mathbf{q} = \tilde{\mathbf{B}}^{-1}\mathbf{x} \tag{1.108}$$

Using the diagonalized form in Eq. (1.104), the required deconvolution is written as

$$\mathbf{q} = \mathbf{W}\boldsymbol{\Lambda}_B^{-1}\mathbf{W}^H\mathbf{x}$$

and in summation form

$$\mathbf{q} = \left(\sum_k \frac{1}{\lambda_k}\mathbf{w}_k\mathbf{w}_k^H\right)\mathbf{x}$$

A particular regularization is achieved by discarding λ_k's which are smaller in modulus than a pre-assigned threshold. Another approach, also using the diagonalized form of the circulant matrix is as follows: Define a cost function J in the usual manner

$$J = ||\tilde{\mathbf{B}}\mathbf{q} - \mathbf{x}||^2 + \mu||\mathbf{q}||^2$$

where μ is the tradeoff or ridge regression parameter of the problem (also the hyperparameter). J is minimized by taking derivatives and equating to zero. The latter yields the following system of normal equations

$$(\tilde{\mathbf{B}}^H\tilde{\mathbf{B}} + \mu\mathbf{I})\mathbf{q} = \tilde{\mathbf{B}}^H\mathbf{x}$$

Replacing the circulant matrix, $\tilde{\mathbf{B}}$, by its spectral expansion

$$(\mathbf{W}\boldsymbol{\Lambda}_B\boldsymbol{\Lambda}_B^H\mathbf{W}^H + \mu\mathbf{I})\mathbf{q} = \mathbf{W}\boldsymbol{\Lambda}_B^H\mathbf{W}^H\mathbf{x} \tag{1.109}$$

Since $\mu\mathbf{I} = \mu\mathbf{W}\mathbf{I}\mathbf{W}^H$ we can write Eq. (1.109) as

$$\mathbf{W}(\mathbf{\Lambda}_B\mathbf{\Lambda}_B^H + \mu\mathbf{I})\mathbf{W}^H\mathbf{q} = \mathbf{W}\mathbf{\Lambda}_B^H\mathbf{W}^H\mathbf{x} \tag{1.110}$$

Designating the DFT of $\mathbf{q}$ and $\mathbf{x}$ by the vectors $\mathbf{q_f}$ and $\mathbf{x_f}$, we have

$$\mathbf{q_f} = \mathbf{W}^H\mathbf{q}$$

and

$$\mathbf{x_f} = \mathbf{W}^H\mathbf{x}$$

Eq. (1.110) becomes

$$(\mathbf{\Lambda}_B\mathbf{\Lambda}_B^H + \mu\mathbf{I})\mathbf{q_f} = \mathbf{\Lambda}_B^H\mathbf{x_f}$$

or

$$\mathbf{q_f} = (\mathbf{\Lambda}_B\mathbf{\Lambda}_B^H + \mu\mathbf{I})^{-1}\mathbf{\Lambda}_B^H\mathbf{x_f} \tag{1.111}$$

Since the matrix $(\mathbf{\Lambda}_B\mathbf{\Lambda}_B^H + \mu\mathbf{I})$ is diagonal we can easily compute the solution in terms of a spectral division. If the DFT samples of $\mathbf{q_f}$ and $\mathbf{x_f}$ are denoted by Q_k and X_k, respectively, the deconvolution can be expressed as

$$Q_k = \frac{\Lambda^*_{k\,k}}{\Lambda_{k\,k}\Lambda^*_{k\,k} + \mu}X_k$$

We have already shown that the eigenvalues of the circulant matrix are the Fourier coefficient of the signal that one wishes to deconvolve. In this case we have

$$\Lambda_{k\,k} = NB_k$$

and therefore the regularized solution becomes

$$Q_k = \frac{N\,B_k^*}{N^2\,B_kB_k^* + \mu}X_k \tag{1.112}$$

We will meet Eq. (1.112) again quite a few times. The hyperparameter μ, often referred to as the water-level, avoids the amplification of small spectral components in the spectral division.

An interesting aspect of the circulant matrix formulation is that the eigenvalues of the matrix $\tilde{\mathbf{B}}^T\tilde{\mathbf{B}}$ can be easily identified with the spectral power of the wavelet. The problem we have outlined shows something that is very well known, the main source of ill-posedness in deconvolution is the band-limited nature of the seismic wavelet. In other words, small spectral amplitudes correspond to small eigenvalues.

1.9.4.3 Efficient Computation of Large Problems

It often happens in practice that the computational cost associated with a particular algorithm is dominated by a matrix times vector multiplication that requires $O(M^2)$ operations for a $M \times M$ matrix. A particular case is one that we treat in Section 5.4.8.2, where the matrix is Toeplitz. We show, in that section, that the cost of the matrix-vector product can be considerably diminished by augmenting the Toeplitz matrix so that it becomes circulant and exploiting the relationship between a circulant matrix and the FT explored when dealing with Radon transforms in Chapter 5.

1.9.5 Polynomial and FT Wavelet Inversion

As we have seen in Section 1.8.1, the stability of dipole inversion depends crucially on the phase properties of the dipole. Thus, whereas minimum phase dipoles can be inverted causally, maximum phase dipoles must be inverted acausally. Suppose however, that we are interested in inverting a mixed phase wavelet, one containing both minimum and maximum phase dipole components. Clearly, to achieve a stable inverse, we must factor the wavelet into these components and invert each in the proper time direction. One way of doing this is to find all the roots of the wavelet polynomial and form the appropriate components from the roots inside and outside of the unit circle. This is not very propitious in general[12], however, and we look at the problem in some more detail.

Let w_n be the mixed phase wavelet with FT W_k and ZT $W(z)$. Let b_n be the minimum phase equivalent of w_n (we emphasize that b_n is not the same as the minimum phase component of w_n) that we obtain using the Hilbert transform for example (Section 2.3.2). In as much as we can determine the stable causal inverse of b_n, which we designate as b_n^{-1}, by

$$b_n^{-1} = \mathcal{Z}^{-1}\left[\frac{1}{B(z)}\right]$$

so, we can also determine this filter using the FT as

$$b_n^{-1} = \mathcal{F}^{-1}\left[\frac{1}{B_k}\right]$$

an expression that we can write as

$$\begin{aligned} b_n^{-1} &= \mathcal{F}^{-1}\left[\frac{B_k^*}{B_k B_k^*}\right] \\ &= \mathcal{F}^{-1}\left[\frac{B_k^*}{P_k}\right] \end{aligned} \tag{1.113}$$

where P_k is the power spectrum

The FT has no problem in inverting P_k. Zeros in this spectrum are not allowed by virtue of the fact that we have defined b_n to be minimum phase and, consequently, all zeros are outside $|z| = 1$[13]. It now becomes clear why the FT does not care about the wavelet phase for stability. P_k, as we have seen, is identical for w_n and b_n. To obtain the inverse of our mixed phase wavelet, we use Eq. (1.113) and write

$$w_n^{-1} = \mathcal{F}^{-1}\left[\frac{W_k^*}{W_k W_k^*}\right] \tag{1.114}$$

[12]Although recent work by Porsani and Ursin (2000) and Ursin and Porsani (2000) explores this approach in a novel and interesting manner.

[13]The problem of zeros on $|z| = 1$ in deconvolution is treated later when we consider the issue of regularization.

$$= \mathcal{F}^{-1}\left[\frac{W_k^*}{P_k}\right]$$

Because of the intimate relationship between the FT and the ZT, we can use the FT recipe to tackle the ZT approach. Transforming Eq. (1.114) into the z domain, we obtain

$$\begin{aligned} w_n^{-1} &= \mathcal{Z}^{-1}\left[\frac{W(z^{-1})}{W(z)W(z^{-1})}\right] \qquad (1.115)\\ &= \mathcal{Z}^{-1}\left[\frac{W(z^{-1})}{R(z)}\right] \end{aligned}$$

where $R(z)$ is the autocorrelation. As we know, however, $R(z)$ may be factored into minimum and maximum phase components

$$R(z) = W(z)W(z^{-1}) = B(z)B(z^{-1})$$

so that Eq. (1.115) becomes

$$\begin{aligned} w_n^{-1} &= \mathcal{Z}^{-1}\left[\frac{W(z^{-1})}{B(z)B(z^{-1})}\right]\\ &= \mathcal{Z}^{-1}\left[\frac{W(z^{-1})}{B^{-1}(z)B^{-1}(z^{-1})}\right]\\ &= w_{-n} * b_n^{-1} * b_{-n}^{-1} \qquad (1.116) \end{aligned}$$

where b_n^{-1} is the causal minimum phase inverse and b_{-n}^{-1} is the acausal maximum phase inverse. Eq. (1.116) describes the ZT procedure that is required. The steps are:

(1) Obtain the equivalent minimum phase wavelet, b_n.

(2) Compute the stable causal inverse, b_n^{-1}, by polynomial division and inverse z transformation.

(3) Time reverse b_n^{-1} to obtain b_{-n}^{-1} and convolve with b_n^{-1} (or autocorrelate b_n^{-1}).

(4) Convolve the result in step (3) with the time reversed wavelet, w_{-n}.

The reason why the FT does not require such steps is that the FT vector is constructed with positive and negative frequencies and, consequently, with positive (causal) and negative (acausal) time. Stability issues simply do not arise. The ZT approach, on the other hand, has to 'manufacture' -ve time and keep track of it. Clearly, the FT, in this particular application, is cleverer and more efficient. These remarks are not to be construed as suggesting that the ZT approach is not useful. In fact, in applications such as prediction and recursive filtering, for example, issues of stability and minimum phase are absolutely central.

APPENDICES

A.1 Expectations etc.,

Data processing and analysis has much to do with averages. In this regard, we have at our disposal an operator, E, that does averaging for us. The beauty of this operator is that it is linear, which means that

$$E[f+g] = E[f] + E[g]$$

As we know, linearity simplifies life considerably (it does not make it more interesting). To get a feel for E, consider a random experiment which has n possible outcomes, $x_1, x_2, \ldots, x_n$. The random variable, $\mathbf{x}$, which describes the experiment, can take on any of the n outcome values. We perform the experiment a very large number of times, N, and count $m_i, i = 1, n$, the number of outcomes favorable to x_i. In standard fashion, the average outcome is $\bar{x}$ where

$$\begin{aligned} \bar{x} &= \frac{1}{N}(m_1 x_1 + m_2 x_2 + \cdots + m_n x_n) \\ \bar{x} &= \frac{m_1 x_1}{N} + \frac{m_2 x_2}{N} + \cdots + \frac{m_n x_n}{N} \end{aligned}$$

But, the frequency definition of probability tells us that

$$\frac{m_i}{N} \to P_{\mathbf{x}}(x_i) \quad \text{as } N \to \infty$$

Consequently

$$\bar{x} = \sum_{i=1}^{n} x_i P_{\mathbf{x}}(x_i)$$

We write $E[\mathbf{x}] = \bar{x}$ where we introduce the expectation operator. Hence

$$E[\mathbf{x}] = \sum_i x_i P_{\mathbf{x}}(x_i)$$

and for the continuous case

$$E[\mathbf{x}] = \int x p_{\mathbf{x}}(x) dx$$

The extension to functions of random variables follows naturally. If, for example, $\mathbf{z} = g(\mathbf{x}, \mathbf{y})$

$$\begin{aligned} E[\mathbf{z}] &= \int z p_{\mathbf{z}}(z) dz \\ &= \int g(x, y) p_{\mathbf{x},\mathbf{y}}(x, y) dx dy \end{aligned}$$

As the simplest of examples of the use of E, consider the expression for the variance which is defined as

$$\sigma_x^2 = E[(\mathbf{x} - \mu_x)^2] = \int (x - \mu_x)^2 p_{\mathbf{x}}(x) dx$$

where μ_x is the mean of the associated pdf. But, using the linear properties of $E[\cdot]$, we have

$$\begin{aligned} E[(\mathbf{x} - \mu_x)^2] &= E[\mathbf{x}^2 - 2\mu_x + \mu_x^2] \\ &= E[\mathbf{x}^2] - 2\mu_x E[\mathbf{x}] + E[\mu_x^2] \\ &= \bar{\mathbf{x}}^2 - 2\mu_x^2 + \mu_x^2 \\ \sigma_x^2 &= \bar{\mathbf{x}}^2 - \mu_x^2 \end{aligned}$$

which expresses the well known relationship between variance and the 2nd moment.

Certain relationships involving E are particularly useful, and we employ them throughout the book. Of significance in the present context are relationships that are applicable to linear functions of random variables. For rv's $\mathbf{x}$ and $\mathbf{y}$, we have

$$\begin{aligned} E[a\mathbf{x} + b\mathbf{y}] &= aE[\mathbf{x}] + bE[\mathbf{y}] \\ \text{var}[a\mathbf{x} + b\mathbf{y}] &= a^2\text{var}[\mathbf{x}] + b^2\text{var}[\mathbf{y}] + 2ab\text{cov}[\mathbf{x}, \mathbf{y}] \end{aligned}$$

More generally

$$\begin{aligned} E\left[\sum_i a_i \mathbf{x}_i\right] &= \sum_i a_i E[\mathbf{x}_i] \\ \text{var}\left[\sum_i a_i \mathbf{x}_i\right] &= \sum_i a_i^2 \text{var}[\mathbf{x}_i] + \sum_i \sum_{\substack{j \\ i \neq j}} a_i \mathbf{x}_j \text{cov}[\mathbf{x}_i, \mathbf{x}_j] \end{aligned}$$

Also, if

$$\mathbf{y}_1 = \sum_i a_i \mathbf{x}_i$$

and

$$\mathbf{y}_2 = \sum_i b_i \mathbf{x}_i$$

a useful relationship is that

$$\text{cov}(\mathbf{y}_1, \mathbf{y}_2) = E[\mathbf{y}_1 \mathbf{y}_2] - E[\mathbf{y}_1] E[\mathbf{y}_2]$$

It is important to realize that, for the general nonlinear case, it is always true that

$$E[\mathbf{xy}] = E[\mathbf{x}]E[\mathbf{y}] + \text{cov}(\mathbf{x}, \mathbf{y})$$

A.1.1 The Covariance Matrix

We use second order statistics of our random vector, $\mathbf{x}$, constantly. Here they are.

- The mean of $\mathbf{x}$ is also a random vector defined by

$$\bar{\mathbf{x}} = E[\mathbf{x}]$$

- The correlation matrix, $\mathbf{R}_{\mathbf{xx}}$, is defined as

$$\mathbf{R}_{\mathbf{xx}} = E[\mathbf{xx}]^T$$

- We are usually concerned with second order statistics of zero mean processes which are characterized by covariance matrices, $\mathbf{C}_{\mathbf{xx}}$, where

$$\mathbf{C}_{\mathbf{xx}} = E[(\mathbf{x} - \bar{\mathbf{x}})(\mathbf{x} - \bar{\mathbf{x}})^T] = \mathbf{R} - \bar{\mathbf{x}}\bar{\mathbf{x}}^T$$

A.2 Lagrange Multipliers

Optimization involves the imposition of constraints. It is seldom that one is faced with the problem of obtaining the maximum or minimum of a cost function that is not subject to some constraint on the desired model solution. For example, consider a cost function, J, that is proportional to some quadratic form of the required solution vector, $\mathbf{x}$

$$J = \mathbf{x}^T \mathbf{I} \mathbf{x} \tag{A.1}$$

where we have set the usual weighting matrix to $\mathbf{I}$ for simplicity.

As a toy problem, we take a two element vector, $\mathbf{x}^T = [x_1, x_2]$, and we look for the minimum of J in Eq. (A.1). Clearly, with no constraint on $\mathbf{x}$, the solution is $\mathbf{x} = 0$ and is illustrated in Fig. 1.28(a). Suppose, however, that we do know something about our vector, and that is that it must lie on a curve defined by $2x_1 - x_2 = 5$. The question is, how do we impose this constraint? The solution is to build the constraint into a new cost function, L, sometimes designated as the Lagrangian, where we augment J by a constant, λ, multiplied by zero. This may seem counter productive but let us go on. The new cost function is

$$\begin{aligned} L &= J + \lambda(2x_1 - x_2 - 5) \\ &= (x_1^2 + x_2^2) + \lambda(2x_1 - x_2 - 5) \end{aligned} \tag{A.2}$$

where it is clear that the second term, formed from the constraint, is equal to zero and therefore does not alter the value of J. We treat λ as an undetermined multiplier, an unknown to be found, and we look for the optimal $\mathbf{x}$ by differentiating L with respect to $\mathbf{x}$ and λ

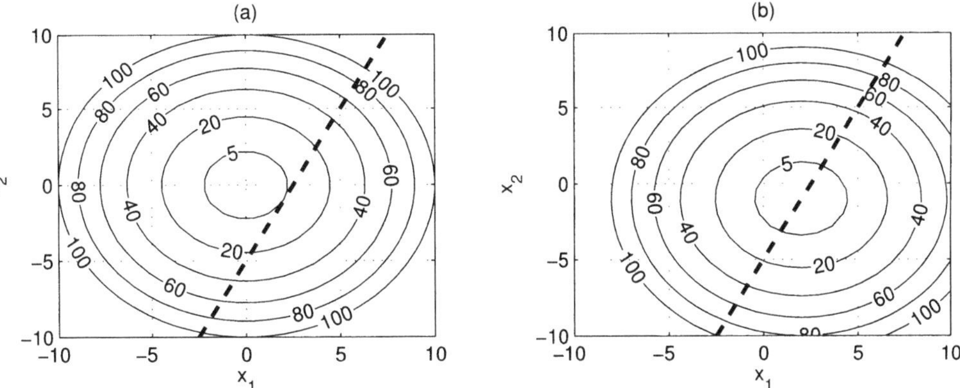

Figure 1.28: Illustrating constraints via Lagrange multipliers. (a) Contour plot of the quadratic surface without constraint. (b) Contour plot of the quadratic surface with constraint. The dashed line indicates the constraint $2x_1 - x_2 = 5$.

$$
\begin{aligned}
\frac{\partial L}{\partial x_1} &= 2x_1 + 2\lambda &= 0 &\quad\Rightarrow & x_1 &= -\lambda \\
\frac{\partial L}{\partial x_2} &= 2x_2 - \lambda &= 0 &\quad\Rightarrow & x_2 &= \lambda/2 \\
\frac{\partial L}{\partial \lambda} &= 2x_1 - x_2 - 5 &= 0 &\quad\Rightarrow & -5/2\lambda &= 5
\end{aligned}
$$

We see that the 3rd equation brought back the constraint, not surprisingly since it was designed to do just that, and we compute

$$\lambda = -2$$

and, by substitution into the first two equations

$$
\begin{aligned}
x_1 &= 2 \\
x_2 &= -1
\end{aligned}
$$

The solution is indicated in Fig. 1.28(b). In general, of course, we have various linear constraints that need be imposed, and we can designate these by $\mathbf{Ax} = \mathbf{f}$. The Lagrangian is now cast in terms of a vector of Lagrange multipliers, $\boldsymbol{\lambda}$, where

$$L = J + \boldsymbol{\lambda}^T(\mathbf{Ax} - \mathbf{f}) \tag{A.3}$$

The solution for $\mathbf{x}$ follow precisely the same steps as outlined for the toy example above.

Chapter 2

Linear Time Series Modelling

2.1 Introduction

Time[1] series modelling attempts to build a model for the available data. The importance of this approach cannot be overestimated. From the very early work of Yule (1927) on the modelling of sunspot numbers, to the present day when we attempt to model, for example, the behavior of multiply reflected energy in the earth, models allow us to understand, and hopefully predict, the behavior of systems that are approximately characterized by the data that we observe. As one particular example, we will examine just how modelling allows the estimation of smooth spectral densities where sample estimates often fail dismally. Another, ubiquitous, time series model that we investigate in this chapter, is one that describes a harmonic process in additive random noise.

We stress the *linear* in the title to this section. Nonlinear models are much more complicated to develop and apply. However, they are often imperative and always very rich in structure. Some discussion of this important topic will be presented in Section 2.8. We begin with a fundamental theorem in time series analysis.

2.2 The Wold Decomposition Theorem

The Wold decomposition theorem (Robinson, 1964) states that, any real valued, stationary stochastic process, x_t, can be decomposed into

$$x_t = u_t + v_t \tag{2.1}$$

where

u_t is a predictable (i.e., deterministic) process

v_t is a non predictable (i.e., stochastic) process

$u_t \perp v_t$ (processes are orthogonal).

[1]and of course this includes space.

Further

$$v_t = \sum_i \beta_i q_{t-i} \qquad (\sum |\beta_i|^2 \leq \infty) \tag{2.2}$$

where $\beta_0 = 1$ (to ensure that $\sigma_v^2 = \sigma_q^2$) and q_t has the properties

$$\begin{aligned} E[q_t] &= 0 \\ E[q_t q_s] &= \sigma_q^2 \delta_{ts} \end{aligned} \tag{2.3}$$

Eq. (2.2) is the representation of a moving average (MA) process and is the fundamental time series model. Ignoring u_t for the moment (let us suppose that it has been filtered out or is zero), (2.2) tells us that any stationary stochastic process may be represented by the convolution of a *source wavelet* (an energy signal, $\sum |\beta_i|^2 \leq \infty$) with a white noise process, q_t (Eq. (2.3)).

2.3 The Moving Average, MA, Model

It is, indeed, often the case that $u_t = 0$ [2] and β_i is a finite energy signal, a wavelet which we designate by b_t. Then, the moving average , MA, model is

$$x_t = \sum_i b_i \, q_{t-i} \tag{2.4}$$

We may view x_t as the response of a filter, with impulse response b_t, to a white noise input. Since b_t is of finite length, we call such filters FIR or finite impulse response filters. Fig (2.1) shows the flow diagram in terms of the unit delay operator z.

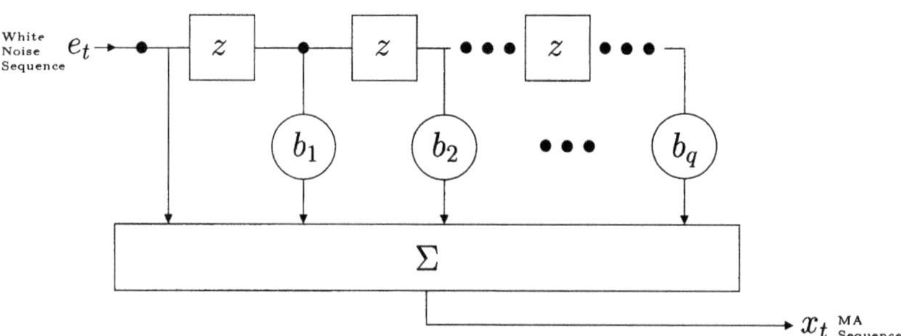

Figure 2.1: MA model flow diagram.

Let us look at the details of the moving average process. Transforming to the z domain and forming the autocovariance [3]

[2]We also, almost always and without loss of generality, assume that the process under consideration, $x(t)$, is zero mean.

[3]Since, as noted above, we generally assume that the process is zero mean, we deal with autocovariances rather than autocorrelations.

$$
\begin{aligned}
R_{xx}(z) &= X(z)\,X(\frac{1}{z}) \\
&= B(z)\,Q(z)\,B(\frac{1}{z})\,Q(\frac{1}{z})
\end{aligned}
$$

Hence

$$R_{xx}(z) = R_{bb}(z)\sigma_q^2 \tag{2.5}$$

In the time domain we have

$$r_k = r_{-k} = \begin{cases} \sigma_q^2 \sum b_t b_{t+k} & 0 \le k \le m \\ 0 & \text{otherwise} \end{cases}$$

The autocovariance of the MA process truncates after m terms. Eq. (2.5) is of central importance in seismic modelling and processing. It says that, given that the model of a seismogram is MA (Eq. (2.4)), where b_t is the seismic wavelet and q_t is the primary reflectivity series of the earth which is assumed white (this is not quite true but good enough at present) then, the autocovariance of the the seismic wavelet is the same, except for a scale factor, as the autocovariance of the seismic trace. The power spectrum of the MA process, which we require for later use, follows immediately

$$P_{xx}(\omega) = \sigma_q^2\,|1 + \sum_{k=1}^{m} b_k\, e^{-i\omega k\ \Delta t}|^2$$

which is in fact the *periodogram* of x_t.

2.3.1 Determining the Coefficients of the MA Model

We define two vectors

$$
\begin{aligned}
\mathbf{q}_t &= [q_t, q_{t-1}, \ldots, q_{t-m}]^T \\
\mathbf{b} &= [b_0, b_1, \ldots, b_m]^T
\end{aligned}
$$

Then

$$x_t = \mathbf{q}_t^T \mathbf{b}$$

We use the l_2 norm in an attempt to estimate $\mathbf{b}$. We know of course that the problem is ill-posed, one equation and two unknowns, but as we will see, the exercise is illuminating. The residual energy we wish to minimize is $E[e_t^2]$ where

$$e_t = x_t - \mathbf{q}_t^T\,\mathbf{b}$$

Minimizing $E[e_t^2]$ with respect to $\mathbf{b}$, we form

$$
\begin{aligned}
\nabla_{\mathbf{b}} E[e_t^2] &= \nabla_{\mathbf{b}} E[(x_t - \mathbf{q}_t^T\mathbf{b})^T(x_t - \mathbf{q}_t^T\mathbf{b})] \\
&= \nabla_{\mathbf{b}} E[x_t^2 - x_t\mathbf{q}_t^T\mathbf{b} - x_t\mathbf{b}^T\mathbf{q}_t + \mathbf{b}^T\mathbf{q}_t\mathbf{q}_t^T\mathbf{b}]
\end{aligned}
$$

where $\nabla_{\mathbf{b}}$ is the partial differential with respect to $\mathbf{b}$. Hence

$$-2E[x_t\mathbf{q}_t] + 2\sigma_q^2\,\mathbf{I}\,\mathbf{b} = 0$$

or

$$\mathbf{b} = \frac{1}{\sigma_q^2} E[x_t \mathbf{q}_t]$$

These are the normal equations for the MA model. The problem is that $E[x_t \mathbf{q}_t]$ is unknown. The only approach possible is using the relationship entailed in Eq. (2.5) and working with the covariance structure of x_t

$$R_{bb}(z) = \frac{1}{\sigma_q^2} R_{xx}(z)$$

and hence

$$|B(f)| = \frac{1}{\sigma_q^2} \sqrt{P_{xx}(f)}$$

Having obtained the amplitude spectrum of the unknown wavelet, $\mathbf{b}$, we need to associate a meaningful phase spectrum. We now remember our discussion of causal systems and refer to Eq. (1.83). We determine the phase spectrum

$$\Phi_B(f) = \frac{1}{\pi} \int \frac{\log |B(f')|}{f' - f} df'$$

and then the FT as

$$B(f) = |B(f)| e^{i\Phi_B(f)}$$

Finally

$$b_t = \mathcal{F}^{-1}[B(f)]$$

As we discussed previously, this phase is the minimum phase function which corresponds to the particular amplitude spectrum. Consequently, we have indeed determined the desired unknown wavelet, solving one equation in two unknowns. Of course, this was accomplished at the expense of two constraints. The ubiquitous constraint of minimum phase and the constraint of the whiteness of the primary reflectivity series. More on this subject later. The approach outlined above is called the Kolmogoroff factorization. Other approaches exist, for example, the double Levinson and autoregressive methods and we will deal with them in due course.

Let us now see how to compute the minimum phase wavelet, $b_t|_{MP}$, using the fast Fourier transform, FFT. This will lead us to the concept of the analytic signal which has an important place in seismic data processing.

2.3.2 Computing the Minimum Phase Wavelet via the FFT

Just as phasors simplify mathematical manipulations in engineering applications, so the analytic signal, $\hat{x}(t)$[4]

$$\hat{x}(t) = x(t) - i\mathcal{H}[x(t)] \tag{2.6}$$

[4]Note that we jump to continuous notation for convenience.

where $\mathcal{H}[\cdot]$ signifies Hilbert transformation, serves a very useful purpose in time series analysis. In seismology, such properties of $\hat{x}(t)$ as the envelope and instantaneous frequency have been used as seismic attributes. Also, this formulation allows the computation of rotation of signals, where the rotation implies a frequency independent phase shift.

Computation of $\hat{x}(t)$ is straight forward. We simply suppress the scaled negative frequency components of $X(f)$. To see this, we write $H(f)$ as the Heaviside function in the frequency domain and form $\hat{x}(t)$ as

$$\hat{x}(t) \Leftrightarrow 2H(f)X(f) \tag{2.7}$$

Since

$$H(f) \Leftrightarrow \frac{1}{2}\delta(t) + \frac{i}{2\pi t}$$

we have, from Eq. (2.7)

$$\begin{aligned}
\hat{x}(t) &= 2[\frac{1}{2}\delta(t) + \frac{i}{2\pi t}] * x(t) \\
&= x(t) - i(\frac{-1}{\pi t}) * x(t') \\
&= x(t) - i\mathcal{H}[x(t)]
\end{aligned} \tag{2.8}$$

We now use Eq. (2.8)[5] to compute $b_t|_{MP}$. Identifying $x(t)$ with $\log|B(f)|$, we write

$$\begin{aligned}
\log|\hat{B}(f)| &= \log|B(f)| - i\mathcal{H}[\log|B(f)|] \\
&= \log|B(f)| - i\Phi_B(f)
\end{aligned}$$

Hence

$$b_t|_{MP} = \mathcal{F}^{-1}\left[\exp(\log|\hat{B})(f)|\right] \tag{2.9}$$

The steps are, therefore:

(1) Form $|B(f)|$.

(2) Take the FFT of $\log|B(f)|$.

  (a) Suppress all negative frequencies.

  (b) Double all positive frequencies except zero frequency.

(3) Take the exponential.

(4) Take the inverse FFT.

We mentioned that another approach to spectral factorization is by means of the double Levinson recursion. This will become clear in the following section, where we explore autoregressive modelling of time series.

[5] From the definition

$$\mathcal{H}[x(t)] = \frac{1}{\pi}\int \frac{x(t')}{t'-t}dt'$$

we see that $\mathcal{H}[x(t)]$ may be written as

$$\mathcal{H}[x(t)] = \frac{-1}{\pi t} * x(t)$$

2.4 The Autoregressive, AR, Model

We have seen that the MA model is expressed in the z-domain as

$$X(z) = B(z)Q(z) \tag{2.10}$$

Let us assume that $B(z)$ is indeed minimum phase. Then, a stable, one-sided inverse, $G(z) = B^{-1}(z)$, exists and, using Eq. (2.10), we write

$$X(z) = G^{-1}(z)\, Q(z)$$

or

$$X(z)\, G(z) = Q(z) \tag{2.11}$$

Since

$$G(z) = B(z)^{-1} \quad \text{and} \quad b_0 = 1$$

we have

$$g_0\, b_0 = 1 \quad \text{therefore} \quad g_0 = 1$$

Since $B(z)$ is a polynomial of finite order, it is clear that $G(z)$ is a polynomial of infinite order. However, since $B(z)$ is minimum phase, $G(z)$ is also minimum phase and, unless $B(z)$ has zeros *very* close to $|z| = 1$, the coefficients of $G(z)$ will decay in amplitude to give an 'effective' polynomial of order p, say. We now transform Eq. (2.11) into the time domain. We have

$$X(z)\, [1 + g_1 z + g_2 z^2 + \cdots + g_p z^p] = Q(z)$$

and therefore

$$x_t + g_q x_{t-1} + g_2 x_{t-2} + \cdots + g_P x_{t-p} = q_t \tag{2.12}$$

Let us define a vector $\mathbf{a}$ with parameters

$$\mathbf{a} = [a_1, a_2, \ldots, a_p]^T \quad \text{such that} \quad a_i = -g_i, \quad i = 1, \ldots, p$$

Eq. (2.12) now becomes

$$x_t = a_1 x_{t-1} + a_2 x_{t-2} + \cdots + a_p x_{t-p} + q_t \tag{2.13}$$

This is the AR(p) representation of x_t. The a_i's are, clearly, *prediction* coefficients. x_t is predicted from p past values with an error q_t. Eq. (2.12) represents the convolution of a filter, $\mathbf{g} = 1 - \mathbf{a}$, with x_t to produce the error series q_t. $\mathbf{g}$ is, consequently, called the prediction error operator or PEO. q_t, as defined by the Wold Decomposition, is a white error series, also called the *innovation*. In as much as the moving average filter is FIR, the AR filter, represented diagrammatically in Fig. 2.2, is IIR (infinite impulse response). The AR process is an example of a feedback filter. The corresponding flow diagram is shown in Fig. 2.3.

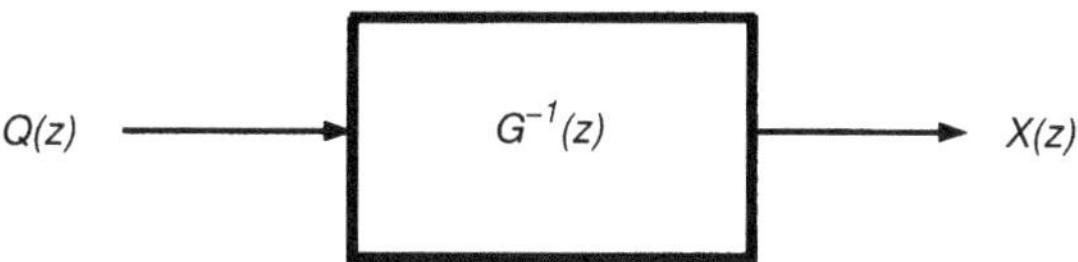

Figure 2.2: The AR IIR filter.

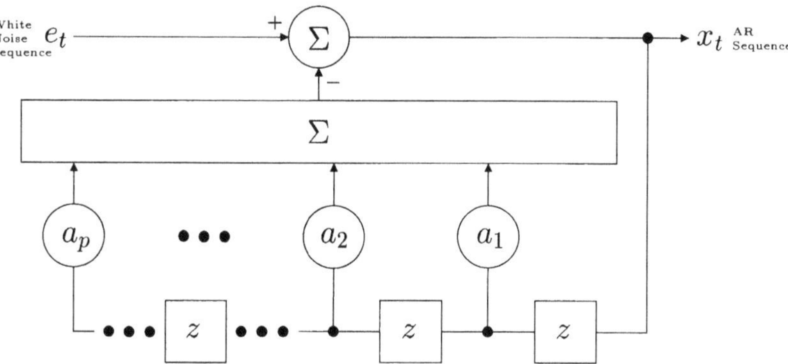

Figure 2.3: AR model flow diagram.

2.4.1 Autocovariance of the AR Process

We now examine the autocovariance of the AR process. This will lead us to some fascinating issues which are involved with this particular model of a time series. We have

$$X(z) = \frac{Q(z)}{G(z)}$$

and therefore, the autocorrelation in the z domain is

$$R_{xx}(z) = \frac{1}{G(z)} \frac{1}{G(1/z)} Q(z)Q(1/z)$$

leading to

$$G(z)R_{xx}(z) = G^{-1}(1/z)\sigma_q^2 \tag{2.14}$$

Transforming to time

$$(1 + g_1 z + \cdots + g_p z^p)(\cdots + r_2 z^{-2} + r_1 z + r_0 + \cdots + r_2 z^2 + \cdots) = \sigma_q^2 (\cdots + x_1 z^{-1} + 1)$$

Equating coefficients of like powers of z

$$\begin{array}{ccccccccc}
r_0 & + & g_1 r_{-1} & + & g_2 r_{-2} & + \cdots + & g_p r_{-p} & = & \sigma_q^2 \\
r_1 & + & g_1 r_0 & + & g_2 r_{-1} & + \cdots + & g_p r_{-p+1} & = & 0 \\
\vdots & & \vdots & & \vdots & & \vdots & & \vdots \\
r_1 & + & g_1 r_0 & + & g_2 r_{-1} & + \cdots + & g_p r_{-p+1} & = & 0
\end{array}$$

In matrix form, since $r_{-j} = r_j$ for real x_t

$$\begin{bmatrix} r_0 & r_1 & \cdots & r_p \\ r_1 & r_0 & \cdots & r_{p-1} \\ \vdots & \ddots & \ddots & \vdots \\ r_p & r_{p-1} & \cdots & r_0 \end{bmatrix} \begin{bmatrix} 1 \\ g_1 \\ \vdots \\ g_p \end{bmatrix} = \begin{bmatrix} \sigma_q^2 \\ 0 \\ \vdots \\ 0 \end{bmatrix} \tag{2.15}$$

or

$$\mathbf{Rg} = \sigma^2 \mathbf{e}_1 \tag{2.16}$$

where $\mathbf{e}_1^T = [1, 0, \ldots, 0]$.

R is, as we have mentioned briefly in Section 1.8.3, a very special matrix. It is symmetric Toeplitz, which means that it is symmetric and equidiagonal. The beauty of this structure is that Norman Levinson (Levinson, 1947) devised an algorithm, the famous Levinson recursion, which solves Eq. (2.16) in $(p+1)^2$ operations rather than conventional methods which require $(p+1)^3$ operations. We will deal with this algorithm a little later. Eqs. (2.15) are called the normal equations for the autoregressive problem and we will derive them again from a different point of view. Since Eq. (2.14) gives rise to

$$R_{xx}(z) = \frac{\sigma_q^2}{R_{gg}(z)}$$

and we can immediately write the autoregressive power spectrum as

$$P_{AR}(\omega) = \frac{\sigma_q^2}{|1 + \sum_k g_k e^{-i\omega k\ \Delta t}|^2} \tag{2.17}$$

We will see later that Eq. (2.17) is equivalent to the *maximum entropy* power spectrum.

2.4.2 Estimating the AR Parameters

Let us, first of all, derive Eq. (2.15) from a different viewpoint. We write Eq. (2.13) as

$$x_t = \mathbf{x}_{t-1}^T \mathbf{a} + q_t$$

where

$$\mathbf{x}_{t-1} = [x_{t-1}, x_{t-2}, \ldots, x_{t-p}]^T$$

Using the expectation operator, $E[\,\cdot\,]$, and performing ensemble averages, we determine **a** in traditional manner, by minimizing the average energy of the prediction error, $E[q_t^2]$.

$$\begin{aligned} E[q_t^2] &= E[(x_t - \mathbf{x}_{t-1}^T \mathbf{a})^T (x_t - \mathbf{x}_{t-1}^t \mathbf{a})] \\ &= E[x_t^2] - 2E[x_t \mathbf{x}_{t-1}^T \mathbf{a}] + E[\mathbf{a}^T \mathbf{x}_{t-1} \mathbf{x}_{t-1}^T \mathbf{a}] \end{aligned} \tag{2.18}$$

Differentiating Eq. (2.18) with respect to **a** and equating to zero, obtains

$$-2E[x_t \mathbf{x}_{t-1}] + 2E[\mathbf{x}_{t-1} \mathbf{x}_{t-1}^T \mathbf{a}] = 0$$

The respective ensemble averages are identified as

$$E[x_t \mathbf{x}_{t-1}^T] = \mathbf{r} = [r_1, r_2, \ldots, r_p]^T$$

where $\mathbf{r}$ is the autocovariance vector, and

$$E[\mathbf{x}_{t-1}\mathbf{x}_{t-1}^T] = \mathbf{R}_p$$

where $\mathbf{R}_p$ is the symmetric, Toeplitz autocovariance matrix of dimension $(p \times p)$. The final result is

$$\mathbf{R}_p \mathbf{a} = \mathbf{r} \tag{2.19}$$

which are the normal equations for $\mathbf{a}$, the vector of AR parameters or, equivalently, the prediction filter. In general, we solve for $\mathbf{g}$, the prediction error operator, or PEO. To do this, we augment Eq. (2.19) by multiplying Eq. (2.13) by x_t and taking expectations

$$E[x_t x_t] = a_1 E[x_{t-1} x_t] + \cdots + a_p E[x_{t-p} x_t] + E[q_t x_t]$$

Therefore

$$r_0 = a_1 r_1 + \cdots + a_p r_p + E[q_t x_t] \tag{2.20}$$

To evaluate the last term, multiply Eq. (2.13) by q_t and take expectations

$$E[q_t x_t] = a_1 E[q_t x_{t-1}] + \cdots + a_p E[q_t x_{t-p}] + E[q_t^2]$$

Since q_t is only correlated with x_t, all the $E[\cdot]$ terms involving q_t and x_t on the right hand side are zero. Finally,

$$E[q_t x_t] = \sigma_q^2 \tag{2.21}$$

and Eq. (2.20) becomes

$$r_0 = a_1 r_1 + \cdots + a_p r_p + \sigma_q^2$$

Augmenting the p normal equations expressed by Eq.s (2.19) with Eq. (2.21)

$$r_0 - \mathbf{r}^T \mathbf{a} \quad = \quad \sigma_q^2$$

and

$$\mathbf{r} - \mathbf{R}_p \mathbf{a} \quad = \quad 0$$

Replacing the vector $1 - \mathbf{a}$ by $\mathbf{g}$ and combining the above two equations, we obtain

$$\mathbf{R}_{p+1} \mathbf{g} \quad = \quad \sigma_q^2 \, \mathbf{e}_1 \tag{2.22}$$

where the Toeplitz matrix is now of order $(p+1 \times p+1)$. From now on, for convenience, we drop the subscript $p+1$.

2.4.3 The Levinson Recursion

We mentioned that the special structure of **R** allows the solution of the normal equations in only $(p+1)^2$ operations. It turns out that all Wiener optimal filtering problems (eg., Section 1.8.3) entail the solution of equations of the form $\mathbf{Rh} = \mathbf{v}$, where $\mathbf{h}$ is the Wiener filter and $\mathbf{v}$ is a vector which depends on the particular quest. The Levinson recursion which we will discuss here will be for the solution of Eq. (2.22), i.e., to obtain the prediction error operator. The general recursion is an extension of the PEO recursion and we will not deal with it here. Interested readers are referred to Robinson and Treitel (2002).

A convenient and illuminating way to derive and explain the Levinson recursion is to assume that we know the solution at order 2 say, and seek the solution at order 3. i.e., we seek g_{13}, g_{23}, g_{33} and P_3 in

$$\begin{bmatrix} r_0 & r_1 & r_2 & r_3 \\ r_1 & r_0 & r_1 & r_2 \\ r_2 & r_1 & r_0 & r_1 \\ r_3 & r_2 & r_1 & r_0 \end{bmatrix} \begin{bmatrix} 1 \\ g_{13} \\ g_{23} \\ g_{33} \end{bmatrix} = \begin{bmatrix} P_3 \\ 0 \\ 0 \\ 0 \end{bmatrix} \tag{2.23}$$

assuming that we know g_{12}, g_{22} and P_2. The middle two equations above are

$$\begin{aligned} r_1 + r_0 g_{13} + r_1 g_{23} + r_2 g_{33} &= 0 \\ r_2 + r_1 g_{13} + r_0 g_{23} + r_1 g_{33} &= 0 \end{aligned}$$

Therefore

$$\begin{bmatrix} r_0 & r_1 \\ r_1 & r_0 \end{bmatrix} \begin{bmatrix} g_{13} \\ g_{23} \end{bmatrix} = -\begin{bmatrix} r_1 \\ r_2 \end{bmatrix} - g_{33} \begin{bmatrix} r_2 \\ r_1 \end{bmatrix}$$

If

$$\begin{bmatrix} r_0 & r_1 \\ r_1 & r_0 \end{bmatrix} = \mathbf{R}_1$$

we can write

$$\mathbf{R}_1 \begin{bmatrix} g_{13} \\ g_{23} \end{bmatrix} = -\begin{bmatrix} r_1 \\ r_2 \end{bmatrix} - g_{33} \begin{bmatrix} r_2 \\ r_1 \end{bmatrix}$$

and consequently

$$\begin{bmatrix} g_{13} \\ g_{23} \end{bmatrix} = -\mathbf{R}_1^{-1} \begin{bmatrix} r_1 \\ r_2 \end{bmatrix} - g_{33} \mathbf{R}_1^{-1} \begin{bmatrix} r_2 \\ r_1 \end{bmatrix}$$

But, for the known order 2, we have

$$\begin{bmatrix} r_0 & r_1 & r_2 \\ r_1 & r_0 & r_1 \\ r_2 & r_1 & r_0 \end{bmatrix} \begin{bmatrix} 1 \\ g_{12} \\ g_{22} \end{bmatrix} = \begin{bmatrix} P_2 \\ 0 \\ 0 \end{bmatrix}$$

and the last two equations give

$$\begin{bmatrix} g_{12} \\ g_{22} \end{bmatrix} = -\mathbf{R}_1^{-1} \begin{bmatrix} r_1 \\ r_2 \end{bmatrix} \tag{2.24}$$

Substituting into Eq. (2.24)

$$\begin{bmatrix} g_{13} \\ g_{23} \end{bmatrix} = \begin{bmatrix} g_{12} \\ g_{22} \end{bmatrix} + g_{33} \begin{bmatrix} g_{22} \\ g_{21} \end{bmatrix} \tag{2.25}$$

In order to find the only unknown, g_{33}, on the right hand side of Eq. (2.25), the Levinson 'trick' is to augment the system by forming a sum of *forward* and *backward* known prediction error operators, i.e., from Eq. (2.25) we write

$$\begin{bmatrix} 1 \\ g_{13} \\ g_{23} \\ g_{33} \end{bmatrix} = \underbrace{\begin{bmatrix} 1 \\ g_{12} \\ g_{22} \\ 0 \end{bmatrix}}_{\text{forward PEO}} + g_{33} \underbrace{\begin{bmatrix} 0 \\ g_{22} \\ g_{12} \\ 1 \end{bmatrix}}_{\text{backward PEO}} \tag{2.26}$$

Premultiplying by $\mathbf{R}_3$, we obtain

$$\mathbf{R}_3 \begin{bmatrix} 1 \\ g_{13} \\ g_{23} \\ g_{33} \end{bmatrix} = \mathbf{R}_3 \left[\begin{pmatrix} 1 \\ g_{12} \\ g_{22} \\ 0 \end{pmatrix} + g_{33} \begin{pmatrix} 0 \\ g_{22} \\ g_{12} \\ 1 \end{pmatrix} \right]$$

$$= \begin{bmatrix} P_2 \\ 0 \\ 0 \\ \Delta_2 \end{bmatrix} + g_{33} \begin{bmatrix} \Delta_2 \\ 0 \\ 0 \\ P_2 \end{bmatrix} = \begin{bmatrix} P_3 \\ 0 \\ 0 \\ 0 \end{bmatrix}$$

The unknown Δ_2 is determined from the last equation in

$$\mathbf{R}_3 \begin{bmatrix} 1 \\ g_{12} \\ g_{22} \\ 0 \end{bmatrix} = \begin{bmatrix} P_2 \\ 0 \\ 0 \\ \Delta_2 \end{bmatrix} \tag{2.27}$$

which is

$$r_3 + g_{12}r_2 + g_{22}r_1 = \Delta_2$$

From the last row of Eq. (2.27), we get

$$\Delta_2 + g_{33}P_2 = 0$$

and therefore

$$g_{33} = -\frac{\Delta_2}{P_2}$$

The updated power is obtained from the first row of Eq. (2.27) as

$$\begin{aligned} P_3 &= P_2 + g_{33}\Delta_2 \\ &= P_2 + g_{33}(-P_2 g_{33}) \\ &= P_2(1 - g_{33}^2) \end{aligned} \tag{2.28}$$

2.4.3.1 Initialization

For the system

$$\begin{bmatrix} r_0 & r_1 \\ r_1 & r_0 \end{bmatrix} \begin{bmatrix} 1 \\ g_{11} \end{bmatrix} = \begin{bmatrix} P_1 \\ 0 \end{bmatrix}$$

we have

$$\begin{aligned} r_1 + r_0 g_{11} &= 0 \qquad g_{11} = -\frac{r_1}{r_0} \\ r_0 + r_1 g_{11} &= P_1 = r_0 - \frac{r_1^2}{r_0} \end{aligned}$$

The recursion now proceeds as outlined above.

2.4.4 The Prediction Error Operator, PEO

The prediction error plays such an important role in many of the aspects which we discuss in this book, that some detail at this stage is worth while. We follow Claerbout (1976), whose style is so graceful. Let us write the AR model in terms of the PEO in matrix form. We do this for the order 2 case considered above for the purpose of continuity. Thus

$$\mathbf{X}_0\mathbf{g}_2^+ = \mathbf{e}_2^+$$

where we define the following quantities. $\mathbf{X}_0$ is the regressor matrix for the problem, defined by

$$\begin{bmatrix} x_0 & 0 & 0 & 0 \\ \vdots & x_0 & 0 & 0 \\ \vdots & \vdots & \ddots & \vdots \\ x_3 & \cdots & x_1 & x_0 \\ \vdots & \vdots & \vdots & \vdots \\ x_{N-1} & \cdots & \cdots & x_{N-4} \\ 0 & x_{N-1} & \vdots & \vdots \\ \vdots & \vdots & \ddots & \vdots \\ 0 & 0 & \cdots & x_{N-1} \end{bmatrix}$$

$(\mathbf{g}_2^+)^T = [1, g_{12}, g_{22}, 0]$ the forward PEO with a zero extension, and $\mathbf{e}_2^+$ is the forward prediction error for order 2. We form the forward mean squared error, mse, defined by E_2^+, as

$$\begin{aligned} E_2^+ &= (\mathbf{e}_2^+)^T \mathbf{e}_2^+ \\ &= (\mathbf{g}_2^+)^T \mathbf{X}_0^T \mathbf{X}_0 \mathbf{g}_2^+ \\ &= (\mathbf{g}_2^+)^T \mathbf{R}_3 \mathbf{g}_2^+ \end{aligned} \tag{2.29}$$

where $\mathbf{R}_3$ is the Toeplitz matrix previously defined. From Eq. (2.27) we can write

$$\mathbf{R}_3 \mathbf{g}_2^+ = \begin{bmatrix} P_2 \\ 0 \\ 0 \\ \Delta_2 \end{bmatrix} \tag{2.30}$$

and substituting into Eq. (2.29)

$$E_2^+ = (\mathbf{g}_2^+)^T \begin{bmatrix} P_2 \\ 0 \\ 0 \\ \Delta_2 \end{bmatrix} = P_2$$

We have thus far considered an error determined by convolution in the forward direction. Let us consider, out of curiosity, what happens if we define a *backward* error, $\mathbf{e}_2^-$, given by

$$\mathbf{X}_0 \mathbf{g}_2^- = \mathbf{e}_2^-$$

where $(\mathbf{g}_2^-)^T = [0, g_{22}, g_{12}, 1]$ is the zero extended backward PEO. Forming the backward mse, E_2^-

$$\begin{aligned} E_2^- &= (\mathbf{e}_2^-)^T \mathbf{e}_2^- \\ &= (\mathbf{g}_2^-)^T \mathbf{X}_0^T \mathbf{X}_0 \mathbf{g}_2^- \\ &= (\mathbf{g}_2^-)^T \mathbf{R}_3 \mathbf{g}_2^- \end{aligned} \tag{2.31}$$

Because of the Toeplitz structure of $\mathbf{R}_3$

$$\mathbf{R}_3 \mathbf{g}_2^- = \begin{bmatrix} \Delta_2 \\ 0 \\ 0 \\ P_2 \end{bmatrix} \tag{2.32}$$

Consequently

$$E_2^- = (\mathbf{g}_2^-)^T \begin{bmatrix} \Delta_2 \\ 0 \\ 0 \\ P_2 \end{bmatrix} = P_2$$

and therefore

$$E_2^- = E_2^+$$

Since the forward and backward errors are equal, one may ask what the point is of introducing the backward error? The answer to this question lies at the core of many different algorithms which employ the forward/backward philosophy. Let us form the error cross power defined by

$$\begin{aligned} E_2^c &= (\mathbf{e}_2^-)^T \mathbf{e}_2^+ \\ &= (\mathbf{g}_2^-)^T \mathbf{X}_0^T \mathbf{X}_0 \mathbf{g}_2^+ \\ &= (\mathbf{g}_2^-)^T \mathbf{R}_3 \mathbf{g}_2^+ \end{aligned}$$

Using Eq. (2.30)

$$E_2^c = (\mathbf{g}_2^-)^T \begin{bmatrix} P_2 \\ 0 \\ 0 \\ \Delta_2 \end{bmatrix} = \Delta_2$$

We have an answer to our question. The unknown parameter Δ, which is required in the Levinson recursion, is computed as the dot product of the forward and backward error vectors. Indeed, using the error vectors, we may write

$$P_2 = (\mathbf{e}_2^+)^T \mathbf{e}_2^+ = (\mathbf{e}_2^-)^T \mathbf{e}_2^-$$

and

$$
\begin{aligned}
g_{33} &= -\frac{\Delta_2}{P_2} = -\frac{(\mathbf{e}_2^+)^T\mathbf{e}_2^-}{(\mathbf{e}_2^+)^T\mathbf{e}_2^+} = -\frac{(\mathbf{e}_2^+)^T\mathbf{e}_2^-}{(\mathbf{e}_2^-)^T\mathbf{e}_2^-} \\
&= -\frac{(\mathbf{e}_2^+)^T\mathbf{e}_2^-}{2[(\mathbf{e}_2^+)^T\mathbf{e}_2^+ + (\mathbf{e}_2^-)^T\mathbf{e}_2^-]}
\end{aligned}
$$

Clearly, one can compute the whole recursion using the concept of forward and backward errors. This is important in various respects. Specifically, as we will show later, the computation of the forward and backward errors in a manner somewhat different to the one described above, lead John Burg to a novel autocovariance estimator which he used in the determination of the maximum entropy power spectrum.

2.4.5 Phase Properties of the PEO

In deriving Eq. (2.16) the assumption was made that $B(z)$ was minimum phase. This implies that $G(z)$ is also minimum phase (since $G(z) = 1/B(z)$). We must, therefore, ensure that **g**, when computed using Eq. (2.16), is minimum delay . In fact, it may be shown that for stationarity of an autoregressive process, **g** *must* be minimum delay.

Before we go on to define the phase properties of the PEO let us write Eq. (2.26) for arbitrary order in z transform notation. Writing the coefficient g_{kk} as c_k we obtain

$$
G_{k+1}(z) = G_k(z) + c_k z^{k+1} G_k(\frac{1}{z}) \tag{2.33}
$$

c_k has a very special significance in seismology which we discuss at some length in Appendix A1. At this stage we refer to it as the kth reflection coefficient. Eq. (2.33) concisely expresses the heart of the famous Levinson recursion.

2.4.5.1 Proof of the Minimum Delay Property of the PEO

Many different proofs of this important property exist, each demonstrating a different facet of the AR model and the Levinson recursion. Here we present one of the earliest proofs and, for the sake of completeness and mathematical interest, we describe two other proofs in Appendix A2.

Considering the system of order 3 of Eq. (2.23) again, we note that since P_3 is obtained as $(\mathbf{e}_3^+)^T\mathbf{e}_3^+$, $P_3 \geq 0$. In fact, the mse will, in general, always be greater then zero (the special case of $P_k = 0$ will be considered later) and we write $P_k > 0 \ \ k = 0, 1, \ldots, p+1$. Since, during the recursion, the powers are updated by means of Eq. (2.28) we can deduce that $|g_{33}| < 1$. This implies that the kth reflection coefficient is bounded by $-1 < c_k < 1$. We now state a theorem due to Rouché which is used in the proof (Marden, 1966).

Theorem 1 (Rouché) *If two polynomials, $P(z)$ and $Q(z)$, satisfy $|P(z)| > |Q(z)|$ on $|z| = 1$, then $P(z)$ and $P(z)+Q(z)$ have the same number of zeroes inside $|z| = 1$.*

In reference to Eq. (2.33), we identify $P(z)$ with $G_k(z)$ and $Q(z)$ with $c_k z^{k+1} G_k(1/z)$. Determining $|Q(z)|$ we have

$$|Q(z)| = |c_k||z^{k+1}||G_k(\frac{1}{z})|$$

Now, $|c_k| < 1, |z_{k+1}| = 1$ and $|G_k(1/z)| = |G_k(z)|$. Consequently $|P(z)| > |Q(z)|$ which implies that $G_k(z)$ and $G_{k+1}(z)$ have the same number of zeroes inside $|z| = 1$. Repeating the argument, we deduce that $G_1(z)$ and $G_0(z)$ have the same number of zeros inside $|z| = 1$. But $G_0(z)$ has no zeroes inside $|z| = 1$ and, consequently, neither do $G_1(z), G_2(z), \ldots, G_p(z)$. $G_p(z)$ is, therefore, minimum phase, which establishes the phase property of the PEO.

Remarks

(1) $c_k = g_{k+1,k+1}$ in Eq. (2.33) is the k-th *reflection coefficient* and is, indeed, the reflection coefficient of a layered earth which may be modelled in terms of the Levinson recursion. This fascinating aspect of the Levinson scheme is explored in Appendix A1.

(2) Since, as we have seen, $G(z) = B^{-1}(z)$, performing the Levinson recursion twice is one way of obtaining the corresponding minimum phase wavelet (more on this later).

(3) We remark that the minimum phase property of the PEO, which is required by the AR model and is imposed by the Levinson recursion, is only true for normal equations which describe the prediction error filter. Normal equations for which the right hand side is different do not, necessarily, enjoy this characteristic.

We have dealt with both moving average and autoregressive models. We now combine these two models and discuss the most general, linear, model which is of autoregressive-moving average, or ARMA, form.

2.5 The Autoregressive Moving Average, ARMA, Model

In as much as the MA is an all zero model and the AR is an all pole model, so the ARMA is a pole and zero model. We write

$$X(z) = \frac{B(z)}{G(z)} Q(z)$$

The difference equation is

$$x_t = \sum_{l=1}^{p} a_l x_{t-l} + q_t + \sum_{n=1}^{m} b_n q_{t-n}$$

where we have written the model as ARMA(p, m). The AR and MA parts are coupled in the autocovariance of the process, but decouple for lags $> m$ and then depend only on the AR coefficients. This is a very useful fact when solving for ARMA coefficients. The flow diagram for the ARMA model, shown in Fig. (2.4), illustrates clearly the process by which x_t is formed as a result of a white input, q_t.

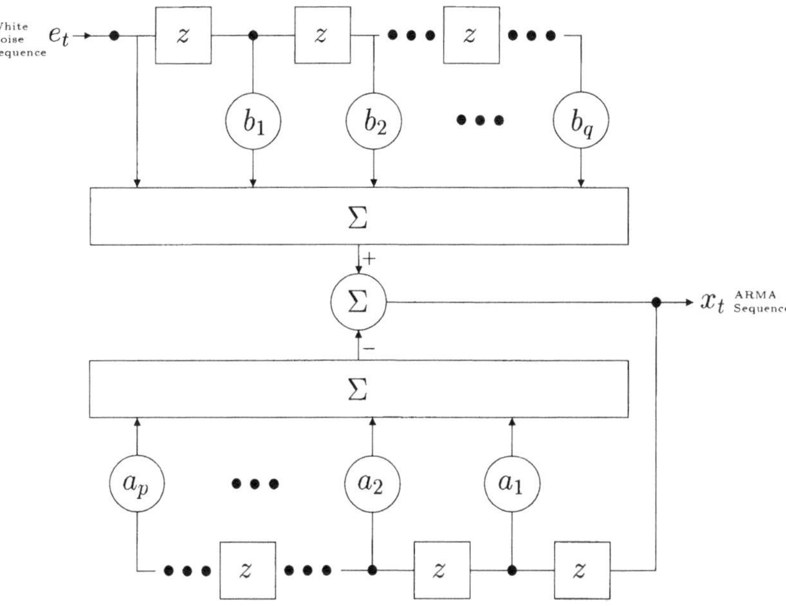

Figure 2.4: ARMA model flow diagram.

We will not deal with details of the ARMA process here for reasons of space and refer the reader to the bible on the subject, the book by Box and Jenkins (1976). We will, however, describe a special ARMA process which has important impact on high resolution properties of estimators, one of the themes of this book.

2.5.1 A Very Special ARMA Process

Consider a harmonic process

$$x_t = \Re[e^{i\omega t}]$$

Then

$$\begin{aligned} x_{t+1} &= \Re[e^{i\omega(t+1)}] \\ &= [\cos\omega t\cos\omega - \sin\omega t\sin\omega] \\ &= x_t\cos\omega - v_t \end{aligned}$$

where

$$v_t = \sin\omega t\sin\omega$$

In similar fashion

$$x_{t-1} = x_t \cos\omega + v_t$$

Forming the sum, we obtain

$$x_{t+1} + x_{t-1} = 2x_t \cos\omega$$

Letting $t+1 \rightarrow t$

$$\begin{aligned} x_t &= 2\cos\omega\, x_{t-1} - x_{t-2} \\ &= a_1 x_{t-1} + a_2 x_{t-2} \end{aligned}$$

where $a_1 = 2\cos\omega$ and $a_2 = -1$

In general, for p harmonics

$$x_t = \sum_{k=1}^{2p} a_k x_{t-k}$$

and represents a special kind of AR process in that the innovation is zero. Now consider the addition of random noise n_t (for which, $E[n_t n_s] = \delta_{ts}\sigma_n^2$)

$$\begin{aligned} y_t &= x_t + n_t \\ &= \sum_{k=1}^{p} a_k s_{t-k} + n_t \end{aligned}$$

for $p/2$ harmonics. Substituting $y_{t-k} = x_{t-k} + n_{t-k}$ into the above equation, we obtain

$$y_t = \sum_k a_k y_{t-k} + n_t + \sum_k a_k n_{t-k}$$

This is a special ARMA(p,p) process with equal AR and MA parameters, an equation first derived by Ulrych and Clayton (1976a). Rewriting the last equation in terms of the PEO, $\mathbf{g}$, and defining

$$\begin{aligned} \mathbf{y}_t &= [y_t, y_{t-1}, \ldots, y_{t-p}]^T \\ \mathbf{n}_t &= [n_t, n_{t-1}, \ldots, n_{t-p}]^T \end{aligned}$$

we obtain

$$\mathbf{y}_t^T \mathbf{g} = \mathbf{n}_t^T \mathbf{g}$$

Premultiplying by $\mathbf{y}_t$ and taking expectations

$$\begin{aligned} E[\mathbf{y}_t\mathbf{y}_t^T]\mathbf{g} &= E[\mathbf{y}_t\mathbf{n}_t^T]\mathbf{g} \\ &= E[(\mathbf{x}_t+\mathbf{n}_t)\mathbf{n}_t^T]\mathbf{g} \\ &= E[\mathbf{x}_t\mathbf{n}_t^T]\mathbf{g} + E[\mathbf{n}_t\mathbf{n}_t^T]\mathbf{g} \end{aligned} \tag{2.34}$$

But, since the ubiquitous assumption is that signal and noise are uncorrelated, $E[\mathbf{x}_t\mathbf{n}_t^T] = 0$. Hence Eq. (2.34) becomes

$$\mathbf{R}\mathbf{g} = \sigma_n^2 \mathbf{I}\,\mathbf{g} \tag{2.35}$$

This is an eigenvalue-eigenvector equation (Ulrych and Clayton, 1976a) . The correct $\mathbf{g}$ corresponds to the minimum eigenvalue which is the noise variance (a proof is given in Section 5.2.8). This eigenvector has all its roots on $|z| = 1$ (Marple, 1987). $\sigma_n^2/|G(z)|^2{}_{z=e^{-i\omega}}$ corresponds to the Pisarenko spectral estimator (Pisarenko, 1973) that we meet again, in some detail, in Section 5.2.8. $\mathbf{R}$ admits the spectral decomposition

$$\mathbf{R} = \mathbf{G}\mathbf{\Lambda}\mathbf{G}^T \tag{2.36}$$

where $\mathbf{G}$ is the matrix of eigenvectors and $\mathbf{\Lambda}$ is the diagonal matrix of eigenvalues.

Let us now designate the PEO which corresponds to $\mathbf{R}$ by $\tilde{\mathbf{g}}$ to distinguish it from $\mathbf{g}$ in Eq. (2.35). Then

$$\mathbf{R}\tilde{\mathbf{g}} = \sigma^2 \mathbf{e}_1$$

and

$$\tilde{\mathbf{g}} = \mathbf{R}^{-1}\sigma^2\mathbf{e}_1$$

But from Eq. (2.36)

$$\begin{aligned} \mathbf{R}^{-1} &= (\mathbf{G}\mathbf{\Lambda}^{-1}\mathbf{G}^T)^{-1} \\ &= (\mathbf{G}^T)^{-1}\mathbf{\Lambda}^{-1}\mathbf{G}^{-1} \\ &= \mathbf{G}\mathbf{\Lambda}^{-1}\mathbf{G}^T \end{aligned}$$

and therefore

$$\tilde{\mathbf{g}} = \sigma^2\mathbf{G}\mathbf{\Lambda}^{-1}\mathbf{G}^{\mathbf{T}}\mathbf{e}_1$$

In summation form

$$\tilde{\mathbf{g}} = \sigma^2 \sum_k \frac{g_{k1}}{\lambda_k}\mathbf{g}_k$$

Therefore, the PEO is the weighted sum of the eigenvectors of R and is most heavily weighted by the minimum eigenvalue[6].

2.6 MA, AR and ARMA Models in Seismic Modelling and Processing

The seismic model is, in general, formulated as

$$x_t = w_t * q_t + n_t \tag{2.37}$$

where

w_t: is the seismic wavelet which is a superposition of earth and instrument response. At this stage we assume all deterministic effects have been removed and, therefore, w_t is the response of the earth to the energy source used.

q_t: is the reflectivity of the earth which consists of all primary reflections, q_t^p, as well as all surface and internal multiples, q_t^m, so that

$$q_t = q_t^p + q_t^m \tag{2.38}$$

n_t: is the additive noise, generally taken to be white with a Gaussian pdf.

Since, in the discussion that follows, we will assume in cavalier fashion that the reflectivity in question is q_t^p, we will, for the sake of notational simplicity, write q_t^p as q_t. The preprocessing step of multiple attenuation is both vital and difficult. We will study this step in more detail later. A most commonly adopted model in applied seismology is that of a horizontally layered, one dimensional earth. This classic model, known as the Goupillaud layered medium (Goupillaud, 1961) (which we will meet again in Chapter 6) has been the subject of many articles over the years. We look at this model and its relationship to the Levinson recursion in more detail in Appendix A1, but at this stage, we consider the associated inverse problem which is the determination of the impedance at depth from the recorded data. In spite of the simplicity of the input model, this problem, as we will see, presents some interesting challenges.

The impedance of the kth layer of our pancake earth is defined as

$$\xi_k = \rho_k \nu_k$$

where ρ_k and ν_k are, respectively, the density and velocity in the kth layer. The relationship between ξ_k and q_k, the reflection coefficient of the kth layer [7], is

$$q_k = \frac{\xi_{k+1} - \xi_k}{\xi_{k+1} + \xi_k}. \tag{2.39}$$

---

[6] We remark that $\dfrac{\sigma^2}{|\tilde{G}(z)|^2}$ is the most highly resolved estimator of the power spectrum for harmonics in white noise.

[7] This is one reason why multiples must be eliminated.

Rearranging Eq. (2.39)

$$\xi_{k+1} = \xi_k(\frac{1+q_k}{1-q_k}) = \xi_1 \prod_{j=1}^{k} (\frac{1+q_j}{1-q_j}). \tag{2.40}$$

We see from Eq. (2.40) that, in order to determine ξ_k, we must determine the reflectivity series, q_t. The only way we can achieve this is by extracting q_t from the data expressed by Eq. (2.37). Two tasks need, therefore, to be performed:

(1) Wavelet estimation: Both the amplitude and *phase* properties of w_t must be determined.

(2) Wavelet deconvolution: w_t must be deconvolved in Eq. (2.37), which is not trivial since w_t is band limited.

Our general time series model is the MA model given by

$$x_t = b_t * u_t \tag{2.41}$$

where u_t is the innovation with statistical properties discussed in Section 2.2. Enders Robinson, in a classical study, Robinson (1957), assumed that Eqs. (2.37) (with $n_t = 0$) and (2.41) may be considered as equivalent, i.e., the assumption is that $q_t \equiv u_t$, which implies that the reflectivity of the Earth (primaries only) is *white*. The assumption that $q_t = q_t^p$ is white is not to be exactly correct. It turns out that q_t is in fact 'blue', which means that it lacks low frequencies. This is however a 2nd order effect and may be compensated for using various schemes.

The second assumption made by Enders Robinson is that w_t is minimum delay. The physical justification for this ubiquitous assumption lies in the Robinson Energy Delay Theorem which can be found in Claerbout (1976). In fact, from a wavelet estimation point of view, as we can see from Eq. (2.37)), we have, at best, one equation and two unknowns, and the minimum phase assumption is the a priori information required to achieve a unique solution. Now we have

$$x_t = w_t * q_t \Leftrightarrow X(z) = W(z)Q(z)$$

Writing $G(z)$ in terms of the wavelet

$$G(z) = W^{-1}(z)$$

and we obtain

$$X(z)G(z) = Q(z)$$

which is precisely the AR model of the seismogram. Solving

$$\mathbf{Rg} = \sigma_q^2 \mathbf{e}_1$$

obtains the PEO, and forming $x_t * g_t = q_t$ returns the *band limited* reflectivity function. The minimum phase wavelet estimate may be obtained in the frequency domain from the inverse of $G(z)$. This is not as straight forward as it sounds, however (please refer back to Section 1.9.4.2), and presents the first of many examples of an inverse problem. The difficulty with the inversion lies in the band limited nature of the wavelet. Inversion in time using a Wiener spiking filter approach will run into problems with the matrix inverse. In frequency, division by zero is the result. To obviate this, we proceed to *regularize* the problem (much more on this in Chapter 4). Here is one way

$$W(f) = \frac{1}{G(f)} \tag{2.42}$$

and multiplying top and bottom by $G^*(z)$

$$W(f) = \frac{G^*(f)}{G^*(f)G(f)} \tag{2.43}$$

A constant, μ, is now added to the denominator to give

$$W(f) = \frac{G^*(f)}{G^*(f)G(f) + \mu} \tag{2.44}$$

μ is called a *hyperparameter* and serves to obviate division by zero. As we have already seen in Chapter 1, Eq. (2.44) involves such matters as discrete versus circular convolution and circulant matrices. We will have much more to say concerning μ, its role in the resolution of estimated models, relationship to inversion in the time domain and other issues in the forthcoming chapters.

2.7 Extended AR Models and Applications

We have explored the AR model in detail in Section 2.4. Specifically, this model, which permits the prediction of a time point, x_k, from previous values, x_{k-1}, x_{k-2}, etc., with a resulting innovation, q_k, is a unit distance or a one step ahead predictor (or filter). For reasons that will become clear in a moment, we now investigate prediction filters with prediction distance equal to α. Designating the filter output as $\hat{x}_{t+\alpha}$, we define an error series in the usual manner

$$\begin{aligned} e_{t+\alpha} &= x_{t+\alpha} - \hat{x}_{t+\alpha} \\ &= x_{t+\alpha} - x_t * a_{\alpha t} \end{aligned}$$

where $a_{\alpha t}$ is the prediction filter for distance α. Transforming to the z domain

$$z^{-\alpha}E(z) = z^{-\alpha}X(z) - X(z)A_\alpha(z)$$

and therefore

$$E(z) = X(z)[1 - z^{\alpha} A_{\alpha}(z)] \tag{2.45}$$

The term $[1 - z^{\alpha} A_{\alpha}(z)]$ is the z transform of the α prediction error operator which we call PEO_{α}. Application of this operator in seismology is termed predictive deconvolution, PD, and was introduced by Peacock and Treitel (1969) in a seminal paper. The objective of this approach may be seen as follows: If we model x_t in term of our usual MA model but allow a predictable sequence, p_t, which may represent periodic multiple reflections, for example, then we may write

$$x_t = w_t * q_t * p_t$$

where, as before, q_t represents our primary reflectivity sequence. Convolving x_t with PEO_{α} should, we hope, if α is properly chosen, obtain $x_t = w_t * q_t$. In other words, the task of PEO_{α} is to remove the predictable part of the signal in x_t. We will see under what circumstances this is possible. Another question that it is important to answer is what effect spiking or predictive deconvolution has upon a nonminimum phase wavelet. In order to this, we need some relationships that exist in the literature. We note first that, since when $\alpha = 1$ the method corresponds to AR modelling, the output of the deconvolution in the case of a minimum phase wavelet, becomes the primary reflectivity series of the subsurface. For $\alpha > 1$, it turns out that prediction filters are related recursively (Ulrych et al., 1973), and this relationship may be used to obtain an interesting connection that exists between PEO_1 and PEO_{α} (Ulrych and Matsuoka, 1991).

2.7.1 A Little Predictive Deconvolution Theory

The method of predictive deconvolution, with prediction distance greater than unity (Peacock and Treitel, 1969), is widely used in the oil industry to eliminate surface multiple reflections from seismic reflection data. When $\alpha = 1$, the method is equivalent, within a scale factor, to the AR PEO of order 1, or PEO_1. As we have seen, when the wavelet is minimum phase, PEO_1 corresponds to the wavelet inverse and the innovation is just the primary reflectivity series. It is important to emphasize that the innovation of the AR process is the reflectivity series of the subsurface only if the seismic wavelet is indeed minimum phase. If this is not the case, an error results which is dependent on the phase spectrum of the wavelet. When $\alpha > 1$, the PD technique is used to attenuate repetitive waveforms of a particular period, with the additional important advantage of controlling the resolution of the deconvolution (Peacock and Treitel, 1969). A brief look at the theoretical development is worth while.

We begin by writing the normal equations for $\mathbf{a}_{\alpha}$, a simple task that follows the lines of the derivation leading to Eq. (2.19). Thus

$$\mathbf{R}\mathbf{a}_{\alpha} = \mathbf{r}_{\alpha} \tag{2.46}$$

where

$$\mathbf{r}_\alpha = [r_{\alpha-1}, r_\alpha, \ldots, r_{\alpha+N-2}]^T$$

for a N point prediction filter

$$\mathbf{a}_\alpha = [a_{\alpha 1}, a_{\alpha 2}, \ldots, a_{\alpha N}]^T$$

In as much as we augmented Eq. (2.19), we can do the same with Eq. (2.46) to obtain normal equations for $\mathbf{g}_\alpha$, the vector expressing PEO$_\alpha$. After some algebra, the equations become

$$\mathbf{R}_L \mathbf{g}_\alpha = \mathbf{p} \tag{2.47}$$

where $\mathbf{R}_L$ is the usual Toeplitz autocovariance matrix, of size $(L \times L), L = \alpha + N - 1$, and

$$\mathbf{g}_\alpha = [\underbrace{1, 0, \ldots, 0}_{\alpha}, \underbrace{-a_{\alpha 1}, -a_{\alpha 2}, \ldots, -a_\alpha}_{N}]^T$$

and

$$\mathbf{p} = [\underbrace{p_1, p_2, \ldots, p_\alpha}_{\alpha}, \underbrace{0, 0, \ldots, 0}_{N}]^T \tag{2.48}$$

$\mathbf{p}$ is a complicated function of the a's and r's (Robinson and Treitel, 2002) but the important point is that it truncates after α terms. This implies that the length of the output wavelet $\leq \alpha$ and, consequently, PEO$_\alpha$ shortens a minimum phase wavelet of length $N + \alpha$ to one of length α. The autocorrelation of the actual output will, consequently, tend to vanish for lags $> \alpha + N$. The length of PEO$_\alpha$ to be used in PD depends on whether short or long period multiples are to be attenuated (relevant discussion can be found in Robinson and Treitel (2002)). Our primary focus here is on the important issue of dependence of the PD output on the phase properties of w_t, the seismic source signature.

2.7.2 The Output of Predictive Deconvolution.

Ulrych and Matsuoka (1991) established a fundamental relationship that we make use of here. Let

$G_1(z)$ be the z transform of the unit delay, PEO$_1$.

$G_\alpha(z)$ be the z transform of PEO$_\alpha$.

$G_1^{-1}(z) \mid_\alpha$ be the z transform of $G_1^{-1}(z)$ truncated to α terms.

Then

$$G_\alpha = G_1^{-1}(z) \mid_\alpha G_1(z). \tag{2.49}$$

We now consider two cases.

Minimum phase wavelet

We have, using Eq. (2.10)

$$X(z) = B(z)Q(z) \tag{2.50}$$

The output of α PD is

$$\begin{aligned} O_\alpha(z) &= G_\alpha(z)X(z) \\ &= G_\alpha(z)B(z)Q(z) \end{aligned}$$

Using Eq. (2.49) we obtain

$$O_\alpha(z) = G_1^{-1}(z) \mid_\alpha G_1(z)B(z)Q(z) \tag{2.51}$$

But, as we have seen repeatedly

$$G_1^{-1}(z) = B(z)$$

and consequently

$$G_1^{-1}(z) \mid_\alpha = B(z) \mid_\alpha$$

Substituting into Eq. (2.51), obtains

$$O_\alpha(z) = B(z) \mid_\alpha Q(z) \tag{2.52}$$

The output is just the desired reflectivity convolved with a truncated minimum phase wavelet.

Nonminimum phase wavelet

In this case, we have

$$X(z) = W(z)Q(z)$$

and, similarly to Eq. (2.51), the output becomes

$$O_\alpha(z) = G_1^{-1}(z)\,|_\alpha\, G_1(z)W(z)Q(z) \tag{2.53}$$

Since $W(z)$ is not minimum phase, $G_1(z)$ is no longer simply its inverse. However, we may write $G_1(z)$ as the inverse of the corresponding minimum phase wavelet, $W_m(z)$, which has the same amplitude spectrum as $W(z)$. Then

$$G_1^{-1}(z)\,|_\alpha = W_m(z)\,|_\alpha$$

and Eq. (2.53) becomes

$$O_\alpha(z) = W_m(z)\,|_\alpha\, G_1(z)W(z)Q(z) \tag{2.54}$$

Considering an isolated wavelet for clarity

$$O_\alpha(z) = W_m(z)\,|_\alpha\, G_1(z)W(z) \tag{2.55}$$

However, since $G_1(z) = W_m^{-1}(z)$, Eq. (2.55) is simplified to

$$O_\alpha(z) = W(z)\,|_\alpha + \Upsilon(z) \tag{2.56}$$

where $\Upsilon(z)$ transforms to $H(t-\alpha)v(t)$ and is some function, which we call a tail function, that is zero for $t \leq \alpha$.

Eq. (2.56), which is illustrated in Fig. 2.5b, shows that the output of PD when applied to a nonminimum phase wavelet, is that wavelet truncated to the gap length followed by a tail, the shape of which depends on the phase properties of the wavelet. Panels (ii) and (iii) in Fig. 2.5b show the results for a spiking filter, $\alpha = 1$, and for a PEO_{10} ($\alpha = 10$) filter, respectively[8]. Fig. 2.5a illustrates that, in the case of a minimum phase wavelet, the tail is absent.

2.7.2.1 Remarks

Since spiking and predictive deconvolution are so widespread in the seismic industry, the shape of the PD output as a function of wavelet phase is an issue of importance. Ulrych and Matsuoka (1991) have developed an interesting relationship (Eq. (2.49)) that can be used to show that only in the case of minimum phase wavelets is the output free of artifacts. When the wavelet is not minimum phase, the deconvolution produces a tail, the shape of which is wavelet phase dependent.

[8]This value of α was chosen for illustration purposes only. In practice, more care is required.

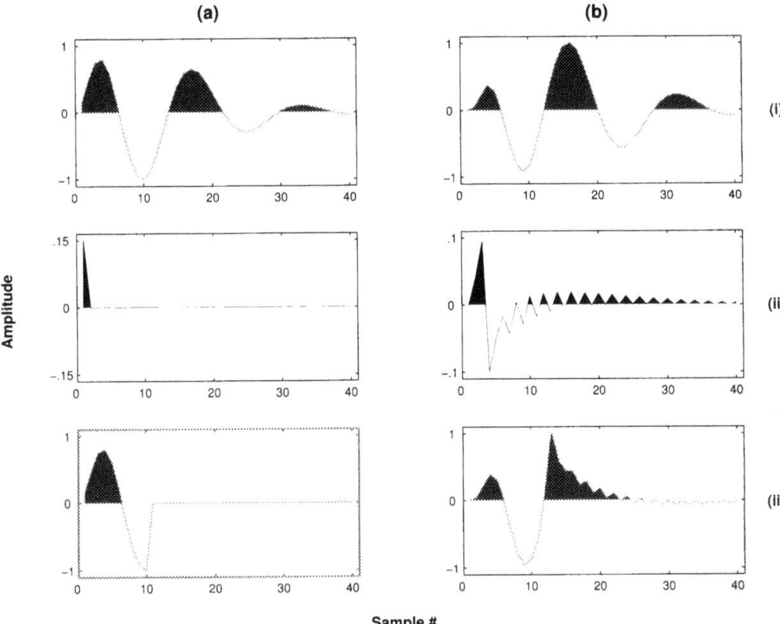

Figure 2.5: Illustrating the effect of PD. (a) Minimum phase wavelet. (b) Mixed phase wavelet. Panels (i) Wavelet. (ii) Output of spiking deconvolution. (iii) Output of PD with $\alpha = 10$.

2.7.3 Summary

To end this section, we summarize the various modelling approaches that we have described.

- The MA model is the general representation of the seismogram with the assumption of a white primary reflectivity. The phase of the wavelet is not an issue in this modelling.

- The AR representation of a seismogram is less general, constraining the seismic wavelet to be of minimum phase. The assumption of a white primary reflectivity is still implicit. The great usefulness of this model lies in its predictive and processing capabilities.

- The extended AR representation is used to model and process certain types of coherent noise in the layered Earth model. It is often used to control the resolution of seismic deconvolution where it is known as the method of predictive deconvolution.

- The ARMA model allows for a colored reflectivity series. The seismic wavelet is modelled as the inverse of the PEO and is assumed minimum phase. A 'special' ARMA model is also the optimal model for harmonics in additive random noise and leads to a time series approach to the derivation of the Pisarenko spectral estimator.

2.8 A Few Words About Nonlinear Time Series

The topic of this chapter has been the modelling of linear time series. We simply cannot end, however, without a brief glimpse into the very complex, difficult and particularly interesting world of nonlinear time series. The important issue is that, often, simple nonlinearities that parameterize a particular model, may be completely misunderstood in a linear description of the process. Since the use of nonlinear time series in our particular field is very limited, our look at this world will be cursory at best. It is true however, that certain transformations that are used in attribute analysis (Taner and Sheriff, 1997), for example, are certainly nonlinear, and some discussion of this topic is in order.

We illustrate our discussion by means of the simplest of nonlinear equations, the canonical logistic equation. In a non dimensional form, we write

$$x_{n+1} = \lambda x_n(1 - x_n) \tag{2.57}$$

where λ is a constant, now called the Feigenbaum parameter, and the limits of which relate to the recursion of x_n and are discussed below. We first encountered this type of equation in an article by R.M. May in Nature in 1976 (May, 1976). Chaos was not then mentioned specifically. Rather, the strange behavior of this equation and possible biological significance, were the issues discussed. In particular, the main focus was the modelling of the pseudo periodic nature of the cicada life-cycle . The logistic equation models, in a very simplistic fashion, the birth and death cycle that we are all involved in. Perhaps this is the reason for its fascination. It is of interest, very briefly, to look at some mathematical highlights of Eq. (2.57) (please see Moon, 1987, for a more comprehensive treatment).

Eq. (2.57) is a particular case of the general mapping

$$x_{n+1} = f(x_n) \tag{2.58}$$

Steady motion, called Period 1 motion, has two fixed, or equilibrium, points, x_e, which are given by the solution of $x_e = \lambda x_e(1 - x_e)$, namely, $x_e = 0$ and $x_e = (\lambda - 1)/\lambda$. The stability of the map i.e., whether the solution moves towards or away from x_e, depends on $f'(x)$, the slope of $f(x)$, at x_e. $|f'(x_e)| < 1$ implies stability and $|f'(x_e)| > 1$ instability. The complete behavior as a function of λ is depicted in Fig. 2.6. For $1 < \lambda < 3$, the solution at the origin is unstable, whereas the solution at $(\lambda - 1)/\lambda$ is stable. At $\lambda = 3$ the solution at the second fixed point becomes unstable and bifurcates at $\lambda = \lambda_1$. The motion, called Period 2 motion, becomes harmonic.
For $3 < \lambda < 4$, Eq. (2.57) exhibits many different multiple periods and *chaotic* motions, where the chaotic behavior is characterized by amplitudes that can adopt any value between ± 1. In particular, period doubling occurs until the value $\lambda_\infty =$ 3.56994... occurs, at which value chaos begins. Fascinatingly, in the range $\lambda_\infty < \lambda < 4$

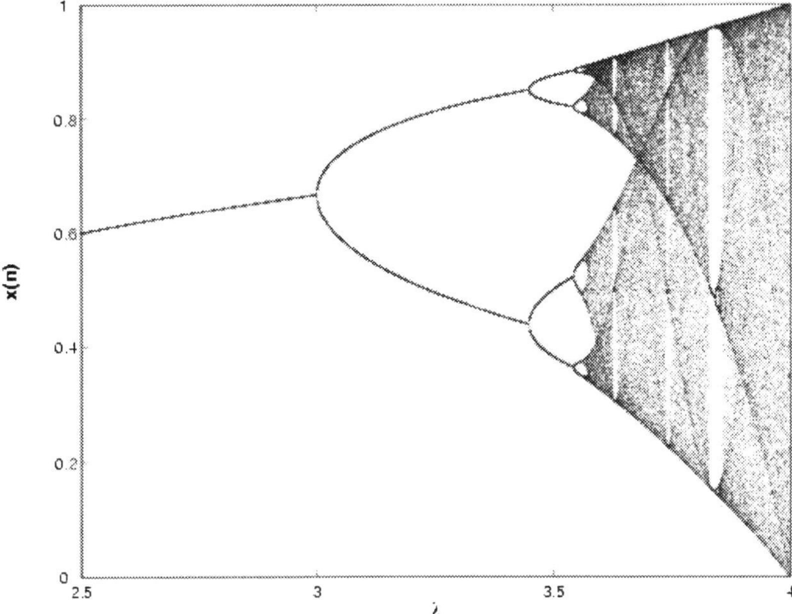

Figure 2.6: Bifurcation diagram for the discrete logistic equation.

windows exist in which the motion is, once again, multi-periodic. The limit, $\lambda = 4$, results from the fact that, at this value, unbounded growth occurs, which is unrealistic. So, $\lambda = 4$ represents the limit of sensible behavior. It is, at this limit, that x_n is white, in the sense that its power spectrum is completely flat.

Let us look at a closeup of the behavior of Eq. (2.57) for a particular value of $\lambda = 3.65$. The actual time series, Fig. 2.7, certainly looks random and does, indeed, have a broadband power spectrum. The return path diagram, also known as a pseudo phase-space map, representing a one-dimensional mapping of x_{n+1} versus x_n, is shown in Fig. 2.8. We see regions forbidden to the path, regions that depend on λ. A value of λ=4 returns a phase-space map that is fully parabolic. The return path diagram has, in this instance, transformed a random time series into a structured curve. There is a deep reason why this has occurred, explained by the principle of embedding.

2.8.1 The Principle of Embedding

The principle of embedding has much importance in the analysis of time series. We note that for the logistic equation, the return path diagram maps x_n against a lagged version, $x_{n+\tau}$, $\tau = 1$. Lags are, of course, the backbone of linear time series analysis and since, as we will outline, embedding is also intimately related to lagged variables, it is likewise, related to signal processing.

We will illustrate our discussion using discrete examples, and note that the concept applies equally to continuous processes. Consider some complex phenomenon that can be modelled by a set of equations that govern the evolution of a particular state of

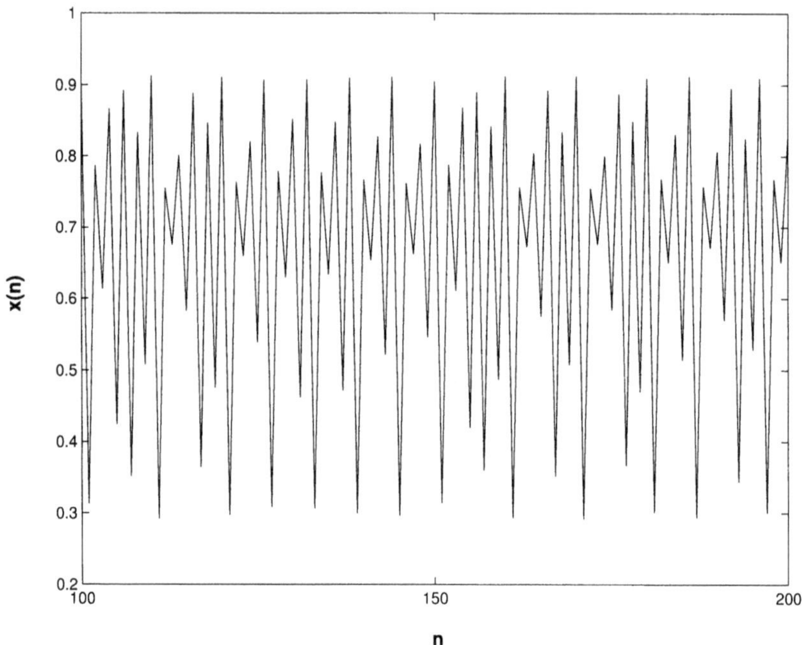

Figure 2.7: The logistic time series for $\lambda = 3.65$.

the system. We do not, however, in general know how many equations, or physical variables, are related to the observed data. Instead, we construct a pseudo phase-space (return map or an embedding space), using delayed measurements of one of the measured quantities. It turns out that the phase-space map contains all the details of the state-space that would have been mapped had we measured all the required variables. As a simple illustration, we consider a modified form of the Henon attractor (Henon, 1976, which is an extension of the quadratic map described by the logistic equation). We modified the standard form of the Henon attractor by introducing a third equation in order to make our embedding example of higher dimension. The governing equations are

$$\begin{aligned} x_{n+1} &= 1 - \alpha x_n^2 + y_n \\ y_{n+1} &= \beta x_n + z_n^2 \\ z_{n+1} &= \gamma y_n^2 * x_n \end{aligned}$$

The strange attractor, illustrated in Fig. 2.9a, has been produced with $\alpha = 0.3$, $\beta = 1.4$ and $\gamma = 1.5$. The state-space diagram, that maps the evolution of the orbit as a function of the two variables, x_n and y_n, is illustrated in Fig. 2.9c on a scale that compares with the embedding-space, Fig. 2.9d, that maps x_n as a function of the lagged variable, x_{n+1}. As can be clearly seen, the two maps are very similar. This

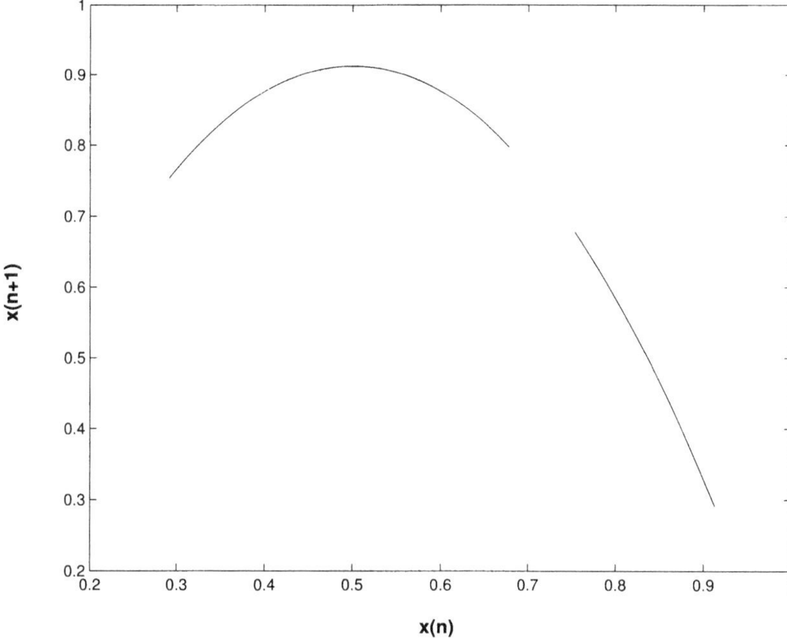

Figure 2.8: The return path diagram for $\lambda = 3.65$.

similarity is not a coincidence. It is a deep property of lagged spaces (Gershenfeld, 1999). In fact, the embedding principle, formulated by Floris Takens (Takens, 1981), tells us that the state-space and the phase-space mappings will, as Gershenfeld (1999) elegantly puts it, '*generically* differ by no more than a smooth invertible change of coordinates ..' (some conditions are specified).

The really important aspect for our work that is embodied by the principle of embedding is that, by utilizing time lags, we can reconstruct the behavior of the complex system, that possibly contains many degrees of freedom that are unknown to us, from the observations of only one of the state variables. Specifically, the steps to be followed are

1. Consider the discrete dynamical system of dimension N

$$\mathbf{x}_{n+1} = \mathbf{f}(\mathbf{x}_n) \tag{2.59}$$

2. which gives rise to the time series

$$\mathbf{x}_n = [x_n, x_{n+1}, x_{n+2}, \ldots]$$

3. We choose a time lag, τ, and an embedding dimension, M, and construct the system

$$\mathbf{y}_n = [y1_n, y2_n, \ldots, yM_n]$$

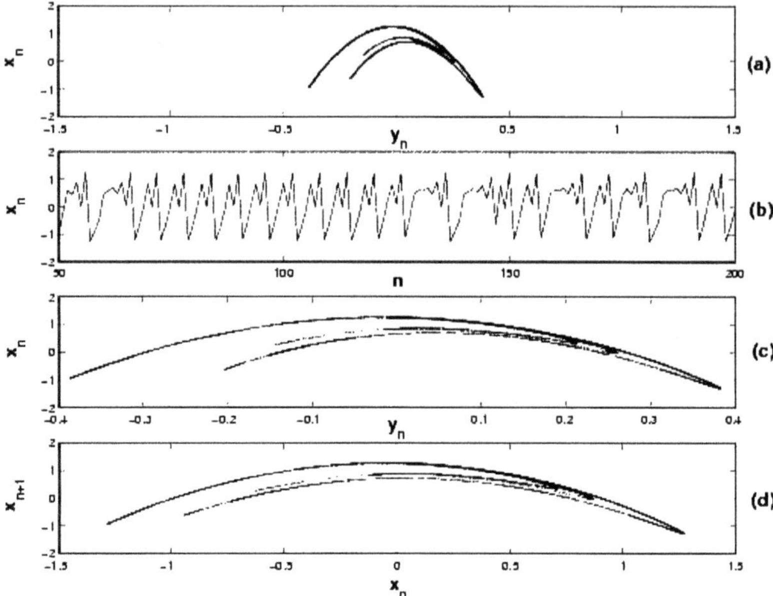

Figure 2.9: A modified Henon attractor. (a) State-space x_n, y_n and z_n. (b) The time series, x_n. As (a) but different y_n scale. (d) Embedding phase-space.

where

$$yk_n = x_{n+k\tau}, \quad k = 1, 2, \ldots, M$$

4. The Takens embedding theorem tells us that, for almost all choices of τ, $\{\mathbf{y}_n\}$ will be an embedding of $\{\mathbf{x}_n\}$, providing that $M > 2N$, thus guaranteeing that the topologies of $\{\mathbf{y}_n\}$ and $\{\mathbf{x}_n\}$ are qualitatively alike.

2.8.2 Summary

We have, quite briefly, ventured into the fascinating world of nonlinear time series. The cursory nature of our treatment is due to the fact that such series are not often utilized in the fields that we explore in this book. We will, however, investigate the very special role of time lags in Chapter 6 when we deal with the topic of blind deconvolution. In particular, we will relate the relationship of lagged time series to conditional probabilities and mutual information, information theoretic quantities that serve in lieu of correlation functions that are not applicable to nonlinear problems.

APPENDICES

A.1 Levinson's Recursion and Reflection Coefficients

As stated before, Levinson's recursion is not only a method of solving Toeplitz normal equations efficiently. The recursion has a physical significance which is particularly well illustrated by considering the propagation of elastic energy in a 1D earth. This is a classic problem in geophysics and has been studied by many authors over the years. In its original form the problem is known as the Goupillaud layered medium (Goupillaud, 1961). In this model it is assumed that the earth can be modelled by a sequence of horizontal layers, in which the vertical two-way time in each layer, T, equals the data sampling interval. It is also assumed that a basic reflection seismic geometry exists where N layers fall below a perfectly reflecting free surface which is interface 0. We have defined the reflection coefficient, q_k, as the amplitude ratio between the downward reflected and upward incident wave at interface k. $-q_k$, consequently, describes the amplitude ratio between the upward reflected and the corresponding downward incident wave. The free surface is defined by $q_0 = -1$ and the whole subsurface subject to the conditions above is described by the reflection coefficients $[q_1, q_2, \ldots, q_N]$. Let us first of all derive the forward problem, i.e. how the surface seismogram is related to the reflection coefficients. We follow the approach of Claerbout (1976) and we stress only the salient points in the derivation.

A.1.1 Theoretical Summary

As shown by Claerbout (1976), the propagation of up and down going waves in the stack of layers may be viewed in terms of a *layer matrix* which describes the relationship between these quantities across a layer interface. Specifically, let us designate the up and downgoing waves at the top of layer k by U_k and D_k respectively. Then, the downward extrapolated waves to layer $k+1$ are given by

$$\begin{bmatrix} U \\ D \end{bmatrix}_{k+1} = \frac{1}{z^{1/2} t_k} \begin{bmatrix} 1 & q_k z \\ q_k & z \end{bmatrix} \begin{bmatrix} U \\ D \end{bmatrix}_k$$

where $z = e^{-i\omega T}$ is the operator that delays by T (multiplying by $z^{1/2}$ is equivalent to delaying by $T/2$, which is the travel time across a layer) and t_k is the transmission coefficient across the kth layer.

In extrapolating from layer j to layer $j+k$, say, we need to multiply k layer matrices together. The general form of such a product is

$$\frac{1}{z^{k/2} \prod_{j=1}^{k} t_j} \begin{bmatrix} F_k(z) & z^k G_k(z^{-1}) \\ G_k(z) & z^k F_k(z^{-1}) \end{bmatrix} \tag{A.1}$$

where the polynomials $F_k(z)$ and $G_k(z)$ are built up in the following manner

$$F_k(z) = F_{k-1}(z) + q_k z G_{k-1}(z) \quad (A.2)$$
$$(A.3)$$

and

$$G_k(z) = q_k F_{k-1}(z) + z G_{k-1}(z) \quad (A.4)$$

with $F_0(z) = 1$ and $G_0(z) = 0$.

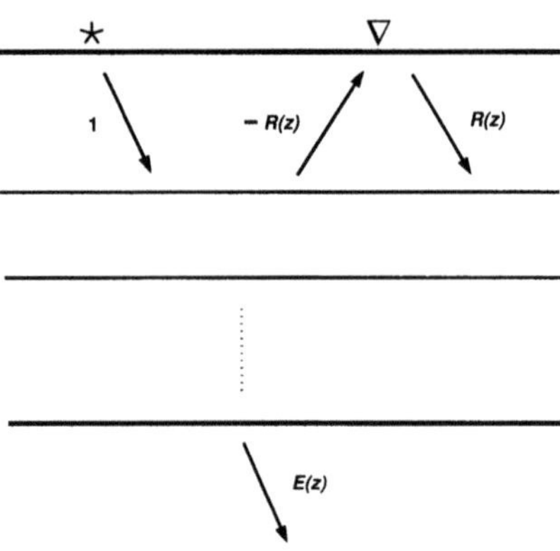

Figure 2.10: The Goupillaud layered model for seismic reflection.

We now consider the layered earth reflection geometry shown in Fig. 2.10. We have the initial downgoing impulse, the scattered energy incident from below, $-R(z)$, and the energy reflected from the surface which, since $q_0 = -1$, is $R(z)$[9]. The energy which escapes from the stack of layers is $E(z)$. Using Eq. (A.1) we have, for the N layers

$$\begin{bmatrix} 0 \\ E(z) \end{bmatrix} = \frac{1}{z^{N/2} \prod_{j=1}^{N} t_j} \begin{bmatrix} F_N(z) & z^N G_N(z^{-1}) \\ G_N(z) & z^N F_N(z^{-1}) \end{bmatrix} \begin{bmatrix} -R(z) \\ 1 + R(z) \end{bmatrix} \quad (A.5)$$

Solving Eq. (A.5) for $R(z)$ we obtain

$$\begin{aligned} R(z) &= \frac{z^N G_N(z^{-1})}{F(z) - z^N G(z^{-1})} \\ &= \frac{z^N G_N(z^{-1})}{A(z)} \end{aligned} \quad (A.6)$$

where

[9]We are considering the marine case.

$$A(z) = F(z) - z^N G(z^{-1}) \tag{A.7}$$

From consideration of the stability of R(z), it follows that $A(z)$ is a minimum phase polynomial.

In as much as the polynomials $F(z)$ and $G(z)$ may be determined recursively, $A(z)$ is also related in a recursive manner. Specifically, from Eqs. (A.2), (A.4) and (A.7) we obtain

$$A_k(z) = A_{k-1}(z) - q_k z^k A_{k-1}(z^{-1}) \tag{A.8}$$

which is the famous Levinson recursion for the prediction error operator which we have often met before. In particular, we have

$$A_k(z) = 1 + a_{k1}z + a_{k2}z^2 + \ldots + a_{kk}z^k$$

and

$$A_0(z) = 1$$

We are interested in the inverse problem of determining the reflection coefficients from a knowledge of the waveform recorded on the surface. i.e., given $R(z)$, we wish to determine $[q_1, q_2, \ldots, q_N]$. Following Hubral (1978), let us assume that we have already recovered the first k reflection coefficients, $[q_1, q_2, \ldots, q_k]$ from the observed reflection seismogram. Using these k reflection coefficients, we can compute the corresponding reflection series due to the k-layer model by means of Eq. (A.6)

$$\frac{z^k G_k(z^{-1})}{A(z)} = R(z) = r_1 z + r_2 z^2 + \ldots r_k z^k + \tilde{r}_{k+1} z^{k+1} + \ldots \tag{A.9}$$

Clearly, the first k reflected pulses $[r_1, r_2, \ldots, r_k]$ obtained in this way must agree with the first k observed pulses. Pulse $\tilde{r}_{k+1}$ will not agree with the actual observed pulse r_{k+1}, however, by virtue of the underlying reflector. The difference between these two pulses is simply given by the contribution of the primary reflected pulse from the $(k+1)$th interface. Specifically

$$r_{k+1} = \tilde{r}_{k+1} + q_{k+1} \prod_{i=1}^{k} (1 - q_i^2)$$

which may be rearranged to give

$$q_{k+1} = \frac{(r_{k+1} - \tilde{r}_{k+1})}{\prod_{i=1}^{k} (1 - q_i^2)} \tag{A.10}$$

We can also obtain an expression for $\tilde{r}_{k+1}$ by multiplying both sides of Eq. (A.9) by $A_k(z)$ and equating coefficients of z^{k+1}. This leads to

$$\tilde{r}_{k+1} = -(a_{k1}r_k + a_{k2}r_{k-1} + \ldots + a_{kk}r_1) \tag{A.11}$$

The coefficients $[a_{k1}, a_{k2}, \ldots, a_{kk}]$ may be obtained from $[q_1, q_2, \ldots, q_k]$ via the recursion formula given by Eq. (A.8). Using Eqs. (A.10) and (A.11) we can now compute $q(k+1)$ from $[r_1, r_2, \ldots, r_{k+1}]$ and $[a_{k1}, a_{k2}, \ldots, a_{kk}]$. Considering that $q_1 = r_1$ and $A_0(z) = 1$, the full recursion to recover the reflection coefficients from the recorded wave train is as follows:

- Solve the system for $k = 1$ to obtain $q_1 = r_1$.
- Increase k by one at each stage and use the Levinson recursion, Eq. (A.8) to obtain $[a_{k1}, a_{k2}, \ldots, a_{kk}]$.
- Use Eq. (A.11) to compute the predicted reflection $\tilde{r}_{k+1}$.
- Finally, compute q_{k+1} by means of Eq. (A.10).

A.1.2 Summary and Remarks

We have considered the problem of the propagation of elastic waves caused by an impulsive source in a layered earth. The model considered was developed by Goupillaud (1961) and by Baranov and Kunetz (1960) and consists of a special discretization of the medium into artificial homogeneous layers with constant two-way travel time. This classical model in exploration seismology is fully characterized by its reflection coefficients, q_k, formed between the kth and the $k+1$th layers, where the index k corresponds to depth. Since from a certain depth the layers are underlain by a homogeneous half-space, only a finite number of non-vanishing reflection coefficients exist.

The inverse problem, that of determining the reflection coefficients from the impulsive synthetic seismogram, may be formulated in terms of Eq. (A.8) which expresses the Levinson recursion. The Levinson recursion thus turns out to be intimately connected to the *physical* process of wave propagation in a stratified medium where the forward and backward system of equations are related to up and downgoing waves in the medium. In the above discussion, which follows the approach of Hubral (1978), no mention has been made either of Toeplitz forms or of the energy flux theorem due to Kunetz and D'Ecerville (1962) which are required in the usual manner of solving the inverse problem (Claerbout, 1976; Robinson and Treitel, 1978). Of course, the equality of both approaches is evident in Eq. (A.8) and a few words will serve to emphasize this important fact.

We have used the first equation of Eqs. (A.5) to determine $R(z)$. The second equation may be used to obtain an expression for the escaping wave, $E(z)$. Specifically (details may be found in Claerbout, 1976)

$$E(z) = \frac{z^N F_N(z{-}1) + [-G_N(z) + z^N F(z{-}1)]R(z)}{z^{N/2}\prod_{i=1}^{N} t}$$

$$= \frac{z^{N/2} \prod_{i=1}^{N} t'_i}{A_N(z)} \tag{A.12}$$

where $t'_k = 1 - q_k$. Since the medium is lossless, the energy difference between the up and downgoing waves at the top of the medium must equal the energy in the escaping wave. Referring to Fig. 2.10, the downgoing wave is $1 + R(z)$, whereas the upgoing wave is $-R(z)$. Consequently, we can express the equality of energies as

$$Y_1[1 + R(z)][1 + R(z^{-1}) - R(z)R(z^{-1})] = Y_N E(z)E(z^{-1})$$

where Y_1 and Y_N are either admittances or impedances (depending on whether pressure or velocity are being measured) of the respective layers. Therefore

$$1 + R(z) + R(z^{-1}) = \frac{Y_N}{Y_1} E(z)E(z^{-1}) \tag{A.13}$$

This is the Kunetz expression (Kunetz, 1964)

(A.14)

Substituting $E(z)$ from Eq. (A.12) obtains

$$\begin{aligned} 1 + R(z) + R(z^{-1}) &= \frac{Y_N}{Y_1} \frac{z^{N/2} \prod_i t'_i}{A_N(z)} \frac{z^{-N/2} \prod_i t'_i}{A_N(z^{-1})} \\ &= \frac{K}{A_N(z)} A_N(z^{-1}) \end{aligned} \tag{A.15}$$

(A.16)

where K is a constant that actually turns out to be $K = \prod_i t'_i t_i$

Now, multiplying through by $A(z)$, results in

(A.17)

$$[1 + R(z) + R(z^{-1})]A(z) = \frac{K}{A(z^{-1})}$$

Remember that $A(z)$ is minimum phase and, therefore, $A(z^-1)$ may be stably expanded as a series in $z^{-k}, k = 1, 2, \ldots, N$. By identifying coefficients of non-negative powers of z the following set of equations result

$$\begin{bmatrix} r_0 & r_1 & \ldots & r_N \\ r_1 & r_0 & \ldots & r_{N-1} \\ \vdots & \vdots & & \vdots \\ r_N & r_{N-1} & \ldots & r_0 \end{bmatrix} \begin{bmatrix} 1 \\ a_{N1} \\ \vdots \\ -q_N \end{bmatrix} = \begin{bmatrix} Y_N \\ 0 \\ \vdots \\ 0 \end{bmatrix} \tag{A.18}$$

$-q_N$ has been inserted as the highest coefficient of $A(z)$ as justified by the definition of $A(z)$ in terms of $F(z)$ and $G(z)$ expressed by Eqs. (A.2), (A.4) and (A.8). Eq. (A.18) expresses the remarkable fact that $1+R(z)$, the downgoing wave in the top medium, is one side of an autocorrelation function. Since the matrix is symmetric Toeplitz, $A(z)$ and consequently q_N, may be immediately determined using the Levinson recursion. The backward Levinson recursion, which is obtained from Eq. (A.8) by determining $A_{N-1}(z-1)$ and substituting it back and which has the form

$$A_{N-1} = \frac{1}{1-q_N^2}[a_N(z) - q_N z^N A_N(z-1)]$$

is now used to recursively determine all the remaining reflection coefficients. This procedure is, of course, identical to the first approach described above, but since the two procedures have been arrived at from quite different starting points, they offer different but complementary insight into this basic problem in waveform propagation. The fundamental point is that the Levinson recursion is not only a fast algorithm for solving Toeplitz normal equations, it is also an expression of the underlying physics of the layered earth.

A.2 Minimum Phase Property of the PEO

Many different proofs of the minimum phase property of the PEO associated with linear prediction exist in the literature (Stoica and Nehorai, 1987; Robinson, 1967; Pakula and Kay, 1983). Each proof is interesting in that it demonstrates yet another facet of the linear prediction problem. We present two quite different proofs in this Appendix. The first proof is based on the spectral decomposition of the autocovariance matrix and on the properties of the associated eigenvectors (Ulrych and Treitel, 1991). The second proof, due to John Burg (personal communication), does not implicitly consider the form of the eigenvectors but shows the relationship between minimum phase, Toeplitz structure and the positive definite property of the autocovariance matrix.

A.2.1 PROOF I

This proof may be found in a slightly different form in (Ulrych and Treitel, 1991)[10] and we repeat it here for convenience, completeness and because it fits so well with all that we have explored in this and the previous chapter.

A.2.1.1 Eigenvectors of Doubly Symmetric Matrices

We consider a $p \times p$ doubly symmetric matrix $\mathbf{Q}$. A doubly symmetric (also known as persymmetric and symmetric centrosymmetric) matrix is one that is symmetric about both the main and secondary diagonal. Two properties that are needed and which are presented in Cantoni and Butler (1976) and Strang (1988) are:

[10]There is rather a charming story associated with this proof, which the curious reader could obtain by writing to ulrych@eos.ubc.ca .

Property 1: If $\mathbf{Q}$ has distinct eigenvalues then $\mathbf{Q}$ has $Int[(p+1)/2]$ symmetric and $Int[p/2]$ skew symmetric eigenvectors that span the eigenspace of $\mathbf{Q}$. $Int[\cdot]$ indicates the integer part of $[\cdot]$. These eigenvectors are unique to within a multiplicative constant and are mutually orthonormal.

Property 2: If an eigenvalue of $\mathbf{Q}$ has multiplicity greater than one $\mathbf{Q}$ still has a complete set of orthonormal eigenvectors. The eigenvectors corresponding to repeated eigenvalues need not necessarily be symmetric or skew symmetric, but they may be chosen to be so, (Makhoul, 1981). An example is useful at this stage. The matrix

$$\begin{bmatrix} 1 & r & r \\ r & 1 & r \\ r & r & 1 \end{bmatrix} \quad \text{has the eigenvalues} \quad \begin{cases} \lambda_1 = \lambda_2 = 1 - r \\ \\ \lambda_3 = 1 + 2r \end{cases}$$

The eigenvectors $\mathbf{u}_1$ and $\mathbf{u}_2$ may be chosen to be

$$\mathbf{u}_1 = \frac{1}{\sqrt{6}}[1,\ 1,\ -2]^T \quad \text{and} \quad \mathbf{u}_2 = \frac{1}{\sqrt{2}}[1,\ -1,\ 0]^T.$$

However they may also be chosen to be symmetric and skew symmetric as follows

$$\mathbf{u}_1 = \frac{1}{\sqrt{6}}[1,\ -2,\ 1]^T \quad \text{and} \quad \mathbf{u}_2 = \frac{1}{\sqrt{2}}[1,\ 0,\ -1]^T.$$

A.2.1.2 Spectral decomposition

Let $\mathbf{U}$ be the eigenvector matrix with column eigenvectors $\mathbf{u}_i = [u_{i1}, u_{i2}, \ldots, u_{ip}]^T$, $\mathbf{u}_i^T\mathbf{u}_i = 1$ and $\mathbf{\Lambda}$ be the diagonal eigenvalue matrix. Then

$$\mathbf{Q} = \mathbf{U\Lambda U}^T$$

and

$$\mathbf{Q}^{-1} = \mathbf{U\Lambda}^{-1}\mathbf{U}^T. \tag{A.19}$$

Consider the normal equations for the prediction error operator

$$\mathbf{Qg}_p = \sigma^2\mathbf{i} \tag{A.20}$$

where

$$\begin{aligned} \mathbf{g}_p &= [1, g_{1p}, \ldots, g_{pp}]^T \\ \mathbf{i} &= [1, 0, \ldots, 0]^T \end{aligned}$$

and σ^2 is the prediction error variance.

From Eqs. (A.19) and (A.20) it follows that

$$\mathbf{g}_p = \sigma^2 \mathbf{U}\mathbf{\Lambda}^{-1}\mathbf{U}^T\mathbf{i} \tag{A.21}$$

Evaluating the first and last elements of $\mathbf{g}$ given by Eq. (A.21) we obtain respectively

$$1 = \sigma^2 \left[\frac{u_{11}^2}{\lambda_1} + \frac{u_{21}^2}{\lambda_2} + \ldots + \frac{u_{p1}^2}{\lambda_p}\right] \tag{A.22}$$

and

$$g_{pp} = \sigma^2 \left[\frac{u_{11}u_{1p}}{\lambda_1} + \frac{u_{21}u_{2p}}{\lambda_2} + \ldots + \frac{u_{p1}u_{pp}}{\lambda_p}\right] \tag{A.23}$$

Using Properties 1 and 2 we have

$$u_{1p} = \pm u_{11} \quad u_{2p} = \pm u_{21} \quad \ldots \quad u_{pp} = \pm u_{p1}$$

where the $\pm$ depends on whether the eigenvector is symmetric or skew symmetric. Substituting into Eq. (A.23)

$$g_{pp} = \sigma^2 \left[\pm\frac{u_{11}^2}{\lambda_1} \pm \frac{u_{21}^2}{\lambda_2} \pm \ldots \pm \frac{u_{p1}^2}{\lambda_p}\right] \tag{A.24}$$

We now assume that $\mathbf{Q}$ is positive definite so that $\lambda_i > 0$ for $i = 1, 2, \ldots, p$. Since the u_{i1}'s enter as their squares it follows from Eq. (A.22) that

$$\frac{1}{\sigma^2} = \frac{u_{11}^2}{\lambda_1} + \frac{u_{21}^2}{\lambda_2} + \ldots + \frac{u_{p1}^2}{\lambda_p} \tag{A.25}$$

is a sum of strictly positive terms. From Eq. (A.24), it follows that

$$\frac{g_{pp}}{\sigma^2} = \pm\frac{u_{11}^2}{\lambda_1} \pm \frac{u_{21}^2}{\lambda_2} \pm \ldots \pm \frac{u_{p1}^2}{\lambda_p}. \tag{A.26}$$

If all terms in Eq. (A.26) were associated with the minus sign, Eqs. (A.25) and (A.26) would give

$$\frac{g_{pp}}{\sigma^2} = -\frac{1}{\sigma^2} \; or \; g_{pp} = -1$$

Because merely some (i.e.,*Int*[$p/2$]) of the terms in Eq. (A.26) are negative, we have

$$-1 < g_{pp} < +1 \tag{A.27}$$

In fact, this must be so, since the last term of the prediction error filter is the negative of the pth reflection coefficient, (see e.g. Robinson, 1966), which must be ≤ 1 in magnitude.

A.2.1.3 Minimum phase property

Thus far we have assumed that $\mathbf{Q}$ is a doubly symmetric positive definite matrix. We now also assume that $\mathbf{Q}$ is Toeplitz. Defining

$$G_k(z) = \sum_i g_{ki} z^i, \quad k = 0, 1, \ldots, p$$

where $g_{k0} = 1$

we reproduce the famous Levinson recursion, Eq. (2.33), as

$$G_k(z) = G_{k-1}(z) + g_{kk} z^k G_{k-1}(z^{-1}) \quad k = 1, 2, \ldots, p \tag{A.28}$$

Since all submatrices of $\mathbf{Q}$ are also symmetric Toeplitz it follows from Eq. (A.27) that the coefficients g_{kk} $k = 1, 2, \ldots, p$ in Eq. (A.28) share the property that $|g_{kk}| < 1$.

At this stage it is customary to invoke Rouché's theorem (Marden, 1966) to prove, using Eq. (A.28), that all the zeros of $G_p(z)$ lie outside the unit circle and hence, by definition, $G_p(z)$ is minimum phase. Ulrych and Treitel (1991) adopted a more direct approach. Evaluating Eq. (A.28) on $|z| = 1$ where $z = exp(-i\omega)$

$$\begin{aligned} G_k(\omega) &= |G_{k-1}(\omega)|e^{i\phi(\omega)} + g_{kk}e^{-i\omega k}|G_{k-1}(\omega)|e^{-i\phi(\omega)} \\ &= G_{k-1}(\omega)\left[1 + g_{kk}e^{i\theta(\omega)}\right] \\ &= G_{k-1}(\omega)H_{k-1}(\omega) \end{aligned} \tag{A.29}$$

where $\theta(\omega) = -\omega k - 2\phi(\omega)$.

We now examine the properties of the phase spectrum of $H_{k-1}(\omega)$, $\angle\{H_{k-1}(\omega)\}$, expressed by

$$\angle\{H_{k-1}(\omega)\} = \tan^{-1}\frac{g_{kk}\sin\theta(\omega)}{1 + g_{kk}\cos\theta(\omega)} \quad -\pi \le \omega \le \pi. \tag{A.30}$$

Since $|g_{kk}| < 1$, $|\angle\{H_{k-1}(\omega)\}| < \pi/2$. Hence, as ω varies from 0 to 2π, the vector $H_{k-1}(\omega)$ traces out a curve in the complex plane which never touches or encloses the origin. $H_{k-1}(\omega)$ $k = 1, 2, \ldots, p$ are, consequently, minimum phase (Robinson et al., 1978). (Actually, as is easily verified, $H_{k-1}(\omega)$ is not only minimum phase, it is also positive real.) Since $G_0(\omega) = 1$ is clearly minimum phase, it follows from Eq. (A.29) that $G_k(\omega)$ $k = 0, 1, \ldots, p$ are minimum phase. Specifically $\mathbf{g}_p$, computed using Eq. (A.20), where $\mathbf{Q}$ is a symmetric, positive definite Toeplitz matrix, is minimum phase.

A.2.2 PROOF II

This proof is due to John Burg[11].

[11] John wrote this proof for TJU on a serviette in a cafeteria in Laramie, Wyoming, in around 1982.

Repeating here the normal equations for the PEO, Eq. (2.22)

$$\mathbf{R}_{p+1}\mathbf{g} = \sigma_q^2 \mathbf{e}_1$$

with the quantities as previously defined. We consider a system of order p=3, for simplicity, which we write as

$$\begin{bmatrix} r_0 & r_1^* & \dots & r_3^* \\ r_1 & r_0 & \dots & r_2^* \\ \vdots & \vdots & \ddots & \vdots \\ r_3 & r_2 & \dots & r_0 \end{bmatrix} \begin{bmatrix} 1 \\ g_{13} \\ g_{23} \\ g_{33} \end{bmatrix} = \begin{bmatrix} P_3 \\ 0 \\ \vdots \\ 0 \end{bmatrix} \tag{A.31}$$

(A.32)

or

$$\mathbf{R}_3\mathbf{g}_3 = P_3\mathbf{e}_1 \tag{A.33}$$

$\mathbf{R}_3$ may be Hermitian, but we consider a real matrix for simplicity and without loss of generality. We define

$$\begin{aligned} G(z) &= 1 + g_{13}z + g_{23}z^2 + g_{33}z^3 \\ &= (1+\alpha z)(1 + b_1 z + b_2 z^2) \end{aligned} \tag{A.34}$$

where we have factored out any one of the three roots of $G(z)$, and α and the $b's$ are complex coefficients. In the time domain

$$\mathbf{g}_3 = \begin{bmatrix} 1 & 0 \\ b_1 & 1 \\ b_2 & b_1 \\ 0 & b_2 \end{bmatrix} \begin{bmatrix} 1 \\ \alpha \end{bmatrix} = \mathbf{B}\boldsymbol{\alpha} \tag{A.35}$$

where $\mathbf{B}$ and $\boldsymbol{\alpha}$ are defined above. From Eqs. (A.33) and (A.35) we have

$$\mathbf{R}_3\mathbf{B}\boldsymbol{\alpha} = P_3\mathbf{e}_1$$

Premultiplying by $\mathbf{B}^H$ (^{H} signifies the Hermitian transpose)

$$\begin{aligned} \mathbf{B}^H\mathbf{R}_3\mathbf{B}\boldsymbol{\alpha} &= \mathbf{B}^H P_3\mathbf{e}_1 \\ &= P_3\mathbf{e}_1 \end{aligned} \tag{A.36}$$

Defining the matrix $\mathbf{S}$

$$\mathbf{S} = \mathbf{B}^H\mathbf{R}_3\mathbf{B} = \begin{bmatrix} s_0 & s_1 \\ s_1 & s_0 \end{bmatrix}$$

we see that, since $\mathbf{R}_3$ is an autocovariance matrix, $\mathbf{R}_3$ is positive definite and, consequently, $\mathbf{S}$ is positive definite. Therefore

$$\det \mathbf{S} > 0 \quad \text{which implies} \quad |s_1| < |s_0| \tag{A.37}$$

Since, from Eq. (A.36)

$$\alpha = -\frac{s_1}{s_0}$$

it follows from Eq. (A.37) that $|\alpha| < 1$.

Consequently, $1 + \alpha z$ has a root outside of $|z| = 1$. But $1 + \alpha z$, in Eq. (A.34), was any factor of $G(z)$.

Consequently, $G(z)$ is minimum phase.

A.2.2.1 Discussion

We have shown, using the properties of the eigenvectors of doubly symmetric matrices and by factorization à la Burg, that the prediction error operator which is computed from normal equations of Toeplitz form is minimum phase. A requirement is that the Toeplitz matrix be positive definite. It is interesting to note that the special properties of the eigenvectors which correspond to the minimum and maximum eigenvalues, namely that the zeros of these eigenvectors lie on the unit circle, were not required in the first proof. Of particular interest is the fact that the property expressed by Eq. (A.27) holds in the general case of positive definite doubly symmetric matrices.

Chapter 3

Information Theory and Relevant Issues

3.1 Introduction

Information theory is a vast subject. For this reason we do not pretend here to do anything more than to introduce the subject. We apply information principles to various issues of interest, and these are dealt with in some more detail. This chapter begins with a view of entropy, introduces the Kullback-Leibler information measure and applies this measure to the development of the AIC that was mentioned in Chapter 2. We also introduce the concept of mutual information and illustrate its use in application to independent component analysis.

3.2 Entropy in Time Series Analysis

This section introduces the concept of entropy, its relationship to probability and discusses the central importance of the principle of maximum entropy, often referred to as MaxEnt, in time series analysis, spectral estimation and the inversion of underdetermined problems. We begin with some general principles which make up the backbone of much of the discussion in this section. In particular, we return to some aspects of probability density functions and their relationship to the modelling of stochastic processes.

3.2.1 Some Basic Considerations

Following the discussion in Section 1.2.2 that introduced the concepts of random variables, random processes, ensembles etc., we now consider the issue of an univariate random process which we designate by $\mathbf{x}_t$, $t = 1, 2, \ldots, N$, in more detail. As is well known, $\mathbf{x}_t$ may be represented in terms of a complete set of orthogonal functions, $\{\phi_k(t)\}$, as

$$\mathbf{x}_t = \sum_k \mathbf{x}_k \phi_k(t) \tag{3.1}$$

Here, $\mathbf{x}_1, \mathbf{x}_2, \ldots, \mathbf{x}_N$ are random variables which, for each sample function of the random process, assume certain values. The random process $\mathbf{x}_t$ can therefore be represented by a point $\boldsymbol{x}(\mathbf{x}_1, \mathbf{x}_2 \ldots, \mathbf{x}_N)$ in the N dimensional hyperspace defined by the basis functions $\{\phi_k(t)\}$, which assumes a certain value, $\mathbf{x}$, for each sample function. The random process, which consists of an ensemble of sample functions may, therefore, be represented by an ensemble of random points in hyperspace. The process is completely specified by the statistics of the point $\boldsymbol{x}$ and therefore by the joint probability density function of the variables $\mathbf{x}_1, \mathbf{x}_2, \ldots, \mathbf{x}_N$ which is denoted by $p_{\boldsymbol{x}}(\mathbf{x}) = p(\mathbf{x}) = p_{\mathbf{x}_1,\mathbf{x}_2,\ldots,\mathbf{x}_N}(\mathbf{x}_1, \mathbf{x}_2, \ldots, \mathbf{x}_N)$. The density of points in hyperspace is directly proportional to $p(\mathbf{x})$.

Probability density functions play a central role in much of what we do and it is elucidating to consider the form of $p(\mathbf{x})$ in more detail for a particular distribution. Specifically, and for many excellent reasons, we choose the Gaussian (or normal) distribution, a distribution which is a pivotal one in statistical analysis.

We define $\mathbf{C_{xx}}$ to be the covariance matrix of the process $\mathbf{x}_t$, with elements given by

$$\rho_{ij} = E[(\mathbf{x}_i - \overline{\mathbf{x}}_i)(\mathbf{x}_j - \overline{\mathbf{x}}_j)]$$

where $E[\,\cdot\,]$ is the expectation operator and $\overline{\mathbf{x}}_i$ is the mean value of the random variable $\mathbf{x}_i$. Under the Gaussian assumption, $p(\mathbf{x})$ becomes

$$p(\mathbf{x}) = \frac{1}{(2\pi)^{N/2}\sqrt{|\mathbf{C_{xx}}|}} \exp\left[-\frac{1}{2}(\mathbf{x} - \overline{\mathbf{x}})^T \mathbf{C}_{\mathbf{xx}}^{-1}(\mathbf{x} - \overline{\mathbf{x}})\right] \tag{3.2}$$

where $|\mathbf{C_{xx}}|$ is the determinant of $\mathbf{C_{xx}}$. Often, we will assume either that the random process is zero mean or that the mean has been removed. In this case, Eq. (3.2) becomes

$$p(\mathbf{x}) = \frac{1}{(2\pi)^{N/2}\sqrt{|\mathbf{C_{xx}}|}} exp\left(-\frac{1}{2}\mathbf{x}^T \mathbf{C}_{\mathbf{xx}}^{-1}\mathbf{x}\right) \tag{3.3}$$

It is important to note the form of Eq. (3.3) , in particular, the quadratic form $\mathbf{x}^T \mathbf{C}_{\mathbf{xx}}^{-1}\mathbf{x}$. We will have many occasions to use this equation.

3.2.2 Entropy and Things

We occupy a probabilistic world, a world where all decisions, inferences, predictions etc. are associated with a measure of uncertainty. The issue of quantifying uncertainty was addressed by Shannon (1948) in a seminal paper who showed that uncertainty is related to entropy. We explore this issue here without going into details of the derivation.

Uncertainty may be understood intuitively in the following manner. Let us assume that we are concerned with the issue of predicting the outcomes of two random events, x_i, $i = 1, 2$, each one associated with a probability p_i (we write p_i rather than P_i for discrete probabilities for convenience). In one experiment, the two probabilities are $p_1 = 0.5$, $p_2 = 0.5$ and in another, the probabilities are $p_1 = 0.01$, $p_2 = 0.99$. Clearly, the outcome of the first event is much more unpredictable, and therefore uncertain, than the outcome of the second event. Uncertainty is proportional to the inverse of the probability of the event occurring. If we denote the uncertainty of an outcome x_i by H_i, we can say, therefore, that

$$H_i \propto \frac{1}{p_i}$$

Since we wish uncertainties of zero and infinity to be associated with probabilities of one and zero, respectively, we rewrite the above equation as

$$H_i \propto \log \frac{1}{p_i} \tag{3.4}$$

Let us now suppose that we wish to quantify an average uncertainty associated with an experiment which has N outcomes. Clearly, this average uncertainty H, also called the entropy, is given by the expected value of H_i or

$$H = -\sum_i p_i \log p_i \tag{3.5}$$

where the equality implies that the base of the logarithm determines the units of the measure. In particular, in Eq. (3.5) , we use the natural logarithm associated with nary units. It may be shown that, given certain requirements which any measure of uncertainty must possess, (see for example Turner and Betts (1974)), the only possible measure is, in fact given by Eq. (3.5).

Shannon was concerned with the transmission and reception of noisy signals and consequently uncertainty arose naturally in this context. Uncertainty, however, is rather an ambiguous term and Shannon sought a different terminology. Apparently, as reported by Kapur (1989), Shannon adopted the term entropy due to a suggestion of Von Neumann who advised its use firstly due to the fact Eq. (3.5) was of the same form as thermodynamic entropy and, secondly, since after 100 years nobody truly understood what entropy was, it was safe to use. In this book, we use entropy in the manner of Jaynes (1982) who formulated a principle which, we believe, is one of the most natural and useful principles in the field of data analysis.

3.2.3 Differential (or Relative) Entropy

Differential entropy is defined by the continuous version of Eq. (3.5). Difficulties arise, however, since one must arbitrarily renormalize the entropy. This is important (Lathi, 1968) and we treat this issue below.

For a discrete random variable $\mathbf{x}$ assuming values $x_1, x_2, \ldots, x_N$ we rewrite Eq. (3.5) explicitly as

$$H(\mathbf{x}) = -\sum_i p(x_i) \log p(x_i) \tag{3.6}$$

Now consider a continuous random variable, $\mathbf{x}_c$, as a limiting form of our $\mathbf{x}$ that assumes discrete values in N intervals $\{(x_k, x_{k+1}]\}_{1 \le k \le N}$ of variable length. If $\Delta_k = x_{k+1} - x_k$, $\mathbf{x}$ assumes a value in the range $(x_k, x_k + \Delta x_k)$ with probability $p(x_k)\Delta x_k$

in the limit as $\Delta x_k \to \Delta x \to 0$. Hence, for a continuous random variable $\mathbf{x}_c$

$$
\begin{aligned}
H(\mathbf{x}_c) &= -\lim_{\Delta x_k \to 0} \sum_k p(x_k)\Delta x_k \log\left(p(x_k)\Delta x_k\right) \\
&= -\lim_{\Delta x_k \to 0} [\sum_k p(x_k)\Delta x_k \log\, p(x_k) - \sum_k p(x_k)\Delta x_k \log\, \Delta x_k] \\
&= -\int p(x) \log\, p(x)\, dx - \lim_{\Delta x \to 0} \log\, \Delta x
\end{aligned}
$$

because of the unimodular constraint. In the limit, as $\Delta x \to 0$, $\log\, \Delta x \to -\infty$. We must therefore consider $-\int p(x) \log p(x)\, dx$ as a *relative* entropy with $-\log\, \Delta x$ serving as a datum or reference. Note also that $\int p(x) \log\, p(x)\, dx$ is *not invariant* under a change of variables and is incorrect dimensionally whenever x has dimensions. It is for this reason of non invariance that the concept of Minimum Relative Entropy, MRE, was introduced by Shore and Johnson (1982), a concept that is important in various applications and which is explored in this chapter.

3.2.4 Multiplicities

A characteristic which is often associated with MaxEnt is that of multiplicity and we will meet this concept again below in the Jaynes Entropy Concentration theorem. Essentially, multiplicity is a measure of the number of ways a particular pdf may be realized. Clearly, choosing a feasible solution with a large multiplicity is very desirable. It turns out, in fact, that the solution with MaxEnt is the one also with maximum multiplicity. Let us examine the relationship between MaxEnt and multiplicity.

We perform N trials of an experiment with M outcomes. The relative nonzero frequencies associated with the outcomes are f_k, $k = 1, \ldots, M$ and the respective counts are then, $n_k = Nf_k$. The multiplicity, W, is the number of ways that the N trials can be arranged to add up to the respective counts and is given by

$$
W = \frac{N!}{n_1! n_2! \ldots n_M!} \tag{3.7}
$$

The relationship we seek may be derived using Stirling's approximation for $N!$ which is

$$
N! \approx \sqrt{2\pi N + 1} N^N e_N
$$

Taking natural logarithms

$$
\log N! \approx (N + \frac{1}{2}) \log N - N + \frac{1}{2} \log(2\pi)
$$

This approximation is quite accurate. For $N = 10$, for example, it is in error only by 0.055% Substituting into $\log W$, where W is given by Eq. (3.7), obtains

$$
\log W = - N \sum_k \frac{n_k}{N} \log \frac{n_k}{N} + \frac{1}{2} \left(\log N - \sum_k \log n_k \right) \tag{3.8}
$$

$$+ \quad \frac{1-M}{2}\log 2\pi \tag{3.9}$$

Defining $H(\mathbf{q})$ to be the entropy associated with the distribution $\mathbf{q}$, $(q_k = n_k/N)$, we see that the first term in equation (3.9) is

$$-N\sum_M q_i \log q_i = NH(\mathbf{q})$$

As N increases, the multiplicity W increases exponentially as $e^{NH(\mathbf{q})}$ and the remaining terms in equation (3.9) very quickly become negligible. Clearly, if we are looking for a distribution that can be realized in the most number of ways, we should maximize the multiplicity. For large N, we should maximize the entropy. Hence the principle of *maximum entropy*, the PME. The PME can be argued in a much more intuitive manner, beginning with the definition of information, as has been done by a number of authors (please see Jaynes, 1982, for a review). The above development allows us to talk about probabilities of solutions, however, in quantitative fashion. It turns out that much of what is central in the development of entropy, mutual information and various other issues, is related to a probabilistic measure known as the Kullback-Leibler, or K-L, information measure.

3.3 The Kullback-Leibler Information Measure

We begin with the definition of this famous measure, introduced in 1951, that may be traced to the beginning of much of what concerns us from an information theoretic point of view. We designate the K-L measure symbolically by $I(1,2)$, where

$$I(1,2) = \int p_1(x)\log\left[\frac{p_1(x)}{p_2(x)}\right]dx \tag{3.10}$$

Eq. (3.10) defines a measure of the 'distance', in probability space, between the two probability density functions, pdf's, $p_1(x)$ and $p_2(x)$ (as always, log indicates the natural logarithm). $p_1(x)$ is taken to be the true pdf and $p_2(x)$ to be the 'best' approximation to $p_1(x)$.

3.3.1 The Kullback-Leibler Measure and Entropy

Let us spend a little time, at this stage, putting into perspective concepts that will form an important part of this book. $I(1,2)$ is intimately related to, and is in fact, a generalization of the Jaynes PME. In turn, the PME has been extended by Shore and Johnson (1980) to the principle of Minimum Cross-Entropy that has become firmly enshrined as the principle of Minimum Relative Entropy, MRE. According to the MRE principle, $p_2(x)$ in Eq. (3.10) is viewed as an *a priori* pdf that expresses our prior knowledge about our model. An estimate $\hat{p_1}(x)$ of $p_1(x)$ is now determined by minimizing $I(1,2)$, the relative entropy between $p_1(x)$ and $p_2(x)$. The PME considers $p_2(x)$ to be uniform, conforming to an absence of prior information (much more about this later), and maximizes $\int p_1(x)\log p_1(x)dx$, the entropy associated with the pdf $p_1(x)$.

3.3.2 The Kullback-Leibler Measure and Likelihood

We now examine the relationship of $I(1,2)$ to the ubiquitous principle of maximum likelihood. Let $p(x_i|\boldsymbol{\theta})$ be the pdf of the observable data x_i $i = 1, 2, \ldots, N$, conditional on the parameter vector $\boldsymbol{\theta} = [\theta_1, \theta_2, \ldots, \theta_M]^T$. The log likelihood, $\mathcal{L}(\boldsymbol{\theta})$ is

$$\mathcal{L}(\boldsymbol{\theta}) = \sum_i \log p(x_i|\boldsymbol{\theta}) \tag{3.11}$$

Since $\mathcal{L}(\boldsymbol{\theta})$ depends on N, we define

$$\mathcal{L}'(\boldsymbol{\theta}) = \frac{1}{N}\mathcal{L}(\boldsymbol{\theta})$$

or

$$= \frac{1}{N}\sum_i \log p(x_i|\boldsymbol{\theta})$$

where $\mathcal{L}'(\boldsymbol{\theta})$ is the average log likelihood. In the limit, as $N \to \infty$

$$\lim_{N\to\infty} \frac{1}{N}\sum_i \log p(x_i|\boldsymbol{\theta}) = E[\log p(x|\boldsymbol{\theta})]$$

and, remembering the definition of $E[\cdot]$ in terms of pdf's, Eq. (A.1), we obtain

$$\frac{1}{N}\sum_i \log p(x_i|\boldsymbol{\theta}) = \int p(x|\boldsymbol{\theta}^*) \log p(x|\boldsymbol{\theta}) dx$$

$\boldsymbol{\theta}^*$ is the vector of true parameters and $p(x|\boldsymbol{\theta}^*)$ is the true pdf, which, of course, is unknown. Identifying $p(x|\boldsymbol{\theta}^*)$ with $p_1(x)$ in Eq. (3.10) and $p(x|\boldsymbol{\theta})$ with $p_2(x)$, we write $I(\boldsymbol{\theta}^*, \boldsymbol{\theta}) = I(1,2)$ where

$$I(\boldsymbol{\theta}^*, \boldsymbol{\theta}) = \int p(x|\boldsymbol{\theta}^*) \log p(x|\boldsymbol{\theta}^*) dx - \int p(x|\boldsymbol{\theta}^*) \log p(x|\boldsymbol{\theta}) dx \tag{3.12}$$

and where we identify the second term in the above equation with $\mathcal{L}'(\boldsymbol{\theta})$, the average log likelihood. Clearly, maximizing the average likelihood is equivalent to minimizing the K-L measure. Hence, the importance of maximum likelihood estimators, MLE's. Their foundation lies in information theory.

3.3.3 Jaynes' Principle of Maximum Entropy

The principle of maximum entropy, PME, expresses a philosophy of dealing with the problem of inference, a problem ubiquitous in all scientific endeavor. Basically, we are faced with the task of making inferences in light of incomplete information. The question is, how do we treat this incompleteness? An example which illustrates the dilemma very well and which, incidentally, is of historical importance, is the problem of estimating the power spectrum of a finite sample of data. As we all know, the conventional approach is to estimate the autocovariance function of the data and

obtain the spectrum by Fourier transformation. In estimating the autocovariance, we are faced with the question of what to assume for the data outside of the known window. The conventional wisdom is to assume a zero extension, an assumption which is obviously extremely constrictive and with deleterious effect. In 1970, John Burg approached the spectral estimation problem by means of the PME and, in so doing, revolutionized this field of data analysis. We state the principle here.
In problems of data analysis and/or inference

Be consistent with the known information and maximally noncommittal with respect to the unknown information

Having formulated this most sound and logical of principles, Jaynes was faced with the problem of expressing it in mathematical terms. Remembering the discussion with relationship to uncertainty, we see that the above principle is consistent with the maximization of our state of ignorance subject to constraints imposed by known information. Maximizing uncertainty appears perhaps to be a paradox. It is not. The solution to a problem of the type illustrated above, which is just an example of a linear inverse problem, is almost always non-unique. In other words, there are an infinity of models satisfying the constraints (e.g., a few autocovariance lags). The uncertainty associated with any of the possible solutions is inversely proportional to the information which we impose. Consequently, the procedure to follow is to seek a solution which, subject to available information, maximizes the uncertainty of the solution and thereby minimizes the influence of unavailable data.

3.3.4 The Jaynes Entropy Concentration Theorem, ECT

We now amplify our discussion concerning MaxEnt by stating, without proof, a beautiful theorem due to Jaynes (1982) which truly encapsulates this paradigm in a formal manner.

3.3.4.1 The Jaynes Entropy Concentration Theorem, ECT

Consider N trials of a random experiment which has M possible results at each trial with entropy

$$H(q_1, q_2, \ldots, q_M) = -\sum_i q_i \log q_i, \quad i = 1, 2, \ldots, M$$

where the frequencies, q_i, are computed from the sample numbers n_i associated with each outcome as $q_i = n_i/N$ with $\sum_i n_i = N$. Consider the subclass $\mathcal{C}$ of all possible outcomes that could be observed in N trials which are compatible with the L linearly independent constraints

$$\sum_i A_{ji} q_i = d_j \quad j = 1, 2, \ldots, L \;\text{ and }\; L < M$$

A certain fraction F of outcomes in $\mathcal{C}$ will have entropy in the range

$$H_{\max} - \Delta H \leq H(q_1, q_2, \ldots, q_M) \leq H_{\max} \tag{3.13}$$

where H_{max} is obtained by maximizing the entropy, $H(q_1, q_2, \ldots, q_M)$, subject to the constraints.

The ECT states that, asymptotically, $2N\Delta H$ is distributed in $\mathcal{C}$ as χ_k^2 with the number of degrees of freedom, $k = M - L - 1$. Specifically, denoting χ_k^2 at the $100P$ percent significance level by $\chi_k^2(P)$, we obtain

$$2N\Delta H = \chi_k^2(1 - F) \tag{3.14}$$

Eq. (3.14) is a remarkable expression. It informs us of the percentage chance that the observed frequency distribution will have an entropy which is outside of the interval in Eq. (3.13). As shown by Jaynes (1982) in a number of examples, for large N, the overwhelming majority of all *possible* distributions will have an entropy which is very close to H_{max}. This is surely a very strong justification for the use of the principle of MaxEnt.

The PME has, traditionally, been used as a principle for the determination of pdf's given information in terms of a few moments of the distribution. We illustrate the above discussion with two very simple examples which serve to put matters into perspective.

3.3.4.2 Example 1. The Famous Die Problem

We consider a die with discrete probabilities $q_i, i = 1, 2, \ldots, 6$ associated with the six faces. Probability reasoning imposes the constraint $\sum_i q_i = 1$. The problem is to find the six probabilities subject to the one constraint. Clearly there are an infinite number of possible solutions and a regularization must be defined. Using the MEP we pose the problem as follows:

$$\text{Maximize } H = -\sum_i q_i \log q_i \quad \text{subject to} \quad \sum_i q_i = 1$$

The solution follows by using a Lagrange multiplier μ and maximizing the Lagrangian, L, where

$$L = H + \mu\left(\sum_i q_i - 1\right)$$

Hence

$$\frac{\partial L}{\partial q_i} = -\log q_i - 1 + \mu = 0$$

resulting in

$$q_i = e^{\mu - 1} \tag{3.15}$$

Using the above, we obtain from the constraint

$$\sum_{i=1}^{6} e^{\mu-1} = 1$$

$$e^{\mu-1} = \tfrac{1}{6}\,. \tag{3.16}$$

Finally, substituting Eq. (3.16) into Eq. (3.15) , obtains the desired result

$$q_i = \frac{1}{6} \tag{3.17}$$

Eq. (3.17) expresses the very reasonable conclusion that, given only the normalizing constraint, the most reasonable assignment for the probabilities is a uniform one. We could, in fact, have deduced this from our intuitive understanding of uncertainty. This result could also have been obtained by minimizing the l_2 norm in this case. This, however, as will become clear from the second example discussed below, is an exception rather than the rule.

Let us now use the PME to solve the problem of estimating the unknown frequencies q_i given the constraint that $\bar{n}$=4.5 (this is not a fair die, for which the value is 3.5). There is also the unimodular constraint that $\sum_i q_i = 1$. We obtain the solution

$$\mathbf{q}_{ME} = (0.0543, 0.0788, 0.1142, 0.1654, 0.2398, 0.3475)$$

with entropy $H_{ME} = 1.6136$. Of course, the ME solution is not the only one possible. Consider another solution, $\mathbf{q}_{AS}$, that also exactly fits our constraints

$$\mathbf{q}_{AS} = (\frac{3}{10}, 0, 0, 0, 0, \frac{7}{10})$$

$\mathbf{q}_{AS}$ has entropy $H_{AS} = 0.6110$. The question that we posed above is, are both solutions equally plausible? Clearly not. Since when does throwing a die result in no 5's coming up? We have deliberately chosen a solution that is so implausible since we wish to stress the point.

But let us look at another possible solution for a more subtle comparison. We obtain this solution, $\mathbf{q}_{MRE}$, by means of the principle of MRE which will be discussed in Section 3.3.1. The $\mathbf{q}_{MRE}$ solution, for a uniform prior with a lower bound of zero and an unconstrained upper bound, is (Ulrych et al., 1990)

$$\mathbf{q}_{MRE} = (0.0712, 0.0851, 0.1059, 0.1399, 0.2062, 0.3918)$$

with entropy $H_{MRE} = 1.6091$.

On the other hand, if the upper bound is constrained to the reasonable value of 1 (Woodbury and Ulrych, 1998), we obtain the MRE solution

$$\mathbf{q}_{MRE} = (0.0678, 0.0826, 0.1055, 0.1453, 0.2232, 0.3755)$$

with entropy $H_{MRE} = 1.6091$. This solution is, interestingly enough, close to the ME solution.

Now the question is, how probable is $\mathbf{q}_{MRE}$ as compared to the ME solution? We can quantify the above results by means of the Jaynes Entropy Concentration Theorem of Section 3.3.4. Specifically, according to the ECT, in an experiment consisting of 1000 trials, 99.99 percent of all outcomes which are allowed by the constraints have entropy in the range $1.609 \leq H \leq 1.614$ (Jaynes, 1982). Although the ME solution is the more probable solution in terms of multiplicities, H_{MRE} is certainly well within the range of the most probable solutions.

We can also compare the multiplicities of the two solutions. For a large number of trials, $N = 1000$ say, we use our approximation to compute the ratio of multiplicities. We have

$$\frac{W_{ME}}{W_{MRE}} = e^{N(H_{ME}-H_{MRE})} = e^{10.2} = 26,903$$

This ratio tells us that for every way that the MRE solution can be realized, the ME solution can be realized in approximately 30,000 ways.

3.3.4.3 Example 2. The *Gull and Newton* Problem

The second compelling reason for using MaxEnt to solve any general linear or linearized inverse problem is beautifully illustrated by this example which is originally due to Gull and Newton (Gull and Newton, 1986) and delightfully fashioned by Paul Fougere (Fougere, 1995) into the form used here which is a summary of Fougere's more detailed presentation.

A poll taken in norther Vermont has shown that 1/4 of the natives were French speaking and 1/4 were ski enthusiasts, where the fractions, 1/4, are relative frequencies. The question to be answered is, how many native Vermonters are both French speaking *and* skiers? To answer this question, we set up the contingency table illustrated in Table (3.1).

	Skiers		$p_{\mathcal{M}}$
French	p	1/4-p	1/4
	1/4-p	p+1/2	3/4
$p_{\mathcal{M}}$	1/4	3/4	

Table 3.1: Contingency table.

We know the marginals of the joint frequency distribution. These have been measured in the survey and are shown in the table as the column and row labelled $p_{\mathcal{M}}$. We now assign p as the joint probability of those who are both French speaking and skiers. The two off-diagonal elements must be $1/4 - p$ and the lower right diagonal element must be $p + 1/2$. Clearly, any value of p in the range $0 \leq p \leq 1/4$ is possible. We will solve the problem using three different norms and compare the results in detail.

3.3.4.4 Shannon Entropy Solution

The total Shannon entropy is

$$H = -\sum_{k=1}^{4} p_k \log p_k$$
$$= -[p \log p + 2(1/4 - p)\log(1/4 - p) + (p + 1/2)\log(p + 1/2)] \quad (3.18)$$

Maximizing the entropy in the usual manner, we obtain

$$\frac{dH}{dp} = 1 + \log p - 2 - 2\log(1/4 - p) + 1 + \log(p + 1/2) = 0$$

or

$$\log\left[\frac{p(p + 1/2)}{(1/4 - p)^2}\right] = 0$$

Clearly, the term in brackets must be equal to 1, yielding

$$p(p + 1/2) = (1/4 - p)^2$$

with the final result

$$p = \frac{1}{6} \quad (3.19)$$

The full contingency table is shown in panel (a) of Table (3.2). This result is exactly what one would expect intuitively if skiing and French speaking are independent characteristics. Since, in the statement of the problem, dependency or lack thereof has not been invoked, we see that the MaxEnt solution has precisely encoded the given information without making any other assumptions. Any correlation existing between the two characteristics would have to be entered into the problem as additional prior information. Here then, is MaxEnt in action. The result is consistent with the available information and maximally noncommittal with respect to unavailable information. Let us see how other approaches compare.

3.3.4.5 Least Squares Solution

We minimize $\sum_{k=1}^{4} p_k^2$. Following the procedure above obtains the result that $p = 0$. The full contingency table is shown in panel (b) of Table (3.2). This result is startling. No Vermonters are both French speaking and skiers. i.e., this approach imposes a negative correlation on the result,

3.3.4.6 Burg Entropy Solution

The quantity $-\sum \log p$ has been given the name, Burg entropy, as a result of its use by John Burg (Burg, 1975) in the problem of estimating the MaxEnt power spectrum.

(a)

$-\sum p_k \log p_k$		**Skiers**	
Shannon Entropy	**French**	1/16	3/16
(Uncorrelated)		3/16	9/16

(b)

$-\sum p_k^2$		**Skiers**	
Least Squares	**French**	0	1/4
(Negative Correlation)		1/4	1/2

(c)

$-\sum \log p_k$		**Skiers**	
Burg Entropy	**French**	1/4	0
(Positive Correlation)		0	3/4

Table 3.2: Comparison of Vermonters results.

pplication of this norm here results in $p = 1/4$, which is the maximum allowed value nd corresponds to a positively correlated result. The full table is shown in panel (c).)nce again, the result obtains more from the observed data than is contained in the ata.

Let us examine the numbers involved in the Gull and Newton example. We take sample size of only 16 Vermonters for simplicity, and compute the counts and the espective multiplicities for the three cases. This has been done for us by Paul Fougere nd the results are as shown in table (3.3)

Norms	Counts	Multiplicity
Shannon Entropy	{ 1 3 3 9 }	1,601,600
Least Squares	{ 0 4 4 8 }	900,900
Burg Entropy	{ 4 0 0 12}	1,820

Table 3.3: Multiplicities for different norms.

Ve see that, for every 9 ways that the least squares solution can be realized, the laxEnt solution can be realized in 16 ways. As a result of the exponential nature f the multiplicity, these ratios become much more accentuated as the number of ndividuals increases. Indeed, for $N = 128$, the ratio of case 1 to case 2 is 9200 to 1. e., the MaxEnt solution is about 9000 times more likely to occur, in this case, than s the least squares solution.

These two simple examples have captured the essence of the MaxEnt philosophy. This is an approach to inference which is consistent with what we know and, in maximal way, oblivious to what we do not know. In the Vermont example, our laxEnt solution returns the intuitively expected uncorrelated result. The other two pproaches which we have investigated return results with maximum positive and naximum negative correlation. The importance of solutions which are free from rtifacts, unknowingly introduced through the topology of the particular regularizer, annot be over emphasized.

3.3.5 The General MaxEnt Solution.

At this stage, we consider the general entropy maximization problem. Our discusion will lead us also to an investigation of the topology of the entropy functional. The general entropy maximization problem is to obtain the probabilities, $\{p_i\}$, by naximizing H subject to $M+1$ constraints which are of the form

$$\sum_i p_i = 1$$

$$\sum_i f_{ki} p_i = c_k \qquad k = 1, 2, \ldots, M \tag{3.20}$$

where c_k are the known constants associated with the variables f_{ik}. It is most convenient to perform the required manipulations in matrix form. To this end we define $\mathbf{p} = [p_1, p_2, \ldots, p_N]^T$ to be the vector of MaxEnt derived probabilities, $\boldsymbol{\lambda}$ is the column vector of M Lagrange multipliers, $\mathbf{u} = [1, 1, \ldots, 1]^T$, $\mathbf{F}$ is the matrix with row vectors $\mathbf{f}_k$ and $\mathbf{c}$ is the vector of known constants. In matrix form, the Lagrangian is written as

$$L = -\mathbf{p}^T \log \mathbf{p} - (\mu - 1)(\mathbf{u}^T \mathbf{p} - 1) - \boldsymbol{\lambda}^T (\mathbf{F}\mathbf{p} - \mathbf{c}) \tag{3.21}$$

where we have written $\mu - 1$ as a Lagrange multiplier for the unimodular constraint for convenience. Solving $\partial L / \partial \mathbf{p} = 0$ for $\mathbf{p}$, we obtain

$$\mathbf{p} = e^{-\mu \mathbf{u}} e^{-\mathbf{F}^T \boldsymbol{\lambda}} \tag{3.22}$$

Substituting into the unimodular constraint leads to

$$\mathbf{u}^T e^{-\mu \mathbf{u}} e^{-\mathbf{F}^T \boldsymbol{\lambda}} = 1 \tag{3.23}$$

We note that Eq. (3.23) relates μ to $\boldsymbol{\lambda}$. Without going into details, it can be shown, following Kapur and Kesavan (1992), that μ is a convex function of $\boldsymbol{\lambda}$. This follows from the fact that the Hessian matrix of μ with respect to $\boldsymbol{\lambda}$ is a covariance matrix and as such is always non-negative definite. The expression for maximum entropy, using Eq. (3.22), becomes

$$\begin{aligned} H_{max} &= -\mathbf{p}^T \log \mathbf{p} \\ &= -\mathbf{p}^T (-\mu \mathbf{u} - \mathbf{F}^T \boldsymbol{\lambda}) \\ &= \mu + \boldsymbol{\lambda}^T \mathbf{c} \end{aligned} \tag{3.24}$$

We see, from Eq. (3.24), that H_{max} is a function of μ, $\boldsymbol{\lambda}$ and $\mathbf{c}$. However, as we have seen, μ is a function of $\boldsymbol{\lambda}$ which in turn is a function of $\mathbf{c}$. Consequently, H_{max} is a function of $\mathbf{c}$. Following Kapur and Kesavan (1992) it may now be simply derived that the matrix of second partial derivatives of H_{max} with respect to $\mathbf{c}$ is negative definite and hence that H_{max} is a concave function of $\mathbf{c}$.

To summarize, it may be shown that, given *linear* constraints, Shannon entropy is a concave function and consequently possesses a solitary maximum value. This point deserves to be stressed. Maximizing the Shannon entropy does not require choosing a suitable initial model which is often the case with nonlinear norms. Further, because of the exponential form of the solution given by Eq. (3.22), the probabilities are guaranteed to be positive, a natural requirement for our problem. It must be pointed out at this stage that the computation of the required Lagrange multipliers is, in general, a nonlinear problem requiring approaches such as the Newton-Raphson iterative technique. More on this subject later.

.3.5.1 Entropic justification of Gaussianity

Ve have seen above, in some detail, the general approach to the maximization of ntropy. In Section 3.3.4.2 we also got our first glimpse of the logic associated with naximum entropy results. We saw that, given only the unimodular constraint, the nost 'reasonable' pdf was the uniform one. We now ask the question, what is the most oncommittal pdf that is associated with the constraint of known variance? This is a articularly important question, since, in most practical cases, all that we do 'know' bout our random signal is, its variance. Many texts outline the calculation that is equired, and we follow one of our favorites, Lathi (1968). It is worth while to outline his development in some detail since, not only is the result of such significance, ut also the path followed serves as a template for later results. For mathematical onvenience, we consider the continuous formulation.

In general, our signals, of whatever type, are zero mean and if not, we can always nake them so. Therefore, we consider here the problem of a known second moment onstraint which we write as σ^2 for generality. i.e., we are given σ^2 and we seek hat $p(x)$ that, while respecting the constraint, is most noncommittal to unavailable nformation. We have the following optimization problem: Maximize the entropy, $H(\mathbf{x}(t))$ where

$$H(\mathbf{x}(t)) = -\int p(x)\log p(x)dx \tag{3.25}$$

iven the constraints

$$\int p(x)dx = 1 \tag{3.26}$$

nd

$$\int x^2 p(x)dx = \sigma^2 \tag{3.27}$$

ollowing the calculus of variation, we define Lagrange multipliers λ_1 and λ_2 and onstraint functions $\phi_1(x,p) = p(x)$ and $\phi_2(x,p) = x^2 p(x)$, and we look for that $p(x)$ vhich satisfies

$$\frac{\partial H(\mathbf{x}(t))}{\partial p(x)} + \lambda_1 \frac{\partial \phi_1(x,p)}{\partial p(x)} + \lambda_2 \frac{\partial \phi_2(x,p)}{\partial p(x)} = 0 \tag{3.28}$$

ubstituting the constraint functions into Eq. (3.28) and performing the differentia-ions, we obtain

$$-(1+\log p(x)) + \lambda_1 + \lambda_2 x^2 = 0$$

olving for $p(x)$

$$p(x) = e^{(\lambda_1 - 1)} e^{\lambda_2 x^2} \tag{3.29}$$

We now substitute the solution of Eq. (3.29) back into the constraint equations, Eqs. (3.26) and (3.27), to obtain the Lagrange multipliers. First, substituting into Eq. (3.26)

$$\begin{aligned}\int e^{(\lambda_1-1)}e^{\lambda_2 x^2}dx &= 2e^{(\lambda_1-1)}\int_0^\infty e^{\lambda_2 x^2}dx \\ &= 2e^{(\lambda_1-1)}\left(\frac{1}{2}\sqrt{\frac{\pi}{-\lambda_2}}\right) \\ &= 1\end{aligned}$$

provided λ_2 is -ve, which gives

$$e^{(\lambda_1-1)} = \sqrt{\frac{-\lambda_2}{\pi}} \tag{3.30}$$

Now, substituting Eqs. (3.29) and (3.30) into Eq. (3.27)

$$\int x^2\sqrt{\frac{-\lambda_2}{\pi}}e^{\lambda_2 x^2}dx = 2\sqrt{\frac{-\lambda_2}{\pi}}\int_0^\infty e^{\lambda_2 x^2}dx = \sigma^2$$

Therefore

$$-\frac{1}{2\lambda_2} = \sigma^2$$

which leads to

$$\lambda_2 = -\frac{1}{2\sigma^2} \tag{3.31}$$

and, by substitution into Eq. (3.30)

$$e^{(\lambda_1-1)} = \frac{1}{2\lambda_2\sigma^2} \tag{3.32}$$

Finally, substituting Eqs. (3.31) and (3.32) into Eq. (3.29), we obtain

$$p(x) = \frac{1}{\sigma\sqrt{2\pi}}e^{-x^2/2\sigma^2} \tag{3.33}$$

This, then, is our final result. Given only the variance of $\mathbf{x}$, the pdf of $\mathbf{x}$ that is most noncommittal with respect to the unknown higher order statistics, is the Gaussian pdf. This is, in our opinion, a beautiful and powerful result. It is precisely the justification of our ubiquitous assumption that the pdf associated with random noise, is Gaussian. Also, of course, it is a result that, given the Central Limit Theorem please see Section 6.7.3.6), is reasonable and expected. We believe together with the late Edwin Jaynes that , for all the reasons explored in this chapter, solutions based on the principle of maximum entropy are eminently reasonable. It is for these reasons that we will use MaxEnt to obtain a priori pdf's required in Bayesian solutions.

3.4 MaxEnt and the Spectral Problem

By far the most famous application of the principle of MaxEnt is that due to John Burg, (Burg, 1975), in the determination of the power spectral density of a stationary random process. The problem is also the first example in our book of the use of information theory in an attempt to obviate the *aperture* inherent in the input data. Basically the issue is this; we are given a finite data set and are required to compute a power spectral estimate. The conventional approach proceeds along two, equivalent, paths.

1. We estimate the autocovariance, $\hat{r}_\tau \quad \tau = 0, 1, 2, \ldots, M$, for our process x_n as

$$\hat{r}_\tau = \sum_{n=1}^{N-1} x_n x_{n+\tau} \tag{3.34}$$

   which is known as the Yule-Walker, YW, estimator. We note immediately, that this estimator implicitly assumes that the data outside the time window of N points are assumed to be **zero**.

   The next step is to produce the power spectrum, $P_{xx}(f)$, by Fourier transformation. i.e.,

$$P_{xx}(f) \Leftrightarrow \hat{r}_\tau$$

2. For the sake of completeness we state without proof, which is left as an exercise for the interested reader, that an equivalent estimate is obtained as

$$P_{xx}(f) = \frac{1}{N} X_k X_k^* \Leftrightarrow \hat{r}_\tau$$

   where $X_k \Leftrightarrow x_n$, the well known periodogram estimator. Note however, that this equivalence holds only when the autocovariance is estimated using the Yule-Walker approach.

We now briefly discuss the application of the PME to the problem of power spectral density estimation, leading to the famous Burg solution.

3.4.1 John Burg's Maximum Entropy Spectrum.

We begin with the Shannon definition of entropy and we use the continuous form for convenience in the derivation.

$$H(x) = -\int p(x) \log p(x) dx \tag{3.35}$$

Since we assume knowledge only of first and second order statistics, we associate with $x(t)$ a Gaussian pdf, $N(\mu, \sigma^2)$. This is an important point that we have discussed at some length in the previous section. The Gaussian form does not imply that the actual pdf is Gaussian, of course. We are merely being true to the spirit of the PME. We use what we know and are non committal with respect to what we do not know. As we have seen, knowledge of the 2nd moment only implies a Gaussian pdf. Anything else would be inconsistent. Integrating (3.35) obtains

$$H(x) = \frac{1}{2}\log\left(2\pi e\sigma^2\right) \tag{3.36}$$

Consider the Fourier series representation of $x(t)$ and assume that the complex Fourier coefficients, $a(f)$, are random variables distributed according to $N(\mu, \sigma^2)$ with

$$E[a(f)] = 0$$

and

$$\begin{aligned} E[a(f)\, a^*(f)] &= \sigma_a^2(f) \\ &= S(f) \end{aligned}$$

Substituting into (3.36)

$$H(f) = \frac{1}{2}\log(2\pi e\sigma^2(f))$$

Burg now maximizes H_B, the average entropy in the bandwidth, where

$$H_B = \frac{1}{2f_N}\int_{-f_N}^{f_N} H(f)df$$

or, in terms of the unknown but desired 'true' spectral density, $S(f)$

$$H_B = \frac{1}{4f_N}\int_{-f_N}^{f_N} \log S(f)\, df + \log\left(2\pi e\right)^{\frac{1}{2}}$$

Clearly we need some constraints. These come from the assumption that we know the first M autocovariances, $r_k, \quad k = 0, 1, \ldots, M$

$$r_k = \int_{-f_N}^{f_N} S(f)\exp(2\pi i k f\Delta t)df \tag{3.37}$$

where Δt is the sampling interval. To maximize (3.37) with (3.37) as constraint, we form

$$\frac{\partial H_B}{\partial r_k} = 0 \qquad \text{for } |k| > M$$

Since

$$S(f) = \Delta t \sum_{k=-\infty}^{\infty} r_k e^{-2i\pi fk\ \Delta t}$$

we have

$$H_B(f) = \frac{1}{4f_N} \int_{-f_N}^{f_N} \log \{\Delta t \sum_k r_k e^{-2i\pi k\ \Delta t f}\}\, df + \text{constant}$$

Hence

$$\begin{aligned} \frac{\partial H_B(f)}{\partial r_k} &= \frac{1}{4f_N} \int \frac{\Delta t \exp(-2i\pi fk\ \Delta t)}{\Delta t \sum r_k \exp(-2i\pi fk\ \Delta t)} df \\ &= \frac{1}{4f_N} \int \frac{\Delta t \exp(-2i\pi fk\ \Delta t)}{S(f)} df \\ &= 0 \quad \text{for } |k| > M \end{aligned}$$

i.e., the Fourier series for $1/S(f)$ truncates for $|k| > M$. Let us expand in the form

$$\frac{1}{S(f)} = \sum_k b_k e^{-2i\pi fk\ \Delta t} \tag{3.38}$$

where $b_k = b^*_{-k}$ for obvious reasons. We now solve for $b_k \Leftrightarrow B(z)$. Since the b_k must be Hermitian (for complex x_t) we can write

$$B(z) = D(z)\, D(z^{-1})$$

where $D(z)$ is minimum phase and $D(z^{-1})$ is maximum phase (Robinson and Treitel, 2002). We now write (3.38) as

$$\begin{aligned} \frac{1}{C(z)} &= D(z)\, D(z^{-1}) \\ &= \frac{G(z)G(z^{-1})}{\sigma^2} \end{aligned} \tag{3.39}$$

where $C(z)$ is the infinitely long 'true' autocovariance function in the z domain and σ^2 is factored out as a scale constant. We rewrite the above equation as

$$C(z)G(z) = \sigma^2 \frac{1}{G(z^{-1}))}$$

and note, from the discussion in Section 1.8.1 that, since $G(z^{-1})$ is minimum phase, its inverse is causal and stable. Taking the z^{-1} transform, we write

$$\begin{aligned}(\cdots + c_2 z^{-2} + c_1 z^{-1} + c_0 + c_1 z &+ c_2 z^2 + \cdots)(1 + g_1 z + g_2 z^2 + \cdots + g_{M-1} z^{M-1}) \\ &= \sigma^2 + \text{ terms in } z^{-j}. \end{aligned} \tag{3.40}$$

Collecting coefficients of equal powers of z and noting that r_k will equal c_k for $|k| = 0, 1, 2, \ldots, M$ so that $S(f)$ is consistent with the constraints, we obtain using Eq. (3.40)

for z^0

$$r_0 + g_1 r_{-1} + g_2 r_{-2} + \cdots + g_M r_M = \sigma^2$$

for z^1

$$r_1 + g_1 r_0 + g_2 r_{-1} + \cdots + g_M r_{M-1} = 0$$

etc., to obtain

$$R\mathbf{g} = \sigma^2 \mathbf{e}_1$$

These are the AR normal equations which we have met before in Section 2.4.1, Eq. (2.16.

3.4.1.1 Remarks

The actual computation of the ME spectrum is identical to that of the AR spectrum in form

$$S_{\text{ME}}(f) = \frac{\sigma^2}{\Delta t |1 + \sum_{k=1}^{M} g_k e^{2i\pi f k \ \Delta t}|^2} \tag{3.41}$$

where the coefficients g_k are identical to the PEO. The actual philosophy is quite different, of course. Another interpretation of the ME spectral estimator has been given by Anderson (1978) who proved that the result obtained by means of Eq. (3.41) is identical to that obtained by first modelling r_k as an AR process, extending to infinity and then taking the Fourier transform.

An interesting and different view which relates these issues is the one developed by Kapur and Kesavan (1992) who ask the question, what is the measure of entropy that, when maximized subject to certain constraints, gives a specified spectral density function, sdf. This is an example of an inverse ME problem. The sdf is $S(f)$ and the

$2M + 1$ constraints, which we assume to be known exactly, are

$$\int_{-1/2}^{1/2} S(f) \exp(-2\pi i f k) df = r_k \qquad k = -M, -M+1, \ldots, 0, \ldots, M-1, M \quad (3.42)$$

As we have seen, the autocovariance constraints, r_k, must in general be estimated from the data, and this introduces much complexity into the problem. This is not the issue here however. Following Kapur and Kesavan (1992), we define the sought after entropy measure to be

$$\int_{-1/2}^{1/2} \phi[S(f)] df \tag{3.43}$$

where $\phi[\cdot]$ is a twice-differentiable concave function. Maximizing expression (3.43) subject to the autocovariance constraints expressed by Eq. (3.42) in the usual manner, we obtain

$$\phi'[S(f)] = \lambda_0 + \sum_{k=-M}^{M} \lambda_k \exp(-2\pi i f k) \tag{3.44}$$

Since the AR sdf, $S(f)$, has the form

$$S(f) = \frac{1}{\lambda_0 + \sum_k \lambda_k \exp(-2\pi i f k)} \tag{3.45}$$

we can see that

$$\phi'[S(f)] \propto \frac{1}{S(f)}$$

and hence, ignoring any additive and multiplicative constants

$$\phi[S(f)] = \log S(f) \tag{3.46}$$

Eq. (3.46) has the form that has become known as the Burg entropy. This derivation complements the development above and stresses the importance of the inverse ME approach.

We mentioned above the complexity introduced by the fact that the r_k must be estimated from the data. This issue deserves a chapter to itself and we only very briefly review the main points here. The YW estimate, given by Eq. (3.34), is a biased but consistent estimate of the true autocovariance. As we have pointed out, it is obtained by insisting that the values of the time series, x_n, are zero for $n > N$. This assumption is clearly not consistent with the whole philosophy of ME where we are attempting to be maximally non committal with respect to unknown data. John Burg's contribution to the spectral problem also involved this issue. As discussed in Section 2.4.4, the Burg autocovariance estimator is based on a PEO that is computed from the data without assuming this zero extension. Ulrych and Bishop

(1975) have compared the YW and Burg autocorrelation estimates as far as resolution and robustness are concerned and the reader is referred to this work and also to the books by Kay (1988) and Marple (1987) that contain a wealth of information on this and many other relevant topics.

3.5 The Akaike Information Criterion, AIC

The computation of the Burg spectral estimator using Eq. (3.41) assumes that the order of the equivalent AR process is known (order 4, for example, for two harmonics with zero additive noise). Of course, in practice, noise is present and the AR order is not known. An approach to order estimation is via an information theoretic criterion known as the Akaike Information Criterion (Akaike, 1973, 1974; Sakamoto et al., 1986) or, simply, the AIC. This criterion is of great importance, however well it works for harmonics in noise. Its importance is due to the information principles that Akaike adopted in deriving the AIC, principles that have been generally recognized as fundamental to the derivation of information measures. This is not the place for an in-depth discussion of these principles. We give abundant references and here we content ourselves, and hopefully the reader, with a description of the salient points. The AIC is a criterion that is intimately related to

- The Kullback-Leibler information measure.
- The principle of maximum entropy.
- The principle of maximum likelihood.
- The Fisher information matrix.

We have already explored the first three items listed above and now we continue with the development of the AIC. We begin with the assumption that the estimated, $\boldsymbol{\theta}$, and true, $\boldsymbol{\theta}^*$, parameter vectors are related by means of

$$\boldsymbol{\theta} = \boldsymbol{\theta}^* + \Delta\boldsymbol{\theta}$$

A Taylor's series expansion of $\log p(x|\boldsymbol{\theta}^* + \Delta\boldsymbol{\theta})$ about $\boldsymbol{\theta}^*$ leads to

$$\log p(x|\boldsymbol{\theta}^*) + \Delta\boldsymbol{\theta}^T \frac{\partial}{\partial\boldsymbol{\theta}} \log p(x|\boldsymbol{\theta})|_{\boldsymbol{\theta}^*} + \frac{1}{2}\Delta\boldsymbol{\theta}^T \frac{\partial^2}{\partial\boldsymbol{\theta}^T\boldsymbol{\theta}} \log p(x|\boldsymbol{\theta})|_{\boldsymbol{\theta}^*}\Delta\boldsymbol{\theta}$$

which, since the second term is zero, results in the approximate expression

$$\log p(x|\boldsymbol{\theta}^* + \Delta\boldsymbol{\theta}) \approx \log p(x|\boldsymbol{\theta}^*) + \frac{1}{2}\Delta\boldsymbol{\theta}^T \frac{\partial^2}{\partial\boldsymbol{\theta}^T\boldsymbol{\theta}} \log p(x|\boldsymbol{\theta})|_{\boldsymbol{\theta}^*}\Delta\boldsymbol{\theta} \qquad (3.47)$$

We now relate the above expression to the K-L measure expressed in Eq. (3.12). Substituting Eq. (3.47) into Eq. (3.12) obtains

$$\begin{aligned} I(\boldsymbol{\theta}^*, \boldsymbol{\theta}) &= I(\boldsymbol{\theta}^*, \boldsymbol{\theta} + \Delta\boldsymbol{\theta}) \\ &\approx \frac{1}{2}\Delta\boldsymbol{\theta}^T \mathbf{C} \Delta\boldsymbol{\theta} \end{aligned}$$

where $\mathbf{C}$ is the Fisher information matrix, defined by

$$\mathbf{C} = E\left[-\frac{\partial^2}{\partial\boldsymbol{\theta}^T\boldsymbol{\theta}} \log p(x|\boldsymbol{\theta})|_{\boldsymbol{\theta}^*}\right] \tag{3.48}$$

We note the following

1. $\frac{1}{2}\Delta\boldsymbol{\theta}^T\mathbf{C}\Delta\boldsymbol{\theta}$ is an approximation to the Kullback-Leibler measure.

2. Since $I(\boldsymbol{\theta}^*, \boldsymbol{\theta})$ depends on the unknown true pdf, Akaike introduced the AIC, that is a robust estimate of $I(\boldsymbol{\theta}^*, \boldsymbol{\theta})$.

We define the following quantities

$$\begin{aligned} W(\boldsymbol{\theta}^*, \boldsymbol{\theta}) &= 2I(\boldsymbol{\theta}^*, \boldsymbol{\theta}^* + \Delta\boldsymbol{\theta}) \\ &= \Delta\boldsymbol{\theta}^T\mathbf{C}\Delta\boldsymbol{\theta} \end{aligned}$$

and

$$||\boldsymbol{\theta}||_C = (\boldsymbol{\theta}^T\mathbf{C}\boldsymbol{\theta})^{\frac{1}{2}}$$

is the norm for $\boldsymbol{\theta} \in \mathcal{D}_M$, the M dimensional model space and defines a measure of distance

Therefore

$$W(\boldsymbol{\theta}^*, \boldsymbol{\theta}) = ||\boldsymbol{\theta} - \boldsymbol{\theta}^*||^2$$

is a [distance]^2

$${}_k\hat{\boldsymbol{\theta}} = \text{the MLE in the space } \mathcal{D}_K$$

a subspace of $\mathcal{D}_M$

$${}_k\hat{\boldsymbol{\theta}^*} = \text{the projection of } \boldsymbol{\theta}^* \text{ in the space } \mathcal{D}_K$$

i.e. $||{}_k\boldsymbol{\theta}^* - \boldsymbol{\theta}^*||_C^2 = \min_{\boldsymbol{\theta}\in\mathcal{D}_K} ||\boldsymbol{\theta} - \boldsymbol{\theta}^*||_C^2$

i.e. ${}_k\hat{\boldsymbol{\theta}}$ is $\perp$ to $\mathcal{D}_K$

The relationships between ${}_k\hat{\boldsymbol{\theta}}, {}_k\boldsymbol{\theta}^*$ and $\boldsymbol{\theta}$ for $M = 3, k = 2$ are shown in Fig. 3.1.

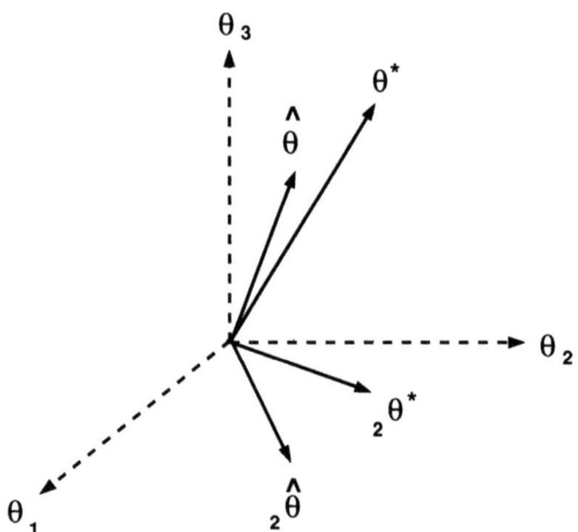

Figure 3.1: Defining the AIC components.

From this figure and the definitions, it follows that

$$
\begin{aligned}
W(\boldsymbol{\theta}^*, {}_k\hat{\boldsymbol{\theta}}) &= ||{}_k\hat{\boldsymbol{\theta}} - \boldsymbol{\theta}^*||_C^2 \\
&= \underbrace{||{}_k\hat{\boldsymbol{\theta}} - {}_k\boldsymbol{\theta}^*||_C^2}_{\text{Parameter estimation errors in model space}} + \underbrace{||{}_k\boldsymbol{\theta}^* - \boldsymbol{\theta}^*||_C^2}_{\text{Goodness of fit of the model}}
\end{aligned}
$$

As we can see, $W(\boldsymbol{\theta}^*, {}_k\hat{\boldsymbol{\theta}})$ is composed of two parts, a trade-off so to speak, between (I) errors that are incurred in the fitting of the parameters and (II) the goodness of fit of the model. As an example, consider the fitting of an AR model. Errors in (I) depend on the number of parameters being fitted. Obviously, the error increases with increasing number. Error in (II) always decreases. This is akin to the fitting of a polynomial. Clearly, the higher the order of the polynomial, the smaller the error of the fit. This does not mean that the fit is correct. As we well know, the overshoot of the high order polynomial between the known data points is a most undesirable effect. Akaike, recognizing this fact, defined the AIC in terms of $W(\boldsymbol{\theta}^*, {}_k\hat{\boldsymbol{\theta}})$ as

$$\text{AIC} = E[NW(\boldsymbol{\theta}^*, {}_k\hat{\boldsymbol{\theta}})] \tag{3.49}$$

where the correct order of the model being fitted is, logically, given by the minimum AIC as a function of the order k. The final form of the AIC that follows from Eq. (3.49) requires much algebra to deduce and, since we have covered the salient points, we leave the interested reader to consult the appropriate references , the publications by Akaike listed above, and the work of Sakamoto et al. (1986) and Matsuoka and

Ulrych (1986). The final form of the measure is

$$\mathrm{AIC}(k) = -2L({}_k\hat{\boldsymbol{\theta}}) + 2k \tag{3.50}$$

$\mathrm{AIC}(k)|_{\min}$ is the optimal compromise between errors in parameter estimation and errors in assigning the model order. Actually, when we look into the details of the AIC, the balance that is struck by Eq. (3.50) in assigning errors is

$-2L({}_k\hat{\boldsymbol{\theta}})$	represents the goodness of fit of the data to the assumed model
k	represents the error that occurs in the estimation of parameters

For normally distributed errors

$$\mathrm{AIC}(k) = N \log s_k^2 + 2k$$

where s_k^2 is the residual sum of squares computed given by $s_k^2 = \frac{1}{N}\sum_i q_i^2$ and q_i is computed from the model $x_t = a_1 x_{t-1} + a_2 x_{t-2} + \ldots + a_k x_{t-k} + q_k$.

3.5.1 Relationship of the AIC to the FPE

A first cousin to the AIC is the FPE, short for the final prediction error. This measure, devised by Akaike prior to the AIC, has been often and successfully used in model identification. An example is the work of Ulrych and Bishop (1975) in its application to the computation of maximum entropy power spectra. It is of interest to compare the two measures. Akaike (1974) has shown that, for normally distributed errors

$$\mathrm{FPE}(k) = \left(1 + \frac{k+1}{N}\right)\sigma_k^2$$

where σ_k^2 is the variance of the innovation at order k. Since an unbiased estimate of the variance is

$$\hat{\sigma}_k^2 = \left(\frac{N}{N-(k+1)}\right) s_k^2$$

the FPE is given by

$$\mathrm{FPE}(k) = \left(\frac{N+(k+1)}{N-(k+1)}\right) s_k^2 \tag{3.51}$$

Taking logarithms in Eq. (3.51) and since the expansion of $\log x$ for $x > 1$ is $\log x = x - x^2/2 + x^3/3 - \ldots$, we have, using only the first term in the expansion[1]

[1] For N=100 and k=9, for example, $\log x = .2007$, whereas $x = .2000$

$$\log\left(\frac{N+(k+1)}{N-(k+1)}\right)s_k^2 \approx \frac{k+1}{N}+\frac{k+1}{N}+\log s_k^2$$
$$\approx 2\frac{(k+1)}{N}+\log s_k^2$$

Therefore

$$\log \mathrm{FPE}(k) \approx \log s_k^2 + 2\frac{(k+1)}{N} \quad \text{for} \quad N >> k$$

Finally

$$N\log \mathrm{FPE}(k) \approx N\log s_k^2 + 2k$$
$$\approx \mathrm{AIC}(k) \tag{3.52}$$

Eq. (3.52) shows that the scaled FPE is equivalent to the AIC when the order of the process is small in comparison to the length of the data. In our experience, although the FPE is a robust measure, the AIC is more highly resolved in the sense that it yields a more defined minimum value. This is at it should be. After all, Akaike developed the AIC as a more general criterion.

We will have much occasion to use the AIC in this book. As an interim example here, consider the important problem of determining whether a time series, x_t, is white. We model x_t as an AR(p) process and compute the AIC(k) as a function of the order $k = 0, 1, \ldots, L$, where $L << N$ is some upper limit. Fig. 3.2 shows the result for a time series ($N = 100$) which is a realization from the logistic equation, Eq. 2.57 Section 2.8, with $\lambda = 4$. As we have mentioned, this value of λ leads to a white process shown in Fig. 3.2d. The minimum value of AIC(k) occurs at $k = 0$ and identifies x_t with q_t, the white innovation. For completeness, Fig. 3.2b illustrates both the periodogram and AR estimates and clearly demonstrates the optimality of the latter estimate. The autocovariance of x_t is shown in Fig. 3.2c.

Other whiteness tests do exist of course, for example the well known Kolmogorov-Smirnov test (Jenkins and Watts, 1969). We have found, however, admittedly with limited experience, that the AIC is superior in its capacity to handle short time sequences.

3.6 Mutual Information and Conditional Entropy

Mutual information and conditional entropy are particularly powerful concepts that we will have occasion to use in different contexts. It turns out, for example, that such processes as blind separation and blind deconvolution can be elegantly formulated in terms of these concepts.

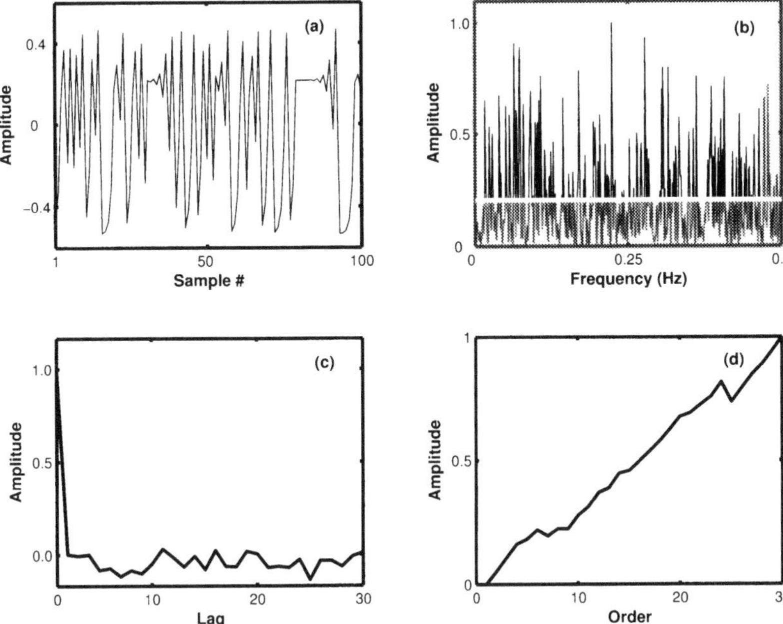

Figure 3.2: Testing for the whiteness of a stochastic process. (a) A realization of the logistic equation with $\lambda = 4$. (b) The periodogram and AR power spectra estimates: thin line is the periodogram, full white line is the AR estimate. (c) The Yule Walker estimate of the autocovariance of the process. (d) The AIC as a function of order k.

3.6.1 Mutual Information

We introduce the concept of mutual information by considering the issue of a signal contaminated with noise. Consider a subsurface source emitting messages, $\mathbf{s}$, that are recorded on the surface as data, $\mathbf{y} = [y_1, y_2, \ldots, y_N]^T$ with probabilities $P_1, P_2, \ldots, P_N$, respectively, $(\sum_i P_i = 1)$. We define the data as being the superposition of some desired signal $\mathbf{s}$ and noise $\mathbf{n}$. The noise is composed of two parts, $\mathbf{n}_c$, coherent noise and $\mathbf{n}_r$, incoherent noise. $\mathbf{n}_{ch}$ is exemplified by such trace to trace coherent events as multiples, air waves, ground roll, etc. $\mathbf{n}_r$ is the random noise that is not coherent from trace to trace.

The information content of a datum y_i is I_i and, by definition $I_i = -\log P_i$. Since the probability of y_i is P_i, the average information per datum received, the entropy of the data, $H(\mathbf{y})$, is defined as in Section 3.2.2

$$\begin{aligned} H(\mathbf{y}) &= E[I] \\ &= \sum_i P_i I_i \\ &= -\sum_i P_i \log P_i \end{aligned}$$

Analytically, it is often more simple to deal with the entropy of a continuous random variable $\mathbf{y}$, where

$$H(\mathbf{y}) = -\int p(x) \log p(x)$$

We reiterate that, although it may appear that the entropy of a continuous variable is infinite, this is not a problem since the entropy is computed *relative* to a reference entropy.

We now ask the question, can the entropy (information content), $H(\mathbf{y})$, of a recorded data set be increased in some manner? For example, given a finite window of data, can prediction outside the window increase $H(\mathbf{y})$? The answer is, of course, no, but to see this more formally, consider the entropy balance associated with a linear system. Since the value of entropy is the same whether expressed in time or frequency, we choose the frequency domain representation for convenience. If $W(f)$ represents the transfer function of a linear filter with input $X(f)$ and output $Y(f)$, the following relationship (Goldman, 1968) can be derived

$$H_Y = H_X + \frac{1}{B}\int_0^B \log |W(f)|^2 df$$

where H_X and H_Y are the input and output entropies/per degree of freedom, respectively.

Entropy power of an ensemble, P_H, is defined in terms of an equivalent ensemble of white noise and, for an ensemble with entropy H, is given by

$$P_H = \frac{1}{2\pi e} e^H$$

Using this definition, the loss of entropy power in a linear system, expressed as a ratio, becomes

$$\frac{P_{H(Y)}}{P_{H(X)}} = e^{\frac{1}{B}\int_0^B \log |W(f)|^2 df}$$

Prediction is a linear process. It does not increase the entropy of the ensemble being predicted. $H(\mathbf{y})$ is all the information that we are ever going to obtain concerning our data set. It can only be increased by means of gathering more data. Our task is not to increase $H(\mathbf{y})$, it is to uncover $H(\mathbf{s})$ given the limitations imposed by $\mathbf{n}$ and by the *finite* amount of data available. A description of our task is elegantly encompassed in the concept of mutual information. Consider our earth to be a channel that carries the message we wish to encode, $\mathbf{s}$. The channel is noisy and we receive $\mathbf{y}$. We now define the mutual information $I(\mathbf{s}, \mathbf{y})$. If the channel were noise free, y_i would faithfully reproduce s_i. The noise introduces a certain uncertainty such that a particular value of $\mathbf{s}$ that has been transmitted lies in the range $(s, s + \Delta s)$. In the limit, as $\Delta s \to 0$, the probability of this event is $p(s)\Delta s$ and the amount of information transmitted is $-\log p(s)\Delta s$. If the value of $\mathbf{y}$ at the geophone is y, and $p(s|y)$ is the conditional probability of $\mathbf{s}$ when $\mathbf{y} = y$, then $p(s|y)\Delta s$ is the probability that $\mathbf{s}$ will lie in the interval $s, s + \Delta s$ when $\mathbf{y} = y$ (providing $\Delta s \to 0$). The uncertainty associated with the interval occupied by s, that is due to the noise, represents a loss of information. Since $-\log p(s)\Delta s$ is the information transmitted and $-\log p(s|y)\Delta s$ is the information lost in the transmission, the net information received is given by

$$I(s,y) = \log \frac{p(s|y)}{p(s)} \tag{3.53}$$

To obtain the average information, denoted by $I(\mathbf{s},\mathbf{y})$, we must average over all values of the ensembles. Thus

$$I(\mathbf{s},\mathbf{y}) = \int\int p(s,y)I(s,y)dsdy \tag{3.54}$$

Following some algebraic steps, we can show that the above equation integrates to

$$I(\mathbf{s},\mathbf{y}) = H(\mathbf{s}) - H(\mathbf{s}|\mathbf{y}) \tag{3.55}$$

where

$$H(\mathbf{s}|\mathbf{y}) = -\int\int p(s,d)\log p(s,y)dsdy \tag{3.56}$$

and represents the average loss of information in the transmission.

An important characteristic of $I(\mathbf{s},\mathbf{y})$ is that it is symmetric. This follows from Eqs. (3.54) and (3.53) which give

$$I(\mathbf{s},\mathbf{y}) = \int\int p(s,y)\log \frac{p(s|y)}{p(s)}dsdy$$

and, using the relation $p(s,y) = p(s|y)p(y)$ we obtain

$$I(\mathbf{s},\mathbf{y}) = \int\int p(s,y)\log \frac{p(s,y)}{p(s)p(y)}dsdy$$

Clearly

$$I(\mathbf{s},\mathbf{y}) = I(\mathbf{y},\mathbf{s}) \tag{3.57}$$

Summarizing, when a particular value of $\mathbf{s}$ is transmitted by the earth and a value y of $\mathbf{y}$ is measured, the average information that is transmitted to us is given by $I(\mathbf{s},\mathbf{y})$. In the process of transmission, the average information that is lost, $H(\mathbf{s}|\mathbf{y})$ is as a result of noise. One of the goals therefore, in signal processing, is to maximize $I(\mathbf{s},\mathbf{y})$ by minimizing the effect of $H(\mathbf{s}|\mathbf{y})$. For this purpose we employ various methods of noise attenuation, some that are described in Chapter 5. An example, at this stage, is illustrative (see Frieden, 2001, for a more thorough discussion).

Suppose we, avid skiers, know from previous experience, that the chance of snow on Whistler mountain in November is 25%. Claude, our powder guide, has 75%

success in predicting snow. The question is, how much information is contained in his Whistler snow prediction. i.e., how far should we trust Claude? We look at the problem in the following way. Define:

$$\text{Message } A \equiv \text{Claude's snow forecast}$$

$$\text{Event } B \equiv \text{occurrence of snow}$$

We have, therefore

$$P(B|A) = 0.75 \quad \text{and} \quad P(B) = 0.25$$

We compute the mutual information by means of a discrete version of Eq. (3.53) which, in this case, becomes

$$I(A,B) = \log \frac{P(B|A)}{P(B)}$$

This quantity represents the information in a message A about an event B. For the assumed probabilities, we obtain $I(A,B) = \log_2(0.75/0.25) = 1.6$ bits.

In order to get a better feeling for what this number represents, we compute the equivalent figure supposing that Claude was a genius and that his forecasts always work, i.e., $P(B|A)$=1. The new value of mutual information becomes $I(A,B) = 2.0$ bits, a small increase in information, even for a perfect track record. If, however, we were to suppose that $P(B)$=1/16, that the chance of snow was, in fact, quite small then, even if Claude was still only 75% correct, $I(A,B) = 4.0$, a quadrupling of information. This example merely serves to demonstrate the relationship that we have stressed all along

$$\text{Information} \propto \text{Uncertainty}$$

3.6.2 Entropy and Aperture

As we have mentioned, one of the goals in signal processing is to maximize the signal to noise ratio in the observed data (after a suitable definition of what is noise in a particular case). We do this not just to make the data 'look nicer' and allow us to see hidden details more clearly, but also to allow the recovery of the model parameters, $\mathbf{m}$, from the recorded data. This ubiquitous task is further complicated by the problem of aperture, in other words, by limitations imposed by the domain of the measured information. In as much as the concept of entropy, as we have seen, plays a fundamental role in denoising, so also, entropy is fundamental to the problem imposed by limited data availability. An excellent illustration is the computation of the power spectrum, $\{P_k\}$ say, from a finite data set, which we call $\{d_t\}$. An estimate of $\{P_k\}$, $\mathbf{p}$, is obtained as the solution to the linear inverse problem (please see Section 4.2)

$$\mathbf{Fp} = \mathbf{r}$$

$\mathbf{F}$ is the known kernel (the Fourier basis functions) and $\mathbf{r}$ is an estimate of the autocovariance that is obtained in some manner from $\{d_t\}$. The problem is under-determined (as all real problems are) and an infinity of solutions are possible. The problem is regularized by the choice of a suitable norm. The l_2 norm (ie., least squares) results in the smallest model, that is the periodogram. This estimate is unbiased but inconsistent. It must be smoothed with the consequent loss in resolution and introduction of bias. As discussed in Section 3.4.1, John Burg suggested the entropy norm which has become so famous in spectral analysis. In reality, what the entropy norm does is equivalent to extending the vector $\mathbf{r}$ by linear prediction, and then taking the Fourier transform. In this process, as we have seen above, no new information is added. The point is that we have mapped the solution into a domain where the support of the solution has been limited. We mention the aperture problem here to stress the importance of information theory in all aspects of data analysis, and as an introduction to the next chapter where we enlarge on the topic of inversion, aperture effects and the importance of norms.

3.6.3 Discussion

Some final thoughts concerning the approach of maximizing entropy are in order. Entropy is a measure of the uncertainty associated with a pdf. Maximizing entropy, consequently, maximizes our uncertainty *concerning what we do not know*. The resultant solutions are, therefore, maximally free of artifacts that occur as a result of, generally, unknowingly, imposing unrealistic constraints. These solutions exhibit parsimony, or equivalently, simplicity. They are devoid of spurious detail. They are what we seek. Examples of unrealistic constraints are to be found everywhere, a particularly common one being the computation of the autocovariance by means of the Yule-Walker estimator, Eq. (3.34). The constraint here is that the time series is zero outside of the known interval. Of course this is not true, and represents in fact, a most brutal termination. Maximizing entropy in such a case, is a mathematical way of replacing the imposed **0**'s by means of **?**'s. Of course, assigning values to the **?**'s depends on the problem at hand.

Chapter 4

The Inverse Problem

4.1 Introduction

It is a truism, in our discipline of geophysics as it is in many other fields, that most of what interests us will, forever, remain hidden from direct observation. The constitution of the inner core, the viscosity of its outer partner, the temperature at a subduction zone and so on. Even if amenable to direct measurement, often the cost of obtaining such measurements is prohibitive (we remember the Mohole, a geopolitical fiasco). We are left, therefore, with the problem of estimating parameters of interest from indirect measurements, measurements on the surface for example, to obtain subsurface parameters. This is the domain of inversion and this is what this chapter is about. Inverse theory is a vast topic and this book does not pretend to encompass this vastness. We do, however cover many topics which are particularly relevant to our quest. We will begin with the simple linear, or linearized, problem and end with an inference formulation using a Bayesian methodology.

4.2 The Linear (or Linearized) Inverse Formulation

In order to formulate the inverse problem, we assume that the forward problem is known (we will have more to say on this subject later). We consider here the discrete linear inverse problem (nonlinear problems may also be tackled using this formulation by first linearizing the equations using Taylor's expansion (as in Section 4.4.2.1). The problem may be stated as follows:
Given a discrete set of measured data, $\mathbf{d}_i, \quad i = 1, 2, \ldots M$, and a known $(M \times N)$ kernel matrix, $\mathbf{G}$, related according to

$$\mathbf{d} = \mathbf{G}\mathbf{m} \tag{4.1}$$

where $\mathbf{m} = [m_1, m_2, \ldots, m_N]^T$ is the vector of unknown "true" model parameters, obtain an estimate, $\hat{\mathbf{m}}$ of $\mathbf{m}$, which satisfies Eq. (4.1).

In most cases of practical interest, inverse problems are underdetermined. The data which are at our disposal exist only at isolated points in data space while the model parameters are, essentially, infinite in number. We are, consequently, justified in assuming that $M < N$ and we are faced with the situation where an infinity of model vectors exist which satisfy the data to within the experimental accuracy. Our inverse problem must, therefore, be regularized. Many different approaches exist which allow one to choose a particular, regularized, model or a particular class of models which satisfy the data. An example is the smallest model, a solution which has the smallest energy. This model is a particularly popular one because often sound physical arguments may be put forward for its adoption, and because it is relatively easy to compute. For these reasons, we consider the smallest, or l_2, inverse solution, and we do so in two different ways.

It is important to note that, for illustrative simplicity, the treatment that follows assumes noise free data. A prior model is not considered at this stage but, of course, we will include both noise and prior models in later discussions.

4.2.1 The Lagrange Approach

We form a cost function, Φ, that consists of two parts, the model norm, and the data constraint

$$\Phi = \mathbf{m}^T \mathbf{I} \mathbf{m} + \boldsymbol{\lambda}^T [\mathbf{G}\mathbf{m} - \mathbf{d}] \tag{4.2}$$

where $\mathbf{I}$ is the identity matrix and $\boldsymbol{\lambda}$ is the vector of Lagrange multipliers (please see Appendix A.2 in Chapter 1 and Section 3.3.5). We have included $\mathbf{I}$ to emphasize that, in general, $\mathbf{I}$ can be replaced by a suitably chosen weighting matrix. Differentiating Φ with respect to $\mathbf{m}$ and setting the result to zero in the usual manner, obtains

$$2\mathbf{m} + \mathbf{G}^T \boldsymbol{\lambda} = 0 \tag{4.3}$$

and therefore

$$\mathbf{m} = -\frac{1}{2}\mathbf{G}^T \boldsymbol{\lambda} \tag{4.4}$$

Substituting Eq. (4.4) into Eq. (4.1), obtains

$$-\frac{1}{2}\mathbf{G}\mathbf{G}^T \boldsymbol{\lambda} = \mathbf{d}$$

yielding

$$\boldsymbol{\lambda} = -2(\mathbf{G}\mathbf{G}^T)^{-1}\mathbf{d} \tag{4.5}$$

Substituting Eq. (4.5) into Eq. (4.4) gives the required solution which we designate $\mathbf{m}_{SVD}$ for reasons that will become apparent

$$\mathbf{m}_{SVD} = \mathbf{G}^T (\mathbf{G}\mathbf{G}^T)^{-1}\mathbf{d} \tag{4.6}$$

$\mathbf{m}_{SVD}$ satisfies the data constraints exactly and is one of the infinity of possible solutions. We now justify the nomenclature. $\mathbf{G}$ admits the singular value decomposition (described in Appendix A.1)

$$\mathbf{G} = \mathbf{U\Sigma V}^T$$

where the quantities have been defined in Eq. (A.7). Substituting for $\mathbf{G}$ in Eq. (4.6), we obtain

$$\begin{aligned}
\mathbf{m}_{SVD} &= \mathbf{V\Sigma}^T\mathbf{U}^T(\mathbf{U\Sigma V}^T\mathbf{V\Sigma}^T\mathbf{U}^T)^{-1}\mathbf{d} \\
&= \mathbf{V\Sigma}^T\mathbf{U}^T(\mathbf{U\Sigma\Sigma}^T\mathbf{U}^T)^{-1}\mathbf{d} \\
&= \mathbf{V\Sigma}^T\mathbf{U}^T(\mathbf{U}^T)^{-1}(\mathbf{\Sigma\Sigma}^T)^{-1}\mathbf{U}^{-1}\mathbf{d} \\
&= \mathbf{V\Sigma}^T\mathbf{U}^T\mathbf{U}(\mathbf{\Sigma\Sigma}^T)^{-1}\mathbf{U}^T\mathbf{d} \\
&= \mathbf{V\Sigma}^T(\mathbf{\Sigma\Sigma}^T)^{-1}\mathbf{U}^T\mathbf{d} \\
&= \mathbf{V\Sigma}^\dagger\mathbf{U}^T\mathbf{d}
\end{aligned}$$

which is just the SVD pseudo inverse solution. $\mathbf{\Sigma}^\dagger$ indicates that, what has been accomplished, was to invert the M (assumed non zero) singular values in the diagonal positions of the first M rows maintaining the zeros in the rest of the matrix. In fact, when using the SVD algorithm, $\mathbf{\Sigma}^\dagger$ is a $M \times M$ diagonal matrix and $\mathbf{V}$ is $N \times M$ where the last $N - M$ columns are discarded.

We conclude that the Lagrange multiplier constraint leads to the truncated SVD solution for the model vector. The Lagrange vector, as is clear from Eq. (4.5), is proportional to the data vector weighted by the inverse of the kernel covariance matrix. It must be emphasized that this approach obtains a model that exactly fits the data and, when noise is present, a modified approach is required.

4.2.2 The Hyperparameter Approach

Here, the cost function is also composed of two parts, the model norm stipulating, in this case, the minimum length constraint, and the data misfit constraint. The latter constraint is imposed in a global manner, implying that the recovered model need not fit the data exactly. Thus

$$\Phi = \mathbf{m}^T\mathbf{Im} + \mu[\mathbf{Gm} - \mathbf{d}]^T[\mathbf{Gm} - \mathbf{d}] \tag{4.7}$$

where $\mathbf{I}$ is the identity matrix and μ is the hyperparameter which controls the trade-off between data misfit and model norm. We have, expanding Eq. (4.7)

$$\Phi = \mathbf{m}^T\mathbf{Im} + \mu[\mathbf{m}^T\mathbf{G}^T\mathbf{Gm} - 2\mathbf{d}^T\mathbf{Gm} + \mathbf{d}^T\mathbf{d}]$$

Minimizing in the usual fashion

$$2\mathbf{m} + 2\mu[\mathbf{G}^T\mathbf{Gm} - \mathbf{G}^T\mathbf{d}] = 0$$

and

$$\mathbf{Im} + \mu\mathbf{G}^T\mathbf{Gm} = \mu\mathbf{G}^T\mathbf{d}$$

Letting $\eta = 1/\mu$ and gathering terms

$$\mathbf{m}[\mathbf{G}^T\mathbf{G} + \eta\mathbf{I}] = \mathbf{G}^T\mathbf{d}$$

finally yielding the solution that we designate as $\mathbf{m}_{LS}$

$$\mathbf{m}_{LS} = [\mathbf{G}^T\mathbf{G} + \eta\mathbf{I}]^{-1}\mathbf{G}^T\mathbf{d} \tag{4.8}$$

Thus, the hyperparameter approach leads to the least squares, LS, solution for an overdetermined system of equations. As is well known, $\eta\mathbf{I}$ is required in the general LS solution to prevent $(\mathbf{G}^T\mathbf{G})^{-1} \to \infty$ should $\mathbf{G}$ be poorly conditioned. In the present case, there is no doubt that $\eta\mathbf{I}$ is required since $\mathbf{G}$ is, at most, of rank M.

4.2.3 A Hybrid Approach

In this section we consider imposing both a data misfit and a Lagrangian constraint. We will, at the outset, specify the constraint that will be imposed via a vector of Lagrangian multipliers as $\mathbf{Am} = \mathbf{q}$, where $\mathbf{A}$ and $\mathbf{q}$ are specified according to some a priori information. Our cost function becomes‘

$$\Phi = \mathbf{m}^T\mathbf{Im} + \mu[\mathbf{Gm} - \mathbf{d}]^T[\mathbf{Gm} - \mathbf{d}] + \boldsymbol{\lambda}^T(\mathbf{Am} - \mathbf{q}) \tag{4.9}$$

giving rise to

$$2\mathbf{m} + 2\mu[\mathbf{G}^T\mathbf{Gm} - \mathbf{G}^T\mathbf{d}] + \mathbf{A}^T\boldsymbol{\lambda} = 0 \tag{4.10}$$

or

$$2[\mathbf{I} + \mu\mathbf{G}\mathbf{G}^T]\mathbf{m} = 2\mu\mathbf{G}^T\mathbf{d} - \mathbf{A}^T\boldsymbol{\lambda}$$

Letting $\eta = 1/\mu$ as before, rewriting $\boldsymbol{\lambda}$ as $2\boldsymbol{\lambda}$ and gathering terms, we obtain

$$\begin{aligned} \mathbf{m} &= [\mathbf{G}^T\mathbf{G} + \eta\mathbf{I}]^{-1}[\mathbf{G}^T\mathbf{d} - \eta\mathbf{A}^T\boldsymbol{\lambda}] \\ &= \mathbf{S}^{-1}[\mathbf{G}^T\mathbf{d} - \eta\mathbf{A}^T\boldsymbol{\lambda}] \end{aligned} \tag{4.11}$$

where

$$\mathbf{S} = [\mathbf{G}^T\mathbf{G} + \eta\mathbf{I}]$$

Substitution into the Lagrange equation, obtains

$$\mathbf{AS}^{-1}[\mathbf{G}^T\mathbf{d} - \eta\mathbf{A}^T\boldsymbol{\lambda}] = \mathbf{q}$$

and solving for $\boldsymbol{\lambda}$ ($\eta \neq 0$)

$$\boldsymbol{\lambda} = \eta^{-1}[\mathbf{A}\mathbf{S}^{-1}\mathbf{A}^T]^{-1}[\mathbf{A}\mathbf{S}^{-1}\mathbf{G}^T\mathbf{d} - \mathbf{q}] \tag{4.12}$$

Substituting Eq. (4.12) into Eq. (4.11) yields

$$\begin{aligned} \mathbf{m} &= \mathbf{S}^{-1}\mathbf{G}^T\mathbf{d} - \mathbf{S}^{-1}\eta\mathbf{A}^T\eta^{-1}[\mathbf{A}\mathbf{S}^{-1}\mathbf{A}^T]^{-1}[\mathbf{A}\mathbf{S}^{-1}\mathbf{G}^T\mathbf{d} - \mathbf{q}] \\ &= \mathbf{m}_{LS} - \mathbf{S}^{-1}\mathbf{A}^T[\mathbf{A}\mathbf{S}^{-1}\mathbf{A}^T]^{-1}[\mathbf{A}\mathbf{S}^{-1}\mathbf{G}^T\mathbf{d} - \mathbf{q}] \end{aligned} \tag{4.13}$$

and with the designation $\mathbf{m}_{LLS}$, we have

$$\mathbf{m}_{LLS} = \mathbf{m}_{LS} - \Delta\mathbf{m} \tag{4.14}$$

The reader may wonder what happens when we set $\mathbf{A} = \mathbf{G}$ and $\mathbf{q} = \mathbf{d}$ in Eq. (4.13). We then have

$$\begin{aligned} \Delta\mathbf{m} &= \mathbf{S}^{-1}\mathbf{G}^T[\mathbf{G}\mathbf{S}^{-1}\mathbf{G}^T]^{-1}[\mathbf{G}\mathbf{S}^{-1}\mathbf{G}^T - \mathbf{I}]\mathbf{d} \\ &= \mathbf{S}^{-1}\mathbf{G}^T[\mathbf{I} - (\mathbf{G}\mathbf{S}^{-1}\mathbf{G}^T)^{-1}]\mathbf{d} \end{aligned}$$

and the hybrid model becomes

$$\mathbf{m}_{LLS} = \mathbf{S}^{-1}\mathbf{G}^T\mathbf{B}^{-1}\mathbf{d} \tag{4.15}$$

where

$$\mathbf{B} = \mathbf{G}[\mathbf{G}^T\mathbf{G} + \eta\mathbf{I}]^{-1}\mathbf{G}^T \tag{4.16}$$

In fact, it can be shown with recourse to the SVD decomposition of $\mathbf{G}$ that, in this particular case, and not surprisingly in view of the Lagrange constraint, $\mathbf{m}_{LLS} = \mathbf{m}_{SVD}$.

4.2.4 A Toy Example

The beauty of toy examples, used throughout this book, is that they teach us much about the general problem in a very simple setting (please see Scales and Tenorio, 2001). At this stage, a $(M = 2) \times (N = 3)$ toy example is illuminating. We consider the following toy components

$$\mathbf{G} = \begin{bmatrix} 2 & 0 & 1 \\ 1 & 2 & 1 \end{bmatrix}$$

$$\mathbf{m} = \begin{bmatrix} 1 & 0 & 1 \end{bmatrix}^T$$

$$\mathbf{d} = \begin{bmatrix} 3 & 2 \end{bmatrix}^T$$

and we evaluate and tabulate the various model estimators that we have discussed.

- The estimator $\mathbf{m}_{SVD}$

$$\mathbf{m}_{SVD} = \begin{bmatrix} 1.1905 & 0.0952 & 0.6190 \end{bmatrix}^T$$

$$\mathbf{d}_{SVD} = \begin{bmatrix} 3 & 2 \end{bmatrix}^T$$

Remark

Since the (2×2) matrix $\mathbf{GG}^T$ is full rank, in this case, the SVD solution fits the data exactly (not what we want for noisy data).

- The estimator $\mathbf{m}_{LS}$

$\eta = 0.1$

$$\mathbf{m}_{LS} = \begin{bmatrix} 1.1669 & 0.1085 & 0.6106 \end{bmatrix}^T$$

$$\mathbf{d}_{LS} = \begin{bmatrix} 2.9444 & 1.9946 \end{bmatrix}^T$$

Remark

We have set $\eta = 0.1$, arbitrarily, assuming noisy data. As expected, the solution does not fit the data exactly. We will consider issues concerned with the choice of η later.

- The estimator $\mathbf{m}_{LLS}$

We have chosen as a priori information, for illustration purpose only of course, knowledge of the mean model vector. i.e., $\bar{\mathbf{m}}_{LLS} = 2/3$. This implies $\mathbf{A} = [\frac{1}{3}\ \frac{1}{3}\ \frac{1}{3}]$ and $\mathbf{q} = 2$.

$$\mathbf{m}_{LLS} = \begin{bmatrix} 1.0492 & 0.0656 & 0.8852 \end{bmatrix}^T$$

$$\mathbf{d}_{LLS} = \begin{bmatrix} 2.9836 & 2.0656 \end{bmatrix}^T$$

Remark

$\bar{\mathbf{m}}_{LLS}$ does, indeed, equal 2. The model does not, as expected, fit the data.

This simple example illustrates well many of the salient points concerning the inverse problem. The Lagrange constraint is a 'hard' constraint, in the sense that it forces the model vector to assume the constraint's desire. This property is often very useful, for example in the determination of pdf's from moment constraints, but too severe when the measured data contain noise, which is always the case. The hyperparameter data misfit constraint adjusts the solution, to some extent, for the presence of noise, but we will deal much more fully with this aspect in a later section. The hybrid approach is potentially very useful if we have strong a priori information.

We hope that, at this stage, the reader has gained some appreciation and liking for what is a passion of ours. Everyone who is involved in inverse matters has his or her own favorite methodology. Ours is a probabilistic one. We now delve into this topic in some detail.

4.2.5 Total Least Squares

We have examined the essential method of least squares in some detail in Chapter 2 and in Section 4.2. In this section we wish to extend our discussion to the method of total least squares, TLS, that finds very important application in various disciplines. The history of TLS is quite long and under different guises. Van Huffel and Vandewalle (1991) tell us that the method is known in the statistical literature as *orthogonal regression* and was explored as early as 1901 by Pearson (1901). The name "total least squares" was probably introduced by Golub and Van Loan (1980). The method first came to the attention of one of us (TJU) during his career as an isotoper. Here, the problem was to compute a line fit to variations in the isotopic ratios, Pb^{207}/Pb^{204} vs. Pb^{206}/Pb^{204}. The problem is that classical LS allows for errors in one of the variables only and, clearly, in this case errors, and correlated errors to boot, occur in both. The correct approach, for the isotope community, was first published by York (1980).

The basic difference between LS and TLS can be best illustrated by the exercise of fitting a line to a series of noisy measurements. Specifically, we wish to fit a line relationship, $y = mx + c$, to our known observations, y_i and x_i, $i = 1, 2, \ldots, N$. The conventional LS approach to determine the slope and intercept parameters, m and c, is to consider the x_i to be error free and the estimate of the parameter vector, $\mathbf{b} = [m, c]^T$ is determined by minimizing $\sum_i (\delta y_i)^2$. In other words, the cost function to be minimized is

$$\phi_y = \sum_i (y_i - mx_i - c)^2$$

This case is illustrated in Fig. 4.1a. In many cases, and certainly in the isotopic ratio problem mentioned above, errors δx_i exist and must be taken into account. Assuming for the moment that δy_i and δx_i are i.i.d. with common variance, a little algebra shows that the sum of squared distances is

$$D_{xy} = \frac{\sum_i (y_i - mx_i - c)^2}{1 + m^2}$$

and the cost function to be minimized is, therefore

$$\phi_{xy} = \sum_i (y_i - mx_i - c)^2$$

subject to the constraint

$$1 + m^2 = k$$

where k is some constant. Formulating in terms of a Lagrange multiplier, λ, we have

$$\text{minimize } [\phi_{xy} + \lambda(1 + m^2 - k)] \tag{4.17}$$

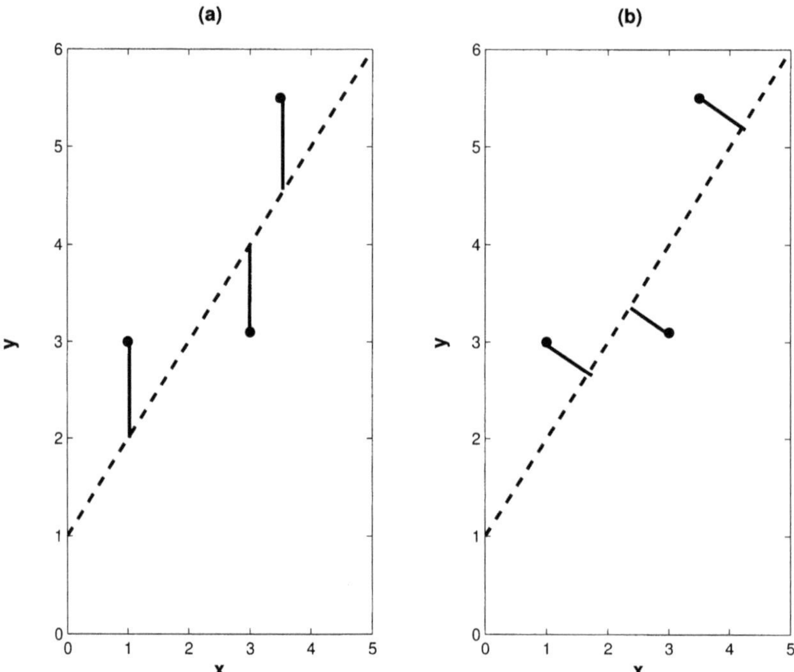

Figure 4.1: Conventional and total least squares. (a) Conventional LS. (b) Total LS.

This entails the minimization of $\sum_i (p_i)^2$, where p_i is the distance from the point to the line as shown in Fig. 4.1b, and the solution will turn out to be an eigen problem with eigenvalue λ. This is the TLS problem which we will now formalize by considering the TLS solution to the fitting of a set of points to a M dimensional plane[1]. This problem is treated by Hamilton (1964) and we follow the author's lucid presentation here, in simplified form.

4.2.5.1 The TLS Solution

A plane in M dimensions is defined by

$$\mathbf{b}^T \mathbf{x}_i + c = 0 \tag{4.18}$$

where $\mathbf{b}$ is a vector of coefficients and $\mathbf{x}_i$, $i = 1, 2, \ldots, N$ represents a particular point in M space. Since errors are associated with all variables, the distance from a point to the plane, assuming i.i.d. and equal variances for the moment, is given by

$$p_i = \frac{\mathbf{b}^T \mathbf{x}_i + c}{(\mathbf{b}^T \mathbf{b})^{1/2}}, \quad i = 1, 2, \ldots, N \tag{4.19}$$

[1] $M = 2$ corresponds to the line fit that we have illustrated in Fig. 4.1.

We now define the augmented quantities

$$\mathbf{y}_i = \begin{bmatrix} \mathbf{x}_i \\ 1 \end{bmatrix}$$

$$\mathbf{Y} = [\mathbf{y}_1\ \mathbf{y}_2\ \cdots\ \mathbf{y}_N]$$

$$\mathbf{a} = [\mathbf{b}^T\ c]^T$$

and the $(N+1) \times (N+1)$ matrix

$$\mathbf{J} = \begin{bmatrix} \mathbf{I} & \mathbf{0} \\ \mathbf{0}^T & 0 \end{bmatrix}$$

where $\mathbf{0}$ is a N element zero vector. Eq. (4.18) can now be written as

$$\mathbf{a}^T\mathbf{y}_i = 0$$

and Eq. (4.19) becomes

$$p_i = \frac{\mathbf{a}^T\mathbf{y}_i}{(\mathbf{a}^T\mathbf{J}\mathbf{a})^{1/2}}$$

We now introduce a weight matrix $\mathbf{W}$, to be defined, that is required when the errors are not i.i.d. Imposing the normalization constraint

$$\mathbf{b}^T\mathbf{b} = \mathbf{a}^T\mathbf{J}\mathbf{a} = 1 \tag{4.20}$$

in the usual manner, we form the cost function

$$\Phi = \mathbf{p}^T\mathbf{W}\mathbf{p} - \lambda(\mathbf{a}^T\mathbf{J}\mathbf{a} - 1) \tag{4.21}$$

where $\mathbf{p} = [p_1, p_2, \ldots, p_N]^T$.

Φ in Eq. (4.21) is minimized with respect to $\mathbf{a}$ in standard fashion to yield

$$(\mathbf{Y}\mathbf{W}\mathbf{Y}^T - \lambda\mathbf{J})\mathbf{a} = 0$$

Setting

$$\mathbf{Y}\mathbf{W}\mathbf{Y}^T = \mathbf{Q} \tag{4.22}$$

we have

$$(\mathbf{Q} - \lambda\mathbf{J})\mathbf{a} = 0 \tag{4.23}$$

Eq. (4.23) is a generalized eigenvalue problem and, in this case, the required solution is the eigenvector corresponding to the minimum eigenvalue, λ. It may sometimes be convenient to rewrite Eq. (4.23) as

$$(\mathbf{J} - \frac{1}{\lambda}\mathbf{Q})\mathbf{a} = 0$$

in which case the solution vector is the eigenvector corresponding to the maximum eigenvalue of

$$\mathbf{Q}^{-1}\mathbf{J}\mathbf{a} = \beta\mathbf{a} \tag{4.24}$$

assuming of course, as we do from here on, that $\mathbf{Q}^{-1}$ exists.

4.2.5.2 Computing the Weight Matrix

The weight matrix in LS is, in general, associated with the inverse of the covariance matrix of the observations. Consequently, defining $\mathbf{C_{pp}}$ to be the covariance matrix of the distances p_i defined in Eq. (4.19), we obtain

$$\mathbf{W} = \mathbf{C}_{\mathbf{pp}}^{-1} \tag{4.25}$$

In order to evaluate W, however, we must relate the unknown $\mathbf{C_{pp}}$ to the 'known' (more accurately, estimated) covariance matrix of the observations $\mathbf{C_{xx}}$. Defining a matrix $\mathbf{B}$ as

$$\mathbf{B} = \begin{bmatrix} \mathbf{b}^T & \mathbf{0} & \cdots & \mathbf{0} \\ \mathbf{0} & \mathbf{b}^T & \cdots & \mathbf{0} \\ \vdots & \vdots & \cdots & \vdots \\ \mathbf{0} & \mathbf{0} & \cdots & \mathbf{b}^T \end{bmatrix}$$

and lexicographically ordering the $\mathbf{x}_i$ into a vector $\mathbf{x}$, we can write Eq. (4.19) (making use of the constraint in Eq. (4.20)) as

$$\mathbf{p} = \mathbf{B}\mathbf{x} + \mathbf{c} \tag{4.26}$$

where $\mathbf{c} = [c, c, \ldots, c]^T$. We now form the appropriate covariance matrix by taking expectations of $\mathbf{p}\mathbf{p}^T$. With the aid of Eq. (4.26) we have

$$\begin{aligned} \mathbf{C_{pp}} &= E[\mathbf{p}\mathbf{p}^T] \\ &= E[\mathbf{B}\mathbf{x}\mathbf{x}^T\mathbf{B}^T] \\ &= \mathbf{B}\mathbf{C_{xx}}\mathbf{B}^T \end{aligned} \tag{4.27}$$

We note that $\mathbf{W}$ depends on the parameters of the plane that we are seeking and the solution must be an iterative one. In consequence, the appropriate eigenvector in Eq. (4.23) may no longer correspond to the minimum eigenvalue (or the maximum eigenvalue when Eq. (4.24) is used). The correct procedure is to choose that eigenvector that corresponds to the minimum $\mathbf{p}^T\mathbf{W}\mathbf{p}$ (see Eq. (4.21)) computed for each eigenvalue (Hamilton, 1964).

4.2.5.3 Parameter Covariance Matrix

The statistics of the derived parameters are contained in the parameter covariance matrix, $\mathbf{C_{aa}}$. To this end, since

$$\mathbf{p} = \mathbf{Y}^T\mathbf{a}$$

left multiplying by $\mathbf{YW}$ obtains

$$\mathbf{YWp} = \mathbf{YWY}^T\mathbf{a}$$

and, by virtue of Eq. (4.22)

$$\mathbf{YWp} = \mathbf{Qa}$$

yielding

$$\mathbf{a} = \mathbf{Q}^{-1}\mathbf{YWp} \tag{4.28}$$

Using Eq. (4.28), we can write

$$\mathbf{aa}^T = \mathbf{Q}^{-1}\mathbf{YWpp}^T\mathbf{WY}^T\mathbf{Q}^{-1}$$

and taking expectations, we have

$$\mathbf{C_{aa}} = \mathbf{Q}^{-1}\mathbf{YWC_{pp}WY}^T\mathbf{Q}^{-1}$$

Since $\mathbf{W} = \mathbf{C_{pp}^{-1}}$, we finally obtain

$$\begin{aligned}\mathbf{C_{aa}} &= \mathbf{Q}^{-1}\mathbf{YWY}^T\mathbf{Q}^{-1} \\ &= \mathbf{Q}^{-1}\end{aligned} \tag{4.29}$$

The TLS solution for the plane is now complete. Two simple examples, where we fit points to a line and a plane, will serve to illustrate the approach.

4.2.5.4 Simple Examples

In both examples, for reason of comparison with LS results, we illustrate errors in all variables that are uncorrelated. Correlated errors present no difficulty. The first example concerns the fitting of a line to 5 points with errors in both x and y and we show it because the simple geometry is easily visualized. Fig. 4.2 illustrates the results for both the LS and TLS methods. In this case, for this particular error distribution, the MSE computed in the conventional manner is 0.6464 for the LS result and .2432 for the TLS result. Of course, because of the small number of points being fitted, a different realization would produce results that might not favor the TLS approach. In general however, as pointed out by Van Huffel and Vandewalle (1991), simulations have shown the improved accuracy of TLS with respect to LS when $N >> M$.

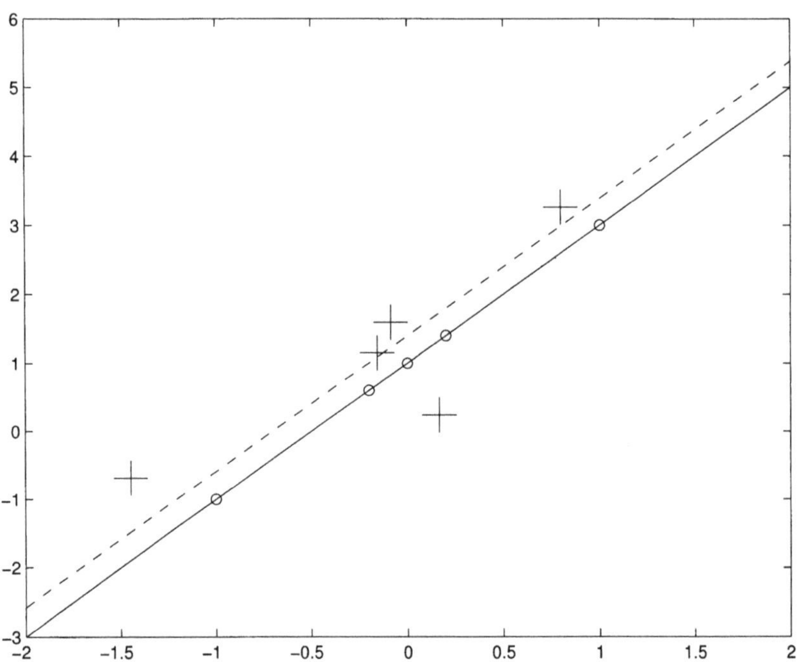

Figure 4.2: Fitting to a line with LS and TLS. — actual model. $-\cdot$ LS solution. $--$ TLS solution.

The second example entails the fitting of a plane to 5 points where errors occur in each of the three variables. The results are shown in Fig. 4.3 and, in this particular case, the respective MSE's are 2.6588 for LS and 2.9394 for TLS. The same remarks regarding these values that were made for the first example apply in this case as well. The interesting point that arises in this particular example is that, as discussed above, the correct choice of the eigenvector was not determined by the maximum eigenvalue, but by the minimum value of $\mathbf{p}^T\mathbf{W}\mathbf{p}$.

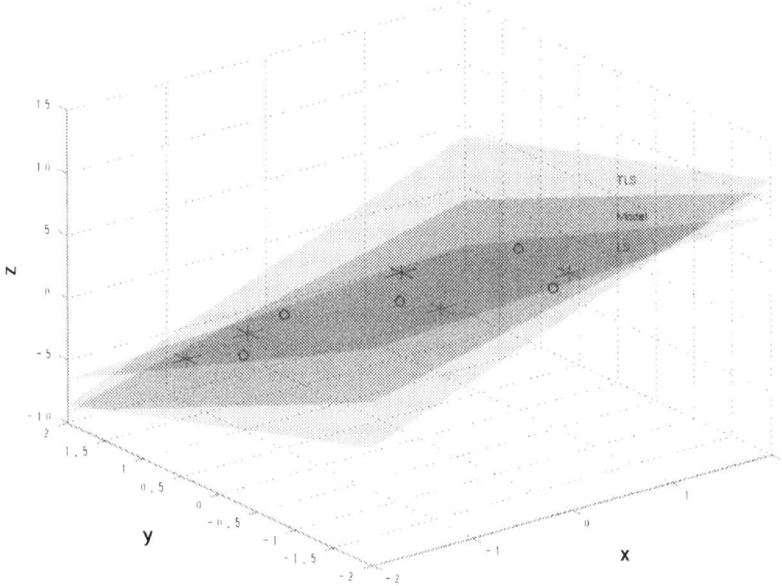

Figure 4.3: Fitting to a plane with LS and TLS.

.2.6 The General TLS Problem

Ve have thus far, for the sake of simplicity and hopefully clarity, illustrated the TLS pproach by considering the fitting of noisy observations to a M dimensional plane. Ve have used a geometrical approach by means of which the perpendicular distance f a point to the hyperplane is defined and the sum of all such distances is minimized. 'his view of TLS is known in the literature as *orthogonal least squares fitting* (Späth, 986). Since the scope of TLS is indeed large, it is advantageous to introduce another iew of TLS that makes other problems more amenable to the TLS model. The pproach we wish to present is that of Van Huffel and Vandewalle (1991) (based on he classical work of Golub and Van Loan (1980)) and makes very general use of one f our very favorite decompositions, SVD, described in Appendix A.1. Specifically, nd using the same notation as in Eq. (4.1), we consider the general linear inverse roblem that is (restating for clarity), given

$$\mathbf{Gm} = \mathbf{d} \tag{4.30}$$

here $\mathbf{m} \in \mathbf{R}^N$ is the vector of unknown "true" model parameters, $\mathbf{d} \in \mathbf{R}^M$ is he vector of measured data and $\mathbf{G} \in \mathbf{R}^{M \times N}$ is a known kernel matrix, obtain an stimate, $\hat{\mathbf{m}}$ of $\mathbf{m}$, which is consistent with Eq. (4.30). Whereas, in our treatment f this problem thus far, $\mathbf{G}$ has been assumed known without error, we now obviate his assumption[2]. It turns out, as discussed by Van Huffel and Vandewalle (1991), hat the TLS solution to this inverse problem, under different conditions, applies

[2] In the line fitting problem, $y = mx + c$, that we have considered above, this is precisely equivalent) assuming that both y and x contain errors.

to the hyperplane problem that we have discussed, and also embraces such worthy issues as Pisarenko harmonic decomposition (Section 2.5), noisy autoregression and ARMA processes (Sections 2.5 and 5.2.8) and a host of other applications. There are, essentially, two cases of interest to us in TLS applications, over and underdetermined sets of linear equations (evendetermined problems will also surface). We will deal with each case separately, but first, we expand on our treatment of the SVD that we present in Appendix A.1, by formulating the problem in a manner suitable for TLS.

4.2.6.1 SVD for TLS

Since we are interested in formulating the problem so that both $\mathbf{G}$ and $\mathbf{d}$ contain errors, we consider the SVD of the augmented matrix defined below.

$$[\mathbf{G}\ \ \mathbf{d}] = \mathbf{U}\boldsymbol{\Sigma}\mathbf{V}^T \tag{4.31}$$

where $[\mathbf{G}\ \ \mathbf{d}]$ is $M \times (N+1)$ and $M > N$ for the overdetermined case[3]. The singular vector and singular value matrices may be defined in terms of partitioned matrices as follows:

The matrix of left singular vectors

$$\mathbf{U} = [\mathbf{U}_1\ \ \mathbf{U}_2]$$

where

$$\mathbf{U}_1 = [\mathbf{u}_1, \mathbf{u}_2, \ldots, \mathbf{u}_N] \text{ and } \mathbf{U}_2 = [\mathbf{u}_{N+1}, \mathbf{u}_{N+2}, \ldots, \mathbf{u}_M]$$

the matrix of right singular vectors

$$\mathbf{V} = \begin{bmatrix} \mathbf{V}_{11} & \mathbf{V}_{12} \\ \mathbf{V}_{21} & \mathbf{V}_{22} \end{bmatrix} \begin{matrix} N \\ 1 \end{matrix}$$
$$\qquad\qquad N \quad 1$$

and the matrix of singular values

$$\boldsymbol{\Sigma} = \begin{bmatrix} \boldsymbol{\Sigma}_1 & \mathbf{0} \\ \mathbf{0} & \boldsymbol{\Sigma}_2 \end{bmatrix}$$

where

$$\begin{aligned} \boldsymbol{\Sigma}_1 &= \operatorname{diag}[\sigma_1, \sigma_2, \ldots, \sigma_N] \ \in \mathbf{R}^{N \times N} \\ \boldsymbol{\Sigma}_2 &= \operatorname{diag}[\sigma_{N+1}, \sigma_{N+2}, \ldots, \sigma_M]^T \ \in \mathbf{R}^{(M-N) \times 1} \end{aligned}$$

[3]In general, the augmented matrix is $[\mathbf{G}\ \ \mathbf{D}]$ where $\mathbf{D}$ is $M \times L$, but we will only consider the case $L = 1$.

4.2.6.2 SVD Solution for TLS - Overdetermined Case ($M > N$)

First, some notation. Since quite a bit of algebra will be involved in our discussion, it is convenient to introduce a condensed notation. Specifically, we will denote all vectors and matrices which apply to the LS solution by $\tilde{\mathbf{v}}$ and $\tilde{\mathbf{V}}$ for example, and when describing the TLS solution by $\hat{\mathbf{v}}$ and $\hat{\mathbf{V}}$ respectively.

We have already met the LS solution for an overdetermined system of equations in Eq. (4.8) which we repeat here using the above mentioned notation

$$\begin{aligned} \tilde{\mathbf{m}} &= [\mathbf{G}^T\mathbf{G} + \eta\mathbf{I}]^{-1}\mathbf{G}^T\mathbf{d} \\ &= [\mathbf{G}^T\mathbf{G}]^{-1}\mathbf{G}^T\mathbf{d} \end{aligned} \tag{4.32}$$

where we have set $\eta = 0$ since, for simplicity, we will consider our system of equations to be well conditioned. In other words, the solution represented by Eq. (4.32) is unique and in some 'optimal' manner, solves the overdetermined system

$$\mathbf{Gm} \approx \mathbf{d} \tag{4.33}$$

which has no exact solution. In preparation for the TLS formulation, we restate the LS problem here (Van Huffel and Vandewalle, 1991) as, given the set of equations represented by Eq. (4.33), $\tilde{\mathbf{m}}$ (equivalent to $\mathbf{m}_{LS}$ in Section 4.2.2) is the LS solution vector to the problem

$$\begin{aligned} &\underset{\tilde{\mathbf{d}} \in \mathbf{R}^M}{\text{minimize}} \quad ||\tilde{\mathbf{d}} - \mathbf{d}||_2 \\ &\text{subject to} \quad \tilde{\mathbf{d}} \in R(\mathbf{G}) \end{aligned} \tag{4.34}$$

where $R(\mathbf{A})$ indicates the range, or column space of $\mathbf{A}$.

Remarks

The following points are important.

1. Defining the LS correction to be $\Delta\tilde{\mathbf{d}}$, where $\Delta\tilde{\mathbf{d}} = \mathbf{d} - \tilde{\mathbf{d}}$, $\Delta\tilde{\mathbf{d}}$ is the orthogonal projection of $\mathbf{d}$ onto $R(\mathbf{G})$ and $\Delta\tilde{\mathbf{d}} \perp \mathbf{G}\tilde{\mathbf{m}}$.

2. It is clear from the minimization in Eq. (4.34) that $\mathbf{G}$ is assumed error-free. Errors occur only in the data vector $\mathbf{d}$.

Using an equivalent formulation to the one above for the LS problem, the TLS problem is to obtain a vector $\hat{\mathbf{m}}$ by solving

$$\begin{aligned} &\underset{[\hat{\mathbf{G}} \;\; \hat{\mathbf{d}}] \in \mathbf{R}^{M\times(N+1)}}{\text{minimize}} \quad ||[\mathbf{G} \;\; \mathbf{d}] - [\hat{\mathbf{G}} \;\; \hat{\mathbf{d}}]||_F \\ &\text{subject to} \quad \hat{\mathbf{d}} \in R(\hat{\mathbf{G}}) \end{aligned} \tag{4.35}$$

$||\cdot||_F$ indicates the Frobenius norm which, for an $M \times N$ matrix $\mathbf{A}$ is given by

$$||\mathbf{A}||_F = \sqrt{\sum_{i=1}^{M}\sum_{j=1}^{N} a_{ij}^2} = \sqrt{\text{trace}(\mathbf{A}^T\mathbf{A})}$$

We now make the assumption, common in the TLS literature, that our equations have been so scaled that the errors in $\mathbf{G}$ and $\mathbf{d}$ are Gaussian i.i.d. with zero mean and equal variance. To obtain the TLS solution using Eq. (4.31), we write Eq. (4.33) as

$$[\mathbf{G} \;\; \mathbf{d}][\mathbf{m}^T \;\; -1]^T \approx 0 \tag{4.36}$$

and make the assumption that $\sigma_N > \sigma_{N+1}$. To obtain a solution to this equation, the rank of $[\mathbf{G} \;\; \mathbf{d}]$ must be reduced to N. It turns out, not surprisingly, that the best rank N approximation to $[\mathbf{G} \;\; \mathbf{d}]$, for either the Frobenius or the l_2 norm, is obtained using Eq. (4.31) (Van Huffel and Vandewalle, 1991)

$$[\hat{\mathbf{G}} \;\; \hat{\mathbf{d}}] = \mathbf{U}_1\mathbf{\Sigma}_1[\mathbf{V}_{11}^T \;\; \mathbf{V}_{21}^T] \tag{4.37}$$

The minimal TLS correction is

$$\begin{aligned} [\Delta\hat{\mathbf{G}} \;\; \Delta\hat{\mathbf{d}}] = [\mathbf{G} \;\; \mathbf{d}] - [\hat{\mathbf{G}} \;\; \hat{\mathbf{d}}] &= \mathbf{U}_2\mathbf{\Sigma}_2[\mathbf{V}_{12}^T\mathbf{V}_{22}^T] \\ &= \sigma_{N+1}\mathbf{u}_{N+1}\mathbf{v}_{N+1}^T \end{aligned} \tag{4.38}$$

a matrix of rank 1 with Frobenius norm

$$||[\Delta\hat{\mathbf{G}} \;\; \Delta\hat{\mathbf{d}}]||_F = \sigma_{N+1}$$

The solution, $\hat{\mathbf{m}}$, to the set

$$[\hat{\mathbf{G}} \;\; \hat{\mathbf{d}}][\mathbf{m}^T \;\; -1]^T = 0 \tag{4.39}$$

follows by substituting Eq. (4.37) into Eq. (4.39). We obtain

$$\mathbf{U}_1\mathbf{\Sigma}_1\mathbf{V}_{11}^T\hat{\mathbf{m}} = \mathbf{U}_1\mathbf{\Sigma}_1\mathbf{V}_{21}^T$$

and consequently

$$\hat{\mathbf{m}} = (\mathbf{V}_{11}^T)^{-1}\mathbf{V}_{21}^T = (\mathbf{V}_{21}\mathbf{V}_{11}^{-1})^T \tag{4.40}$$

Another way of viewing the solution is by examining the null space of $[\hat{\mathbf{G}} \;\; \hat{\mathbf{d}}]$. The only singular vector to be found here[4] is $\mathbf{v}_{N+1}$ and $\hat{\mathbf{m}}$ is a scaled version so that the $(N+1)$th component is -1. Thus

$$[\hat{\mathbf{m}} \;\; -1]^T = \frac{-1}{v_{N+1,N+1}}\mathbf{v}_{N+1}$$

[4]We emphasize that we are considering the TLS problem with only one RHS vector.

nd

$$\mathbf{m}_{TLS} = \hat{\mathbf{m}} = -\frac{-1}{v_{N+1,N+1}}[v_{1,N+1}, v_{2,N+1}, \ldots, v_{N,N+1}] \quad (4.41)$$

n keeping with the notation used above, this solution may also be written as

$$\mathbf{m}_{TLS} = \hat{\mathbf{m}} = -\mathbf{V}_{12}\mathbf{V}_{22}^{-1}$$

emarks

he following points further characterize the TLS solution and emphasize the differ-nce between the LS and TLS solutions.

1. In reducing the rank of $[\mathbf{G} \ \ \mathbf{d}]$ to N, we made the assumption that $\sigma_N > \sigma_{N+1}$. If $\sigma_N = \sigma_{N+1}$, the solution to the TLS problem is non unique, and the procedure is to single out a unique solution, $\hat{\mathbf{m}}$, that has the minimum l_2 norm. This is also true for the underdetermined case, when $M < N$. In this case (when $\mathbf{G}$ is sufficiently well conditioned), the rank of $[\hat{\mathbf{G}} \ \ \hat{\mathbf{d}}]$ is N and $[\hat{\mathbf{G}} \ \ \hat{\mathbf{d}}] = [\mathbf{G} \ \ \mathbf{d}]$. The TLS solution is one that we have already dealt with in Section 4.2.1 where we discussed the LS solution, $\mathbf{m}_{SVD}$, to the underdetermined inverse problem. Thus, in this case, $\hat{\mathbf{m}} = \mathbf{m}_{SVD}$.

2. In as much as $\mathbf{m}_{LS}$ results from the perturbation of $\mathbf{d}$ by $\Delta\tilde{\mathbf{d}}$, $\mathbf{m}_{TLS}$ results from the perturbation of both $\mathbf{G}$ and $\mathbf{d}$ by $\Delta\hat{\mathbf{G}}$ and $\Delta\hat{\mathbf{d}}$, respectively.

3. $\mathbf{m}_{TLS}$ is obtained by projecting $\mathbf{d}$ orthogonally onto $R(\hat{\mathbf{G}})$, the range of the modified kernel matrix $\hat{\mathbf{G}}$, which is obtained by

$$\underset{[\hat{\mathbf{G}} \ \hat{\mathbf{d}}] \in \mathbf{R}^{M\times(N+1)}}{\text{minimize}} \quad ||[\Delta\hat{\mathbf{G}} \ \ \Delta\hat{\mathbf{d}}]||_F$$

$$\text{subject to} \quad \hat{\mathbf{d}} \in R(\hat{\mathbf{G}})$$

and solving $\hat{\mathbf{G}}\mathbf{m}_{TLS} = \hat{\mathbf{d}}$.

.2.6.3 An Illustration

t is worthwhile, at this stage, to illustrate our discussion with a toy example. We magine that the problem is to determine the slownesses, $\mathbf{s}_0$, associated with a 4-ayered pylon. The experiment has produced 5 approximate time measurements and he approximate structure is known from old drawings. We thus have a 'tomographic' xperiment represented by

$$\mathbf{Gm} = \mathbf{d}$$

vhere both $\mathbf{d}$, the vector of transit times and $\mathbf{G}$, the matrix of path lengths, have ssociated errors. For simplicity of illustration, we assume that the errors are i.i.d.

The true quantities, $\mathbf{G}_0, \mathbf{s}_0$ and consequently $\mathbf{d}_0$, are given by

$$\mathbf{G}_0 = \begin{bmatrix} 1.0000 & 2.0000 & 3.0000 & 4.0000 \\ 1.5000 & 2.5000 & 3.0000 & 5.0000 \\ 2.0000 & 3.0000 & 4.0000 & 5.0000 \\ 2.5000 & 4.0000 & 4.5000 & 5.5000 \\ 3.5000 & 4.5000 & 5.5000 & 6.5000 \end{bmatrix}$$

$$\mathbf{s}_0 = \begin{bmatrix} 0.1000 & 0.0500 & 0.3000 & 0.2000 \end{bmatrix}^T$$

and

$$\mathbf{d}_0 = \begin{bmatrix} 1.9000 & 2.1750 & 2.5500 & 2.9000 & 3.5250 \end{bmatrix}^T$$

The approximately known kernel and vector of transit times, $\mathbf{G}$ and $\mathbf{d}$, are given by

$$\mathbf{G} = \begin{bmatrix} 1.6285 & 2.2207 & 3.4140 & 4.3234 \\ 0.8581 & 2.4726 & 2.9983 & 4.7977 \\ 2.1817 & 3.1102 & 4.2304 & 5.9919 \\ 3.2303 & 3.9647 & 4.9090 & 5.1529 \\ 2.9089 & 4.7526 & 5.6465 & 7.2637 \end{bmatrix}$$

and

$$\mathbf{d} = \begin{bmatrix} 2.6997 & 2.2142 & 2.4652 & 3.2240 & 3.5823 \end{bmatrix}^T$$

We consider, initially, unregularized solutions for clarity. For the LS solution, $\mathbf{G}$ is unchanged, but $\mathbf{m}_{LS}$ results from the perturbation of $\mathbf{d}$ by $\Delta\tilde{\mathbf{d}}$. One way of obtaining this perturbation is by substituting $\mathbf{m}_{LS}$ back into the original system of equations. With

$$\mathbf{m}_{LS} = \begin{bmatrix} -1.4414 & -0.4215 & 2.8032 & -0.8252 \end{bmatrix}^T$$

we compute

$$\begin{aligned} \Delta\tilde{\mathbf{d}} &= \mathbf{d} - \mathbf{G}(\mathbf{G}^T\mathbf{G})^{-1}\mathbf{G}^T\mathbf{d} = \mathbf{d} - \tilde{\mathbf{d}} \\ &= \begin{bmatrix} -0.0194 & 0.0472 & 0.0069 & 0.0428 & -0.0558 \end{bmatrix}^T \end{aligned}$$

We have

$$\mathbf{G}\mathbf{m}_{LS} = \tilde{\mathbf{d}}$$

where

$$\tilde{\mathbf{d}} = \begin{bmatrix} 2.7191 & 2.1668 & 2.4583 & 3.1812 & 3.6381 \end{bmatrix}^T$$

Clearly, $\mathbf{m}_{LS}$ makes no sense, but we will worry about this in a moment.

The TLS solution considers both perturbations in $\mathbf{d}$ and in $\mathbf{G}$. Using Eq. (4.37), we obtain

$$[\hat{\mathbf{G}}\ \ \hat{\mathbf{d}}] = \begin{bmatrix} 1.6317 & 2.2216 & 3.4079 & 4.3252 & 2.7019 \\ 0.8526 & 2.4710 & 3.0089 & 4.7945 & 2.2104 \\ 2.1801 & 3.1098 & 4.2335 & 5.9910 & 2.4641 \\ 3.2253 & 3.9632 & 4.9188 & 5.1500 & 3.2205 \\ 2.9156 & 4.7546 & 5.6336 & 7.2675 & 3.5869 \end{bmatrix}$$

with

$$[\Delta\hat{\mathbf{G}}\ \ \Delta\hat{\mathbf{d}}] = \begin{bmatrix} -0.0032 & -0.0009 & 0.0062 & -0.0018 & -0.0022 \\ 0.0055 & 0.0016 & -0.0106 & 0.0031 & 0.0037 \\ 0.0016 & 0.0005 & -0.0031 & 0.0009 & 0.0011 \\ 0.0051 & 0.0015 & -0.0099 & 0.0029 & 0.0035 \\ -0.0067 & -0.0019 & 0.0129 & -0.0038 & -0.0045 \end{bmatrix}$$

The TLS solution (which also makes no sense) is

$$\mathbf{m}_{TLS} = \begin{bmatrix} -1.4649 & -0.4265 & 2.8382 & -0.8399 \end{bmatrix}^T$$

and solves

$$\hat{\mathbf{G}}\mathbf{m} = \hat{\mathbf{d}}$$

We regularize the problem using the approach specified in Eq. (4.8). Since we know $\mathbf{s}_0$ in this synthetic case, we can determine optimal μ's for both the LS and TLS problems. We obtain, for $\mu = 1.1900$ in both cases, the solutions

$$\mathbf{m}_{LS} = \begin{bmatrix} 0.0964 & 0.0929 & 0.03095 & 0.1740 \end{bmatrix}^T$$

and

$$\mathbf{m}_{TLS} = \begin{bmatrix} 0.0963 & 0.0929 & 0.03097 & 0.1739 \end{bmatrix}^T$$

which compare with $\mathbf{s}_0$. The MSE in both cases is the same and the solutions almost identical because the kernel matrix is badly conditioned and a reasonable solution requires regularization. As can be deduced from our discussion, regularization tends to equalize the LS and TLS solutions. The purpose of the above example has been to illustrate the computational steps involved in order to clarify the presentation. We will further consider the importance of the conditioning of the problem in our discussion of TLS for structured matrices below.

4.2.6.4 Extensions of TLS

We have dealt with two approaches to the problem of computing LS parameter estimates when both $\mathbf{G}$ and $\mathbf{d}$ contain errors. We mentioned that the TLS solution described in Section 4.2.5.1 is known in the literature as orthogonal LS fitting (or orthogonal l_2 approximation). The relationship between this approach and the

SVD formalism that we have explored above, which we call TLS-SVD, for convenience, is simple to demonstrate. Thus, the l_2 problem stated formally is, given $\mathbf{Y} = [\mathbf{G} \;\; \mathbf{d}] \in \mathbf{R}^{M \times (N+1)}$, we seek to

$$\underset{\mathbf{a} \in \mathbf{R}^{N+1}}{\text{minimize}} \qquad ||\mathbf{Ya}||_2$$

$$\text{subject to} \qquad \mathbf{a}^T\mathbf{a} = 1$$

As we have seen in Section 4.2.5.1, the solution to this problem is given by the eigenvector of $\mathbf{Y}^T\mathbf{Y}$ corresponding to the smallest eigenvalue. Since this eigenvector is equal to the right singular vector that corresponds to the smallest singular value, we see that the l_2 solution is equivalent to the TLS-SVD solution, $[\hat{\mathbf{m}}^T \;\; -1]^T$, to within a scale factor.

Mixed LS-TLS Problems

In the particular case that we have treated in Section 4.2.5.1, the two solutions differ however since the normalization in Eq. (4.20) is $\mathbf{a}^T\mathbf{J}\mathbf{a}$ and not $\mathbf{a}^T\mathbf{a}$. In fact, the TLS-SVD solution, in this particular case, is not appropriate since the kernel matrix contains a $(1 \times M)$ column vector, $\mathbf{1}^T = (1, 1, \ldots, 1)$, that is perfectly known. TLS problems where some of the columns of $\mathbf{G}$ are known without error are termed mixed LS-TLS problems and we briefly outline here the solution which is due to Golub and discussed by Van Huffel and Vandewalle (1991). The approach is to compute a QR factorization of the L error-free columns and then to solve a TLS-SVD problem of reduced dimension. Partitioning $\mathbf{G}$ into $\mathbf{G}_1$ $(M \times L)$ and $\mathbf{G}_2$ $(M \times (N - L))$, the QR factorization of $[\mathbf{G}_1 \; \mathbf{G}_2 \;\; \mathbf{d}]$ obtains the upper triangular matrix

$$\mathbf{R} = \begin{bmatrix} \mathbf{R}_{11} & \mathbf{R}_{12} & \mathbf{R}_{13} \\ \mathbf{0} & \mathbf{R}_{22} & \mathbf{R}_{23} \end{bmatrix} \begin{matrix} N \\ M - L \end{matrix}$$
$$\qquad L \quad N-L \quad 1$$

The TLS solution of reduced dimension, $\hat{\mathbf{m}}_2$, is obtained in usual fashion from

$$\mathbf{R}_{22}\mathbf{m} \approx \mathbf{R}_{23}$$

and $\hat{\mathbf{m}}_1$ is obtained as the LS solution to

$$\mathbf{R}_{11}\hat{\mathbf{m}}_1 = \mathbf{R}_{13} - \mathbf{R}_{12}\hat{\mathbf{m}}_2$$

yielding the LS-TLS solution, $\hat{\mathbf{m}} = [\hat{\mathbf{m}}_1^T \;\; \hat{\mathbf{m}}_2^T]^T$

For the simple example of the fitting of a line illustrated in Fig. 4.2, where the deviations in both coordinates are with respect to a line defined by slope $b_1 = 2$ and intercept $b_2 = 1$, the obtained parameters are listed in Table 4.1. The solutions differ from each other for reasons that we have outlined above.

Solutions	Slope b_1	Intercept b_2
TLS	2.0187	1.4960
TLS-SVD	1.8144	1.6610
TLS-LS	2.0543	1.5023
LS	1.3812	1.3759

Table 4.1: Various solutions to the line fit of Fig. 4.2. The true values are, $b_1 = 2.0$ and $b_2 = 1.0$

Very often, in solving TLS problems, the matrix $\mathbf{G}$ is structured. For example, in the ubiquitous deconvolution problem, $\mathbf{G}$ assumes a non-symmetric Toeplitz structure that we have met on various occasions (e.g. Section 1.8.4) and which is illustrated below for completeness.

$$\mathbf{G} = \begin{bmatrix} g_1 & & & \\ g_2 & \ddots & \mathbf{0} & \\ \vdots & \ddots & g_1 & \\ g_L & & g_2 & \ddots & \\ & \ddots & \vdots & \ddots & g_1 \\ & & g_L & & g_2 \\ & \mathbf{0} & & \ddots & \vdots \\ & & & & g_L \end{bmatrix} \in \mathbf{R}^{M \times N}$$

Here, we have constructed a convolutional matrix where $[g_1, g_2, \ldots, g_L] = \mathbf{g}^T$ is the seismic wavelet of length L. The deconvolution problem is to estimate the model $\mathbf{m}$ from noisy estimates of $\mathbf{g}$ and noisy measurements $\mathbf{d}$ expressed by $\mathbf{Gm} = \mathbf{d}$. With $\mathbf{m} \in \mathbf{R}^{N\times 1}$, the dimensions of $\mathbf{G}$ are such that $M = N + L - 1$. Clearly, since knowledge of both $\mathbf{G}$ and $\mathbf{d}$ is approximate, the problem of estimating $\mathbf{m}$ should be attacked using TLS. However, the TLS approach that we have explored thus far, is not optimal in this case since the TLS error matrix, $\Delta\hat{\mathbf{G}}$, is unstructured and does not reflect the nature of the actual error matrix that is related only to errors in $\mathbf{g}$. Recently, various authors have considered this problem, amongst whom Rosen et al. (1996) and Van Huffel et al. (1996) have suggested an approach called the Structured

Total Least Norm, STLN, method that we adopt here. The application of STLN to deconvolution has been studied by Mastronardi et al. (2000) and Yalamov and Yuan (2003). We formulate the basic problem in terms of the wavelet matrix $\mathbf{G}$, a correction matrix $\Delta\mathbf{G}$ applied to $\mathbf{G}$, and a correction vector $\Delta\mathbf{d}$ applied to $\mathbf{d}$. We consider

$$(\mathbf{G}+\Delta\mathbf{G})\mathbf{m} = \mathbf{d}+\Delta\mathbf{d} \tag{4.42}$$

Eq. (4.42) is nonlinear with respect to the unknowns which are $\Delta\mathbf{G}, \mathbf{m}$ and $\Delta\mathbf{d}$. The method of solution (Yalamov and Yuan, 2003) is to split the unknowns into two groups, $\mathbf{m}$ in one, $\Delta\mathbf{G}$ and $\Delta\mathbf{d}$ in the other, for example, such that when the parameters in the first group are considered constant, the problem becomes linear with respect to the parameters in the second group. For $\mathbf{m}$ constant, the problem expressed by Eq. (4.42) can be reformulated as

$$\min_{[\Delta\mathbf{d}\ \ \Delta\mathbf{g}]} \quad ||[\Delta\mathbf{d}\ \ \Delta\mathbf{g}]||_2 \tag{4.43}$$

$$\text{subject to} \qquad \mathbf{G}\mathbf{m}+\mathbf{M}\Delta\mathbf{g} = \mathbf{d}+\Delta\mathbf{d}$$

where $\mathbf{M}$ and $\Delta\mathbf{g}$ are chosen so that

$$\mathbf{M}\Delta\mathbf{g} \;=\; \Delta\mathbf{G}\mathbf{m}$$

$\mathbf{M}$ and $\Delta\mathbf{g}$ depend on the structure of $\Delta\mathbf{G}$ which, of course, is the same as the structure of $\mathbf{G}$. In the particular case of concern to us

$$\mathbf{M} = \begin{bmatrix} m_1 & & & \\ m_2 & \ddots & & \mathbf{0} \\ \vdots & \ddots & m_1 & \\ m_N & & m_2 & \ddots \\ & \ddots & \vdots & \ddots & m_1 \\ & & m_N & & m_2 \\ & \mathbf{0} & & \ddots & \vdots \\ & & & & m_N \end{bmatrix} \in \mathbf{R}^{M\times L}$$

and $\Delta = [\Delta g_1, \Delta g_2, \ldots, \Delta g_L]^T$. We rewrite the problem defined by Eq. (4.43) as

$$\min_{[\Delta\mathbf{d}\ \ \Delta\mathbf{g}]} \quad ||[\Delta\mathbf{d}\ \ \Delta\mathbf{g}]||_2 \tag{4.44}$$

$$\text{subject to} \qquad [\mathbf{I}\ \ -\mathbf{M}][\Delta\mathbf{d}\ \ \Delta\mathbf{g}]^T = \mathbf{G}\mathbf{m}-\mathbf{d}$$

and seek $[\Delta\mathbf{d}\ \ \Delta\mathbf{g}]^T$ as the solution of the underdetermined system of equations expressed by Eq. (4.44). Yalamov and Yuan (2003) suggest an interesting approach which is to solve

$$(\mathbf{I}+\mathbf{M}\mathbf{M}^T)\mathbf{y} \;=\; \mathbf{G}\mathbf{m}-\mathbf{d}$$

and, consequently, identify

$$\Delta\mathbf{d} = \mathbf{y}$$

and

$$\Delta\mathbf{g} = -\mathbf{M}^T\mathbf{y}$$

The solution proceeds in an iterative manner. We begin with the LS solution, $\mathbf{m}_{LS}$, as an initial guess. We now solve two LS problems.

[I] **LS problem #1** Given $\mathbf{m}$, solve Eq. (4.44) for $[\Delta\mathbf{d} \;\; \Delta\mathbf{g}]^T$ using the method outlined above.

[II] **LS problem #2** Form the structured $\Delta\mathbf{G}$ from $\Delta\mathbf{g}$ and solve

$$(\mathbf{G} + \Delta\mathbf{G})\mathbf{m} = \mathbf{d} \tag{4.45}$$

to update $\mathbf{m}$. The updated $\Delta\mathbf{d} = (\mathbf{G} + \Delta\mathbf{G})\mathbf{m} - \mathbf{d}$. Forming the new $\mathbf{Gm} - \mathbf{d}$, return to **[I]**.

The iteration is stopped when the change in $\mathbf{m}$ is smaller than some pre-assigned tolerance. Details of the convergence properties of this scheme are discussed by Yalamov and Yuan (2003).

It is interesting to note that this algorithm can also be used for unstructured matrices. In this case, $\mathbf{M}$ assumes the following structure

$$\mathbf{M} = \begin{bmatrix} m_1 \; \dots \; m_N & & & \mathbf{0} \\ & m_1 \; \dots \; m_N & & \\ & & \ddots & \\ \mathbf{0} & & & m_1 \; \dots \; m_N \end{bmatrix} \in \mathbf{R}^{M \times MN}$$

and the vector $\Delta\mathbf{g} \in \mathbf{R}^{MN \times 1}$ is the columnwise lexicographic ordering of $\Delta\mathbf{G}$.

At this stage it is important to stress the pivotal role that the condition number of $\mathbf{G}$ plays in the TLS problem. In the structured case, for example, the initial guess often used is $\mathbf{m}_{LS}$. If the problem is ill-posed, $\mathbf{m}_{LS}$ is computed in the usual regularized manner

$$\mathbf{m}_{LS} = (\mathbf{G}^T\mathbf{G} + \mu\mathbf{I})^{-1}\mathbf{d}$$

This is also true, of course, for the solution of Eq. (4.45) in step **[II]** of the algorithm. The effect of the regularization is to minimize differences between the LS and TLS solutions. We now illustrate this important point with a simple example that is shown in Fig. 4.4. Fig. 4.4a shows the input wavelet, a simple boxcar with additive noise, the input impulse response consisting of a dipole, and the resulting input trace. Panels (b) and (c) shows the results of LS and TLS deconvolution and the respective errors. Since the input wavelet is full-band, $\mathbf{G}^T\mathbf{G}$ is well conditioned and the TLS result is superior to that using the LS approach. The MSE's in this case are 0.1270 for LS and .04129 for TLS. When a band limited wavelet is used, $\mathbf{G}^T\mathbf{G}$ is ill conditioned due to the zeros in the amplitude spectrum and the required regularization tends to smooth out any difference between the two approaches.

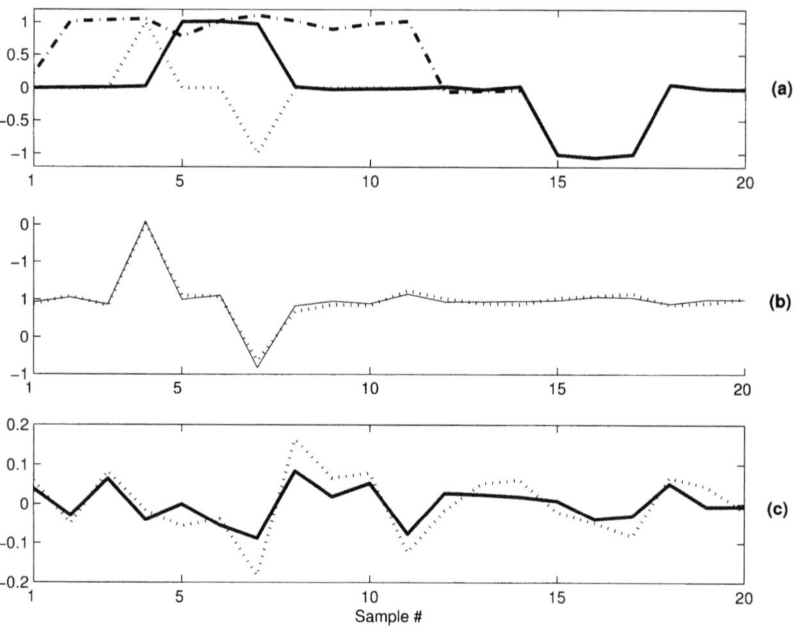

Figure 4.4: Full-band deconvolution using TLS. (a) Input panel. ··· the input impulse response: —· wavelet + noise: — input trace. (b) Output panel. ··· LS model: — TLS model. (c) Error panel. ··· LS error: — TLS error.

—

4.2.6.5 Discussion

The importance of the method of least squares in data processing and inversion cannot be over emphasized. LS is based on the assumption that the kernel matrix, **G**, is known without error. When this assumption is obviated, which is the case in many practical instances, LS is extended to include these errors by means of TLS. We have considered briefly various aspects of the TLS approach. In the first part, using an orthogonal approximation approach, we have shown how errors in all observations may be accounted for in the fitting of a hyperplane. In the following section we showed how the general TLS problem is approached with use of the SVD. Finally, we dealt with extensions to the standard TLS method by considering the problems associated with kernel matrices where some of the columns are perfectly known and with matrices that exhibit a particular structure. Our treatment has, by no means, been exhaustive. TLS, in its many guises and applications, is a very fertile subject and much research is ongoing. Our aim has been to introduce a subject that is well worth exploring.

4.3 Probabilistic Inversion

We have, so far, dealt with the problem of estimating the unknown model from the point of view of inversion. In other words, given the noisy data and the known kernel matrix, obtain the model estimate by inverting the matrix in some manner. As we have seen, this may be a non-unique, or underdetermined, problem. In fact, together with Edwin Jaynes[5] we also believe that a problem that has been well posed is underdetermined and therefore, by its very nature, requires regularization. The previous section has presented, very briefly and simply, various means to achieve this objective. The literature dealing with the regularization of inverse problems is vast, one of the most famous treaties of this subject being that of Tikhonov (Tikhonov, 1963). We will not attempt to review this literature here because such a review is outside of the scope of this book. We will, however, hopefully do some justice to that part of the literature that deals with regularization through the concept of inference.

A word here about inference and inversion. Sivia (1996), in a seminal book, intuitively defines inference as the process of inductive logic or plausible reasoning. Given a set of observations, the task is to unearth the possible causes that might have given them birth. This is a problem of inference simply because that is the best we can do, we use reason, the observed data and the always present prior information. As Sivia (1996) also points out, we reserve the right to change our inference as new data come to light. Inference has to do with probabilities and hence this approach is intimately connected with probability distributions and probability theory.

Our philosophy presented here is based on the ideas expressed by Jaynes (1986) and Shore and Johnson (1980) and may be generally referred to as the Bayesian/ Maximum Entropy approach. Formally expressed and to emphasize the point we have made above, the basic premise is that inversion is an inference problem, where we mean by inference the quantitative use of probability theory to infer a model from specific information about expected values of the model. This definition of inference includes the Bayesian approach as well as the maximum entropy, ME, or the more general minimum relative entropy, MRE, approach.

On the face of it, at least, the inference approach to model estimation appears quite different to that described in Section 4.2 where no mention of probabilities is to be found. We do not wish to overstress this point however. In our opinion, inversion by means of the l_2 norm also uses plausible reasoning, the reasoning that the model we are looking for conforms to the minimum energy principle. Prior models can also be used, of course, although the manner in which they enter into the estimate is quite different to that using the probabilistic approach, as we discuss in what follows. In any case, inference is what we do here and we begin with a particular approach, Minimum Relative Entropy or MRE, proposed by Shore and Johnson (1980) and developed by Ulrych et al. (1990) and Woodbury and Ulrych (Woodbury and Ulrych, 1998; Woodbury et al., 1998; Woodbury and Ulrych, 2000) in a number of papers in application to hydrology. The justification of the use of the principle of ME and MRE in problems of inference lies in the compelling statement made by Jaynes' entropy concentration theorem (Section 3.3.4).

[5]in a conversation with Ed whom we, as so many others, miss greatly, and will continue to do so

4.4 Minimum Relative Entropy Inversion

As we have emphasized, non-uniqueness is inherent in the solution of our inverse problems, and one approach is to use probability theory to do plausible rather than deductive reasoning. Certainly, as pointed out by Jaynes (1986), the most fundamental theorem of scientific inference which accomplishes this goal is expressed by Bayes' theorem with which we will be much concerned. As a prelude, however, we will describe the method of MRE, which turns out to be a special case of the more general Bayesian approach.

A fundamental role in the MRE method is the use of a priori information which expresses the state of our present knowledge and, in particular, reduces the dimensionality of the solution space. The coupling of the principle of ME with the prior information is accomplished by means of the principle of MRE. This approach allows us to obtain the most objective estimate consistent with the data and which is guided to this estimate by any prior information which we might wish to take into account. In our view, the a priori information acts as a lens which focuses the solution into a particular region of the solution space and consequently allows an efficient and essential exploration of the inherent non-uniqueness.

4.4.1 Introduction to MRE

In this section we will briefly review the theoretical aspects of the Bayesian/Maximum Entropy approach which are embodied in the MRE algorithm. We consider a physical and perhaps nonlinear system described by

$$\mathbf{g}(\mathbf{m}) = \mathbf{d} \tag{4.46}$$

where $\mathbf{m}$ is a N parameter model vector and $\mathbf{d} = [d_1, d_2, \ldots, d_M]^T$ are the measured data, assumed to be noise free for the moment. In order to invert such a nonlinear equation, we consider local linearization and iteration. Specifically, if $\mathbf{m}_0$ is some initial model and $\Delta\mathbf{m} = \mathbf{m} - \mathbf{m}_0$, we use Taylor's expansion to obtain

$$\mathbf{g}(\mathbf{m}_0 + \Delta\mathbf{m}) = \mathbf{m}_0 + \mathbf{G}\Delta\mathbf{m} + O(|\Delta\mathbf{m}|^2) \tag{4.47}$$

where $\mathbf{G}$ is the matrix of Fréchet derivatives of $\mathbf{g}$ at $\mathbf{m}_0$ with respect to the model parameters. We now look to solve

$$\mathbf{G}\Delta\mathbf{m} = \mathbf{d} - \mathbf{g}(\mathbf{m}_0) \tag{4.48}$$

It is preferable to rewrite the above equation as

$$\mathbf{G}\mathbf{m} = \mathbf{d} - \mathbf{g}(\mathbf{m}_0) + \mathbf{G}\mathbf{m}_0 \tag{4.49}$$

$$= \tilde{\mathbf{d}} \tag{4.50}$$

We have a linear, underdetermined system to solve, one which is familiar to us from Section 4.2, and we now consider an inference approach to obtain a particular solution.

4.4.1.1 The Bayesian Approach

We outline, briefly, the Bayesian approach to inference which we will meet again in more comprehensive fashion in Section 4.5. Let H and D denote propositions about which we are to reason. In what follows, $P(H|D)$ denotes the probability of H being true conditional on D being true and $\bar{H}$ means H is false. It is particularly important to emphasize that, in Bayesian parameter estimation, probabilities do not represent any measurable property of the parameter but merely our state of knowledge about it. Since no such thing as an absolute probability exists in this system, all probabilities are conditional on some prior information, I. The rules for plausible reasoning are simply the product and sum rules of probability theory which are

$$\begin{aligned} P(H, D|I) &= P(H|I)P(D|H, I) \\ &= P(D|I)P(H|D, I) \end{aligned} \tag{4.51}$$

and

$$P(H|D) + P(\bar{H}|D) = 1$$

where we have written $P(\cdot)$ for prob$(\cdot)$ for convenience. From Eq. (4.51), assuming that $P(D|I) \neq 0$, we obtain Bayes' theorem (or rule)

$$P(H|D, I) = \frac{P(D|H, I)P(H|I)}{P(D|I)} \tag{4.52}$$

Using this fundamental principle of inference we start with the prior probability of the hypothesis H, $P(H|I)$, and obtain the posterior probability, $P(H|D, I)$, given D and the prior information, I.

Since we are going to be dealing with continuous functions for a while, we rewrite Bayes' theorem in its continuous form, which is quite legal to do (see Sivia, 1996), in terms of probability density functions (pdf's). We also stipulate that, for our problem, $\mathbf{m}$ is the hypothesized equivalent of H and $\mathbf{d}$ is the measured data equivalent to D. Then

$$p(\mathbf{m}|\mathbf{d}, I) = \frac{p(\mathbf{d}|\mathbf{m}, I)p(\mathbf{m}|I)}{p(\mathbf{d}|I)} \tag{4.53}$$

and in the form that is most useful for parameter estimation, since $p(\mathbf{d}|I)$ amounts to a scaling

$$p(\mathbf{m}|\mathbf{d}, I) \propto p(\mathbf{d}|\mathbf{m}, I)p(\mathbf{m}|I) \tag{4.54}$$

This beautiful relationship states that, $p(\mathbf{m}|\mathbf{d}, I)$, the probability structure of the model conditional on the observed data and any prior information available, is proportional to the product of two quantities. $p(\mathbf{d}|\mathbf{m}, I)$, the probability of the measured data being observed if the model were correct, called the likelihood, and $p(\mathbf{m}|I)$, the probability of the model prior to data collection. The likelihood function is constructed by computing the noise term, the only uncertain quantity about the measured data, as the difference between the model function and the data. The noise prior probability density function is assigned to be consistent with whatever we know about it and uninformative otherwise. The prior information concerning the model may be incorporated either as informative or uninformative priors, depending on our state of knowledge about the system.

That is all we need to say about Bayes at present, to point out that Eq. (4.54) has it all, so to speak. It infers the pdf of $\mathbf{m}$, which is all the information that we require about $\mathbf{m}$, taking into account the ubiquitous noise and prior information.

4.4.2 MRE Theoretical Details

Similar in spirit to the Bayesian philosophy but different in detail (we will connect MRE and Bayes formally in Section 4.5.9), MRE is an information-theoretic method of problem solving. Its roots lie in probability theory and in an abstract form, it deals with information measures in probability spaces (Kapur and Kesavan, 1992) and follows directly from four basic axioms of consistent inference (Shore and Johnson, 1980).

Consider a system having a set of possible states. Let $\mathbf{x}$ be a state and $q^\dagger(\mathbf{x})$ its multivariate pdf. Assume that $q^\dagger(\mathbf{x})$ exists but is unknown. Let $q^\dagger(\mathbf{x}) \geq 0$ denote a possible pdf such that

$$\int q^\dagger(\mathbf{x})d\mathbf{x} = 1 \qquad (4.55)$$

where the integration is over the full range of random variables $\mathbf{x}$. Note that $\mathbf{x} = [x_1, x_2, \ldots, x_N]^T$ and the integral is a multiple integral over each of the x_i. Suppose an a priori estimate of $q^\dagger(\mathbf{x})$ is $\pi(\mathbf{x})$, and information exists in the form of $j = 1, 2, \ldots, M$ expected value constraints

$$\int q^\dagger(\mathbf{x})f_j(\mathbf{x})d\mathbf{x} = \bar{f}_j \qquad (4.56)$$

where $f_j(\mathbf{x})$ and $\bar{f}_j$ are known. The statistical problem posed here is to determine an estimate, $q(\mathbf{x})$, of $q^\dagger(\mathbf{x})$ based on the information provided. The constraints described by Eq. (4.56) do not uniquely determine $q^\dagger(\mathbf{x})$ but do restrict the allowable densities that $q^\dagger(\mathbf{x})$ could represent. The solution is to minimize $I(q, \pi)$, the Kullback-Leibler or relative entropy measure of the "distance" in probability space, discussed in Section 3.3 and defined below, i.e.,

$$I(q, \pi) = \int q(\mathbf{x}) \log\left[\frac{q(\mathbf{x})}{\pi(\mathbf{x})}\right] d\mathbf{x} \qquad (4.57)$$

subject to the constraints in Eqs. (4.55) and (4.56). The posterior estimate $q(\mathbf{x})$ has the form (Shore and Johnson, 1980; Woodbury and Ulrych, 1998)

$$q(\mathbf{x}) = \pi(\mathbf{x}) \exp\left[-1 - \mu - \sum_{j=1}^{M} \lambda_j f_j(\mathbf{x})\right] \tag{4.58}$$

where μ and the λ_j are Lagrange multipliers determined from Eqs. (4.55) and (4.56). If the final goal is the posterior pdf, Eq. (4.58) is the form of the solution and one has to determine the Lagrange multipliers. For the inverse problem, we are concerned with the expected values of the posterior pdf, $q(\mathbf{x})$, together with the appropriate confidence limits.

We now consider the discrete linear, or linearized (i.e., Eq. (4.50)), inverse problem given, as always, by

$$\mathbf{d} = \mathbf{Gm}$$

where the quantities have been previously defined. It is our goal to obtain an estimate, $\hat{\mathbf{m}}$, of $\mathbf{m}$ which satisfies the above equation and is distinguishable from all other $\hat{\mathbf{m}}$ by a rational criterion. Also, any a priori estimate $\mathbf{s}$ of $\mathbf{m}$ should be included in the inversion scheme. The discretization can also be written as

$$d_j = \sum_{n=1}^{N} g_{jn} m_n \quad j = 1, 2, \ldots, M \tag{4.59}$$

We now denote the expected value of the random vector $\mathbf{m}$ by $\hat{\mathbf{m}}$. That is

$$\hat{\mathbf{m}} = \int_{\mathcal{M}} \mathbf{m} q(\mathbf{m}) d\mathbf{m}$$

where $\int_{\mathcal{M}}$ implies $\int_L^U$, with L and U being, respectively, the lower and upper bounds expressing the a priori allowed values of $\mathbf{m}$, and $q(\mathbf{m})$ the pdf of $\mathbf{m}$. Eq. (4.59) becomes

$$\bar{d}_j = \int_{\mathcal{M}} q(\mathbf{m}) \left[\sum_{n=1}^{N} g_{jn} m_n\right] d\mathbf{m} \tag{4.60}$$

Eq. (4.60) is now in the form of the expected value constraints of Eq. (4.56), where $\bar{d}_j$ corresponds to $\bar{f}_j$ and $\sum_n g_{jn} m_n$ corresponds to $f_j(\mathbf{x})$. Basically, (4.60) is the 'best' estimate of the data in a mean-square sense if the pdf $q(\mathbf{m})$ is known. We then equate $\bar{d}_j$ to the data that we actually observe. However, measurements are usually subject to error so that the strict equality suggested by (4.60) is unrealistic. In a later section we will modify this constraint and allow for measurement error.

We now develop a relationship between the prior $\mathbf{s}$, the model $\mathbf{m}$ and $q(\mathbf{m})$, the pdf to be estimated. Woodbury and Ulrych (1993) deal with the estimation of appropriate prior pdf's for hydrological applications and the essence of their results is briefly repeated here. Specifically, it is acknowledged that hydrological data are such that reasonable upper and lower bounds are obtainable for virtually all model parameters. This information implies that our base level of knowledge is a joint box-car pdf (uniform distribution between an upper and lower bound). Suppose additional information such as a geological interpretation, an informed guess, limited tests or calibrations, becomes available. This "new" prior estimate of the model is defined here as $\mathbf{s}$. Assume $\mathbf{s}$ is the expected value vector of a pdf, $\pi(\mathbf{m})$, which is chosen so that it has minimum relative entropy with respect to a box-car pdf subject to the expected value constraints, s_n. As shown by Woodbury and Ulrych (1993), $\pi(\mathbf{m})$ has the form

$$\pi(\mathbf{m}) = \prod_{n=1}^{N} \frac{-\beta_n}{\exp(-\beta_n U) - 1} \exp(-\beta_n m_n) \tag{4.61}$$

which is a multivariate truncated exponential. The β_n terms are Lagrange multipliers which must be determined from the upper and lower bounds and the expected value constraints (see Woodbury and Ulrych (1993) Appendix B). By definition, this equation satisfies the expected value constraints

$$\int \pi(\mathbf{m})\mathbf{m}d\mathbf{m} = \mathbf{s} \tag{4.62}$$

Thus, unless further constraints are included, the expected values of $\pi(\mathbf{m})$ yield the initial model estimates, $\mathbf{s}$. It is important to note that the above approach of determining $\pi(\mathbf{m})$ is the one which is the most noncommittal with respect to unknown information. Now, in order to deal with the inverse problem, we proceed by minimizing the entropy of $q(\mathbf{m})$ relative to $\pi(\mathbf{m})$ subject to the expected value constraints in Eq. (4.60) and the normalizing constraint

$$\int_{\mathcal{M}} q(\mathbf{m})d\mathbf{m} = 1$$

For M expected value constraints, one obtains

$$q(\mathbf{m}) = \pi(\mathbf{m}) \exp\left[-1 - \mu - \sum_{j=1}^{M} \lambda_j \sum_{n=1}^{N} g_{jn} m_n\right] \tag{4.63}$$

which is the posterior estimate of the pdf $q^{\dagger}(\mathbf{x})$. Defining (Woodbury and Ulrych, 1996)

$$a_n = \beta_n + \sum_{j=1}^{M} \lambda_j g_{jn} \tag{4.64}$$

obtain

$$q(\mathbf{m}) = \prod_n \frac{-a_n}{\exp(-a_n U) - 1} \exp\left[-m_n a_n\right] \tag{4.65}$$

which is a multivariate truncated-exponential pdf. The estimate $\hat{\mathbf{m}}$ is the expected value of Eq. (4.65) and performing the integration

$$\hat{m}_n = \frac{\exp(-a_n U) a_n U + \exp(-a_n U) - 1}{a_n(\exp(-a_n U) - 1)} \tag{4.66}$$

Confidence intervals about the mean value $\hat{\mathbf{m}}$ can be found by obtaining the cumulative distribution function, cdf, for $\mathbf{m}$. Let

$$\int q(\mathbf{x}) d\mathbf{x} = P(\mathbf{m})$$

define the cdf. Carrying out the integration term by term yields

$$P(m_n) = \frac{\exp(-a_n m_n) - 1}{\exp(-a_n U) - 1} \qquad 0 \leq m_n \leq U \tag{4.67}$$

Eq. (4.66) gives an expression for $\hat{\mathbf{m}}$, which is now a function of the unknown Lagrange multipliers, and substituting this expression into Eq. (4.60) yields

$$\bar{d}_j = \sum_{n=1}^{N} g_{jn} \hat{m}_n(\lambda) \tag{4.68}$$

4.4.2.1 Determining the Lagrange Multipliers

Defining

$$F_j(\boldsymbol{\lambda}) = \sum_n g_{jn} \hat{m}_n(\boldsymbol{\lambda}) = \bar{d}_j$$

$\boldsymbol{\lambda} = [\lambda_1, \lambda_2, \ldots, \lambda_M]^T$, the vector of Lagrange multipliers, can be determined using the Newton-Raphson algorithm (see for example, Johnson (1983)). The basic approach proceeds by means of a Taylor expansion. We begin with an initial guess, $\boldsymbol{\lambda}^{(0)}$, and compute approximations by means of

$$\boldsymbol{\lambda}^{(i)} = \boldsymbol{\lambda}^{(i-1)} + \Delta\boldsymbol{\lambda}^{(i)} \tag{4.69}$$

where $\Delta\boldsymbol{\lambda}^{(i)}$ satisfies

$$F_j(\boldsymbol{\lambda}^{(i-1)}) + \sum_l \frac{\partial F_j(\boldsymbol{\lambda}^{(i-1)})}{\partial \lambda_l^{(i-1)}} \Delta\lambda_l^{(i)} = \bar{d}_j$$

In matrix form, we can write

$$\Delta\boldsymbol{\lambda}^{(i)} = \mathbf{F}^{-1}\mathbf{f}$$

where the components of the vector $\mathbf{f}$ are

$$f_j = \bar{d}_j - F_j(\boldsymbol{\lambda}^{(i-1)})$$

and the elements of $\mathbf{F}$ are

$$\begin{aligned} F_{jl} &= \frac{\partial F_j(\boldsymbol{\lambda}^{(i-1)})}{\partial \lambda_l^{(i-1)}} \\ &= -\sum_n g_{jn}\left\{\frac{\partial \hat{m}_n}{\partial a_n} g_{ln}\right\} \end{aligned}$$

where

$$\begin{aligned} \frac{\partial \hat{m}_n}{\partial a_n} &= -\frac{U^2\exp(-a_nU)a_n + U\exp(-a_nU) + U\exp(a_nU)}{a_n(\exp(-a_nU)-1)} \\ &\quad -\frac{\exp(-a_nU)Ua_n + \exp(a_nU) - 1}{a_n^2(\exp(-a_nU)-1)} \\ &\quad +\frac{[\exp(-a_nU)Ua_n + \exp(a_nU) - 1]\,U\exp(-a_nU)}{a_n(\exp(-a_nU)-1)^2} \end{aligned}$$

The iteration is repeated until some convergence criterion is achieved.

4.4.2.2 Confidence Intervals

Since $q(\mathbf{m})$ in Eq. (4.65) is nonGaussian, confidence intervals cannot be easily derived by simply forming the covariance and expanding intervals in the classic way. However, any confidence intervals about the mean value $\hat{\mathbf{m}}$ can be found by first obtaining the cdf for $\mathbf{m}$. Let

$$P(\mathbf{m}) = \int p(\mathbf{x})d\mathbf{x}$$

define the cdf. Carrying out the integration in Eq. (4.65) term by term yields

$$P(m_n) = \frac{\exp(-a_nm_n) - 1}{\exp(-a_nU) - 1} \tag{4.70}$$

Note that if $m_n = U$, $P(U) = 1$. We now determine m_n for some desired probability level, say the 95 percentile. Using Eq. (4.70), we write

$$P(m_n) = c\exp(-a_nm_n) - 1$$

where $c = 1/(\exp(-a_nU) - 1)$ and the expression for m_n becomes

$$m_n = -\frac{\log[\frac{P}{c} + 1]}{a_n}$$

Suppose, for example, that we wish to determine m_n for $P(m_n) = 0.975$, then

$$0.975 = c\exp(-a_n m_n) - 1$$

and

$$m_n = -\frac{\log(b)}{a}$$

where $b = 1.975/c$.

4.4.2.3 The Algorithm

The computation of $\hat{m_n}$ proceeds as follows:

- Modify the input data constraints. Since

$$\bar{\mathbf{d}} = \mathbf{G}\hat{\mathbf{m}}$$

we can obtain

$$\bar{\mathbf{d}} = \mathbf{G}L + \mathbf{G}\hat{\mathbf{m}}_o$$

where $\hat{\mathbf{m}}_o$ is the corresponding model for a zero lower bound with $s_n - L$ as the new prior model. In other words, modify the data to

$$\bar{\mathbf{d}}_L = \bar{\mathbf{d}} - \mathbf{G}L$$

- Determine the Lagrange multipliers λ_j, $j = 1, 2. \ldots, M$ using the Newton Raphson algorithm.

- Compute the final model as

$$\hat{\mathbf{m}} = \hat{\mathbf{m}}_o + L$$

4.4.2.4 Taking Noise Into Account

The uncertainties in the data are taken into account by specifying a weighting matrix, **W**, usually associated with an estimate of the covariance matrix, and considering a global constraint of the form

$$\mathbf{e}^T\mathbf{W}^{-1}\mathbf{e} \;\le\; \epsilon^2 \tag{4.71}$$

where

$$\mathbf{e} \;=\; [e_1, e_2, \ldots, e_M]^T$$

and

$$e_j \;=\; \int_{\mathcal{M}} g_j(u)m(u)du - \tilde{d}_j$$

It can be shown, Johnson and Shore (1984), that the MRE solution, subject to Eq. (4.71), has the same form as the noise free solution but with constraints modified to

$$\int_{\mathcal{M}} g_j(u)m(u)du \;-\; \frac{\epsilon\lambda_j{}'}{(\boldsymbol{\lambda}^T\mathbf{W}^{-1}\boldsymbol{\lambda})^{1/2}} = d_j$$

where $\boldsymbol{\lambda}' = \mathbf{W}^{-1}\boldsymbol{\lambda}$ and $\boldsymbol{\lambda} = [\lambda_1, \lambda_2, \ldots, \lambda_M]^T$ is the vector of Lagrange multipliers in the unconstrained case. The important point is that the functional form of the solution with uncertain data constraints is not changed and a simple modification in the Newton-Raphson algorithm used to compute the $\boldsymbol{\lambda}$s is all that is required.

4.4.3 Generalized Inverse Approach

For comparison purposes, and to complement and amplify Section 4.2.1, we extend here the generalized inverse approach expressed by Eq. (4.6) by incorporating a prior model. The cost function to be minimized becomes

$$\Phi \;=\; (\mathbf{m}-\mathbf{s})^T(\mathbf{m}-\mathbf{s}) \;-\; \boldsymbol{\lambda}^T(\mathbf{Gm}-\mathbf{d})$$

and leads to the solution

$$\hat{\mathbf{m}} \;=\; \mathbf{G}^T(\mathbf{GG}^T)^{-1}(\mathbf{d}-\mathbf{Gs}) + \mathbf{s} \tag{4.72}$$

Eq. (4.72) may be written as

$$\hat{\mathbf{m}} \;=\; \hat{\mathbf{m}}_0 - \mathbf{Qs} + \mathbf{s}$$

where

$$\mathbf{Q} \;=\; \mathbf{G}^T(\mathbf{GG}^T)^{-1}\mathbf{G}$$

As can be readily seen from Eq. (4.73), the prior information plays a distinctly different role in determining the smallest model than it does in the MRE solution. In fact,

Interestingly enough, under certain conditions the smallest solution is transparent to the prior, as may be seen by considering the case when the prior model is chosen to be one of the rows of the kernel matrix $\mathbf{G}$[6]. Denoting the jth row of $\mathbf{G}$ by $\mathbf{g}_j$ and substituting $\mathbf{s} = \mathbf{g}_j$ in Eq. (4.73) yields

$$\hat{\mathbf{m}} = \mathbf{m}_0 - \mathbf{Q}\mathbf{g}_j + \mathbf{g}_j$$

It may be shown, after some algebraic manipulation and the use of $\mathbf{G} = \mathbf{U}\mathbf{\Sigma}\mathbf{V}^T$, that $\mathbf{Q} = \mathbf{V}\mathbf{V}^T$. Forming the matrix product $\mathbf{Q}\mathbf{G}^T$, one obtains

$$\mathbf{Q}\mathbf{G}^T = \mathbf{V}\mathbf{V}^T\mathbf{V}\mathbf{\Sigma}^T\mathbf{U}^T = \mathbf{V}\mathbf{\Sigma}^T\mathbf{U}^T = \mathbf{G}^T$$

Consequently

$$\mathbf{Q}\mathbf{g}_j = \mathbf{g}_j$$

and therefore, using Eq. (4.73)

$$\hat{\mathbf{m}} = \hat{\mathbf{m}}_0$$

showing that the prior model has had no effect on the solution. Other interesting aspects of the role of the prior in the two approaches to inverse problems will be illustrated in the examples below.

4.4.4 Applications of MRE

We show several examples of the application of MRE. The first is application to bandlimited extrapolation of a simple data set and is meant only to stress the enhanced resolution properties of MRE solutions and the role of prior information. The second example is an application to hydrology where we are faced with the problem of plume source reconstruction. Other applications include traveltime transmission tomography (Bassrei, 1990) and diffraction tomography (Lo et al., 1990).

4.4.4.1 Bandlimited Extrapolation

As one of many possible examples of MRE inversion, we consider the universally important topic of bandlimited extrapolation, and specifically, the recovery of very closely spaced impulses from a few values of the Fourier transform. The solution may be formulated as an underdetermined inverse problem. Assumed known, within some error bound, are values of the real and imaginary components of the Fourier transform of the bandlimited seismogram. The kernel functions, $g_j(n)$, in this case are computed as

$$g_j(n) = \cos(2\pi\Delta t\nu_j n) \qquad n = 0, 1, \ldots, N$$

[6]In fact, a prior of this form might be of use in problems such as trace interpolation.

for the real part of the transformed data, and

$$g_j(n) = \sin(2\pi\Delta t\nu_j n) \qquad n = 0, 1, \ldots, N$$

for the imaginary part.

The ν_j are the frequencies in the band of concern and Δt is the sampling interval in time. The prior information is determined as the bandlimited seismogram itself with values smaller than a certain cut-off level set to zero. Such a cut-off level is required since the band-limited seismogram is itself one of the solutions to the inverse problem. The cut-off level is justified by the fact that we are interested in recovering a sparse spike train.

We compare the MRE results with l_2 inversion, using SVD. In this specific case, Eq. (4.56) takes on a particularly simple form. Because of the orthogonality of the kernel functions, $\mathbf{G}\mathbf{G}^T = (N/2)\mathbf{I}$ and consequently, Eq. (4.72) becomes

$$\hat{\mathbf{m}} = \frac{2}{N}\mathbf{G}^T(\mathbf{d} - \mathbf{G}\mathbf{s}) + \mathbf{s}$$

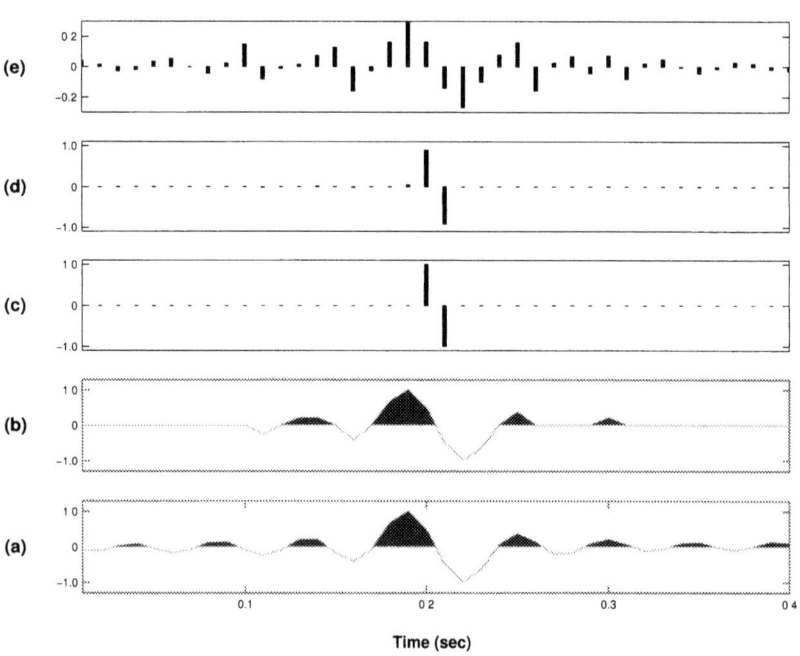

Figure 4.5: MRE extrapolation to full band: smooth prior. (a) Input band-limited model. (b) Prior model. (c) Actual full band model. (d) MRE extrapolated model. (e) SVD extrapolated model.

Fig. 4.5 illustrates both the MRE and l_2 extrapolations. The input trace, Fig. 4.5a, is the 10-40 Hz bandlimited version of the actual spike train shown in Fig. 4.5c,

dipole with impulse spacing of one data interval. 5% (by amplitude) white noise was dded to the frequency band. The kernel functions were computed using nine equally paced frequencies. The prior model is shown in Fig. 4.5b and the deconvolved spike n Fig. 4.5d. Fig. 4.5e shows the SVD result, and clearly illustrates the fact that he MRE approach returns a result with minimum spurious structure. Of course, the mallest norm solution would not be the ideal choice in an application of this type. he comparison is meant only to illustrate the results for two very different norms.

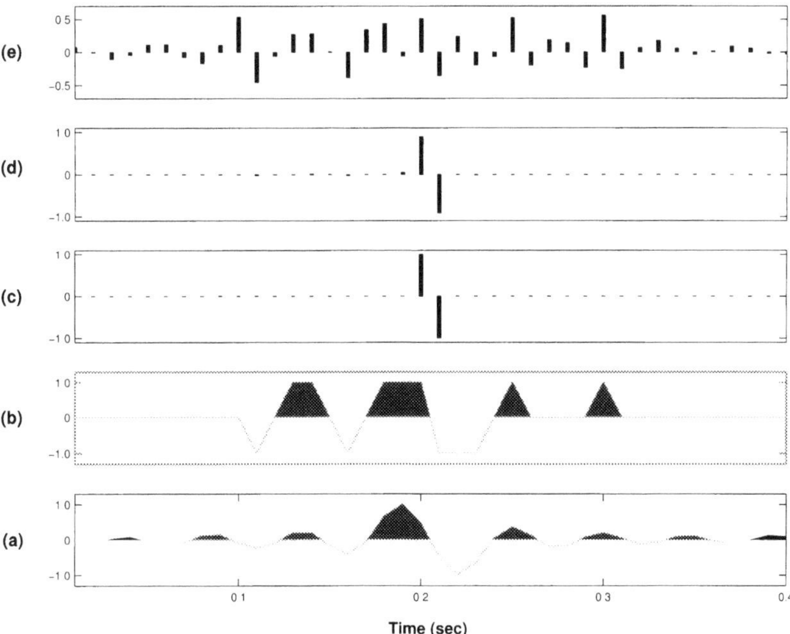

igure 4.6: MRE extrapolation to full band: ± 1 prior. (a) Input band-limited model. b) Prior model. (c) Actual full band model. (d) MRE extrapolated model. (e) SVD xtrapolated model.

Fig. 4.6 explores the sensitivity of the deconvolution to the prior information. he input in this case is the same as in Fig. 4.5a, but the prior is computed as either 1 or -1 depending on the sign of the input data above the cut-off level. As can e clearly seen, the results are very similar, demonstrating that the prior is indeed omfortably 'soft'.

second example, this time consisting of a model where the two spikes have the same olarity, is shown in Fig. 4.7.

.4.4.2 Hydrological Plume Source Reconstruction

challenging issue in groundwater analysis is the recovery of the release history of a roundwater contaminant source. If the groundwater velocity field and dispersion coefficients are known, plume source reconstruction can be categorized as a linear inverse roblem. Wagner (1992) and Skaggs and Kabala (1994) present various solutions to

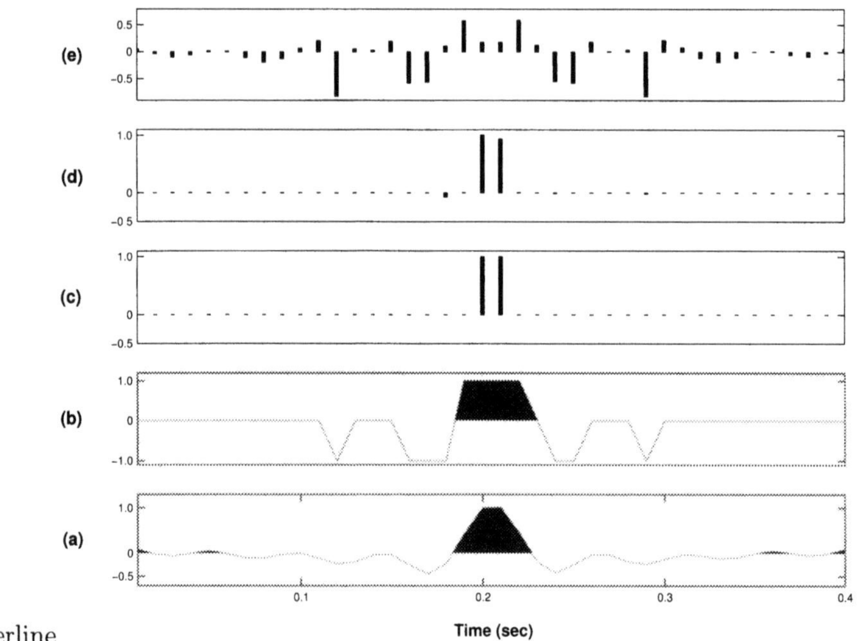

centerline

Figure 4.7: MRE extrapolation to full band: ±1 prior. (a) Input band-limited model. (b) Prior model. (c) Actual full band model. (d) MRE extrapolated model. (e) SVD extrapolated model.

this problem and the reader is referred to these works for a review. As mentioned, Woodbury and Ulrych (1996, 1998) have developed solutions to this problem based on MRE methodology.

Once a plume source history has been developed, future behavior of a plume can be estimated. One advantage of the MRE approach to inversion is that the plume source is characterized by a probability density function. This function can be sampled randomly and then future predictions can be cast in a probabilistic framework. Woodbury and Ulrych (1996) have solved the one dimensional contaminant-source reconstruction problem by means of MRE. Their formulation allows for lower and upper bounds and a prior estimate of the mean value of multiple parameters.

An example of a one-dimensional problem is that of an extremely diffuse pulse given by Skaggs and Kabala (1994), with a dispersion coefficient of $D = 0.5\ \mathrm{m}^2/\mathrm{day}$ and an advective velocity of 1 m/day. Solute is introduced at one boundary according to the source release function

$$C_{\text{in}}(t) = 1.5\exp[-(t-35)^2] + \exp[-(t-42)^2]$$

The solution of the Skaggs and Kabala problem as determined by Woodbury and Ulrych (1996) is illustrated for in Fig. 4.8. The continuous curve represents the true contaminant plume measured at an elapsed time of 50 days. The +'s represent the 30 sample points for the noise free case and the •'s represent data points corrupted with additive normal errors with standard deviation of 0.025.

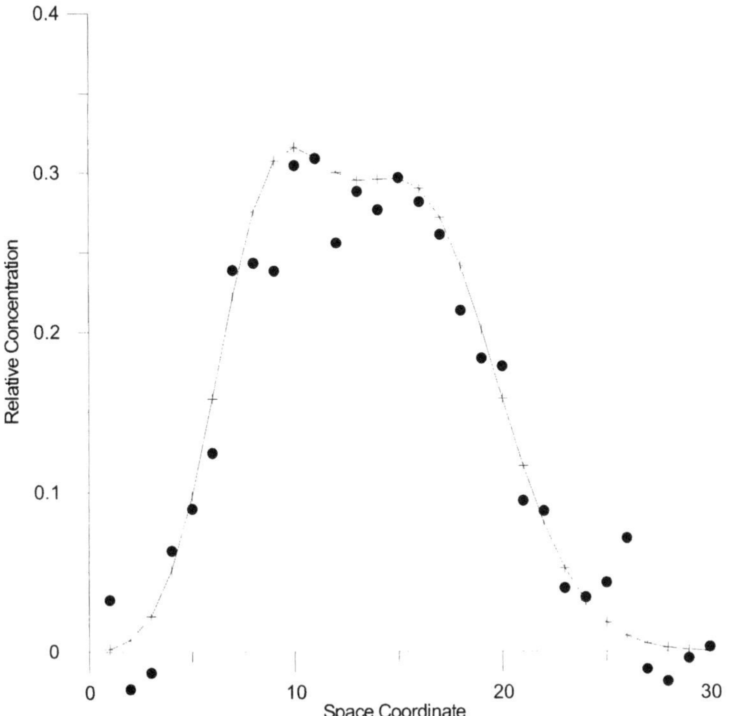

Figure 4.8: Two pulse example data set, showing measured (•) and 'true' data (solid line). + symbols represent discrete samples of noise free data. Solution time of 50 days.

Woodbury and Ulrych (1996) first solved the reconstruction problem by the well known approach of using an l_2 norm with Tikhonov regularization. Here, 400 unknowns were solved by means of a second order regularization coupled with a positivity constraint. The results are depicted in Fig. 4.9
The MRE solution is shown in Fig. 4.10. The two pulses are well reproduced and this result serves to emphasize our thesis, supported by other authors (e.g., Gull and Skilling, 1985), that entropic solutions demonstrate superior resolution compared to conventional l_2 solutions.

4.4.5 Discussion

The MRE approach to inverse problems, introduced by Shore and Johnson (1980) and Shore (1981), following the pioneering work of E. Jaynes in Bayesian/Maximum Entropy methods, has met with considerable success in spectral and image analysis and the hydrological inverse problem. MRE is a method for solving the general underdetermined linear and linearized inverse problem. This approach returns a result which, while consistent with the data, contains the minimum of spurious detail, an attribute that we judge to be of much utility in inverse problems in general. Another

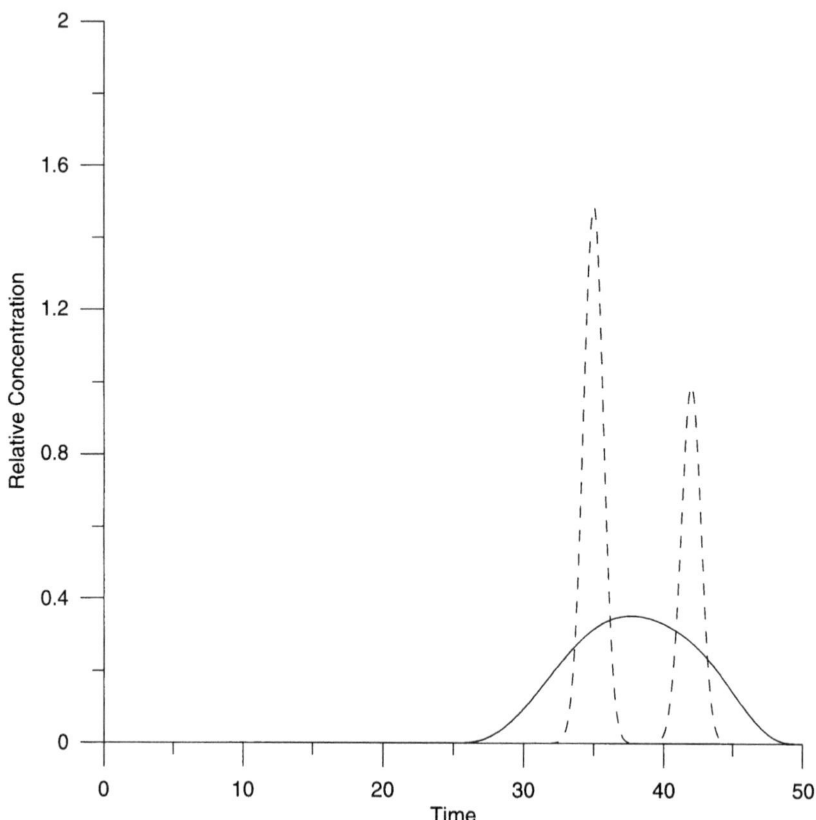

Figure 4.9: l_2 recovery of the two pulse problem. Solid line is the recovered history; dashed line is the 'true' history. 2nd order regularization with 400 unknowns.

important point concerning the role of the prior information in the MRE formalism, is that the use of very different priors promises the efficient exploration of model space. This has been clearly illustrated in our discussion of the famous die problem.

It is important to emphasize that MRE is a methodology which is in keeping with the tradition of Bayesian inference. In other words, MRE model estimates are inferred from prior information and observed data using probability theory. The specific relationship between MRE and Bayesian approaches to inversion will be dealt with in Section 4.5.9

4.5 Bayesian Inference

Bayes is ubiquitous. He is ubiquitous because his theorem describes the way we think. Repeated here again, for convenience, Bayes' theorem (in $\propto$ mode) is

$$P(H|D,I) \;\propto\; P(D|H,I)P(H|I) \tag{4.73}$$

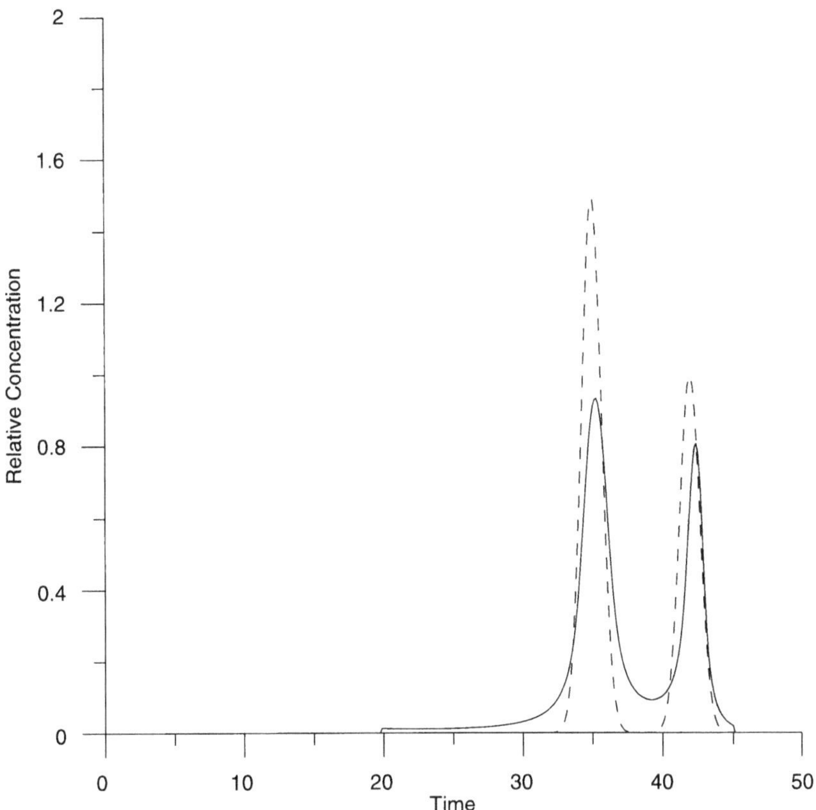

Figure 4.10: MRE recovery of the two pulse problem. Solid line is the recovered history; dashed line is the 'true' history. Box-car prior with lower bound=0 and upper bound=2.0.

with the definitions given in Section 4.4.1.1.

Here is a thought experiment about the use of Eq. (4.73). We live in Vancouver and would like to go skiing, powder skiing, on Whistler mountain, about 120 km. away. Our hypothesis, H, is, 'it is going to snow on Whistler'. We look out at the universe from home and observe D (dark clouds in the sky, a temperature of 2^oC, etc.,). We **always** have some a priori vision, $P(H|I)$, of our hypothesis H, that is conditioned on all the rest of prior information I that we have in mind (e.g., in November, on Whistler the chance of snow ...). $P(D|H,I)$ is the way we evaluate our observations given that our hypothesis is correct (what is the probability of D given that it is snowing on Whistler?) Then, our brain computes a probability distribution for the hypothesis by combining what we observe and what we 'know' and we make our decision, 'better put the snow tires on'.

The Bayes story is very rich, from the personal story of the man, the independent rediscovery of the theorem by Laplace who used it to correctly infer the mass of Saturn, to the frequentist vs. Bayesian controversy. This is not the place for this great story however, and so we launch into the task at hand.

The continuous form of Bayes' theorem is given by Eq. (4.53). For convenience only, we will often omit the I in this equation but we must always be aware of the fact that our pdf's are always conditioned on this property. We write, therefore

$$p(\mathbf{m}|\mathbf{d}) = \frac{p(\mathbf{d}|\mathbf{m})p(\mathbf{m})}{p(\mathbf{d})} \tag{4.74}$$

As already noted (an example of the useful repetition we referred to in the introduction), the denominator is a normalization, does not depend on $\mathbf{m}$ and, for many problems of interest, may be omitted to yield

$$p(\mathbf{m}|\mathbf{d}) \propto p(\mathbf{d}|\mathbf{m})p(\mathbf{m}) \tag{4.75}$$

As we will see in due course however, the denominator which can be cast in the form

$$p(\mathbf{d}) = \int p(\mathbf{d}|\mathbf{m})p(\mathbf{m})d\mathbf{m}$$

to ensure that $p(\mathbf{m}|\mathbf{d})$ integrates to one as all legitimate pdf's should, plays a crucial role in such matters as model estimation using a Bayesian form of the AIC. We now discuss some salient points concerning Eq. (4.75).

- $p(\mathbf{m}|\mathbf{d})$ is the pdf that we desire. It is the distribution of the model parameters posterior to the data $\mathbf{d}$ and is called the posterior or a posteriori pdf.

- $p(\mathbf{d}|\mathbf{m})$ deserves special attention. It is called the likelihood function and has a long and important history in stochastic estimation problems. Before $\mathbf{d}$ has been observed, $p(\mathbf{d}|\mathbf{m})$ represents the pdf that is associated with possible data realizations for a fixed parameter vector. After observation, $p(\mathbf{d}|\mathbf{m})$ has a very different interpretation. It is the likelihood of obtaining the data realization actually observed as a function of the parameter vector $\mathbf{m}$. Some authors call this function $L(\mathbf{d}|\mathbf{m})$ and so do we. It has the same form as $p(\mathbf{d}|\mathbf{m})$ but the interpretation is that it is $\mathbf{d}$ that is fixed and $\mathbf{m}$ is variable.

- It is $p(\mathbf{m})$ that is at the center of any disputes that have arisen concerning the use of Bayes' theorem. It is the prior probability of the model vector and the questions that arise are, how do we assign this pdf and what are the ramification associated with the particular choice? More later but, at this stage, we emphasize that the Bayesian approach is one where the model parameters are considered to be random and a prior probability is, therefore, appropriate if one has a Bayesian disposition.

4.5.1 A Little About Priors

We will have much to say concerning priors throughout this chapter, but, at the outset, we consider the important quest for priors that express complete ignorance so that our estimates will not be biased by uncertain knowledge. We quote from Jeffreys' epic book (Jeffreys, 1939)

> "Our first problem is to find a way of saying that the magnitude of a parameter is unknown, when none of the possible values need special attention. Two rules appear to cover the commonest cases. If the parameter may have any value in a finite range, or form $-\infty$ to $+\infty$, its prior probability should be taken as uniformly distributed. If it arises in such a way that it may conceivably have any value from 0 to ∞, the prior probability of its logarithm should be taken as uniformly distributed."

We have incorporated this quotation for a few reasons. First and foremost, because of Jeffreys' enormous contributions and stature. Second, because the noninformative prior is central in our problem. Finally because the second part of this quotation serves to define what is called the Jeffreys prior. Jeffreys, in considering the prior for the standard error, σ, that can never be negative, invoked the noninformative prior $1/\sigma$ that is in standard use. Certainly, the notion of the uniform prior to describe complete ignorance for a parameter that can assume all values, the so called location parameter, does not require detailed justification (although subtle arguments exist, Scales and Tenorio, 2001). The case for the Jeffreys prior is, perhaps, less obvious. Sivia (1996) gives the following clear justification. Consider a parameter that can assume only positive values and is associated with magnitude, often called a scale parameter. Let σ be, for example, represent standard deviation. We are interested in assigning a pdf, $p(\sigma|I)$, where I represents complete prior ignorance with respect to σ. Clearly, our pdf must be invariant to the units of measurement of σ. For this to be so, it is required that

$$p(\sigma|I)d\sigma = p(\beta\sigma|I)d(\beta\sigma) = p(\beta\sigma|I)\beta d\sigma$$

and for equality to hold, we must have

$$p(\sigma|I) \propto \frac{1}{\sigma} \tag{4.76}$$

that implies

$$p(\log\sigma|I) = \text{constant}$$

thus confirming our intuition that magnitudes involve logarithms.

A somewhat different but interesting derivation may be obtained as follows (Ulrych et al., 2001). Since $\sigma \geq 0$, we write

$$\sigma = e^{\beta}$$

where β is a parameter that can assume all values in $(-\infty, +\infty)$ with uniform probability. In order to find the pdf of σ, we compute the Jacobian of the transformation and write

$$p(\sigma) = \left.\frac{p(\beta)}{e^{\beta}}\right|_{\beta=\log\sigma}$$

Assuming a uniform distribution for β

$$p(\sigma) \propto \frac{1}{\sigma} \tag{4.77}$$

We notice immediately that the above prior does not represent a proper pdf in the sense that it cannot be normalized. We consider this fact in more detail below but, at this stage, we can state with confidence that this characteristic is of little concern and we now have a noninformative prior for many important applications.

4.5.1.1 A Simple Example or Two

We feel that it is very important to simply illustrate the discussion thus far. Our examples are based on some in Lupton (1993). Consider first the almost canonical problem of estimating the mean from N observations drawn from a Gaussian distribution designated by $\mathrm{N}(\mu, \sigma^2)$, where μ is the actual mean and σ^2 the actual variance, assumed known. In order to use Eq. (4.75) we must, first of all, assign a prior density $p(\mathbf{m})$. We do this in the conventional manner by assuming that all mean values, μ, are equally probable and assign a uniform pdf. The likelihood function is just the probability of obtaining the observed sample if we knew that the mean assumed some particular value. If the observations are independent and normally distributed, we obtain

$$p(\mathbf{d}|\mathbf{m}) = \frac{1}{(2\pi)^{N/2}\sigma^N} \prod_i e^{-(d_i-\mu)^2/2\sigma^2} \tag{4.78}$$

Substituting our uniform prior into Eq. (4.75), it is clear that $p(\mathbf{m}|\mathbf{d}) \propto p(\mathbf{d}|\mathbf{m})$. Taking logarithms of $p(\mathbf{m}|\mathbf{d})$ to make life easier, differentiating and setting the resulting expression to zero in the usual manner, obtains the maximum a posteriori, MAP, estimate as

$$\hat{\mu} = \frac{1}{N} \sum_i d_i$$

Lupton (1993) continues with this example and we continue with him. Specifically, we now consider the estimation of the variance, assuming this time that μ is known. With equivalent assumptions on the prior distribution of σ^2, just a little algebra shows that

$$\hat{\sigma}^2 = \frac{1}{N} \sum_i (d_i - \mu)^2 \tag{4.79}$$

Here, however, is the first point of contention. Eq. (4.79) was derived under the hypothesis of a uniform prior. But variance cannot be negative. A possible prior in such a case is the Jeffreys prior that we developed above. Substituting Eq. (4.77) into Eq. (4.75) and proceeding as before yields

$$\hat{\sigma}^2 = \frac{1}{N+1}\sum_i (d_i - \mu)^2 \tag{4.80}$$

'his result may, at first glance, appear somewhat disconcerting. The estimator is, fter all, biased (the denominator is $N+1$ rather than N for known μ). We look at his result in a little more detail later when we introduce the concept of risk. At this tage we itemize and discuss our findings.

- From a Bayesian perspective, the parameter vector is a realization of a random variable. As such, we can associate with it a prior probability that is essential in the application of Bayes' theorem to find the posterior probability. Once we have computed $p(\mathbf{m}|\mathbf{d})$ we have, in fact, computed everything we wish to know about the model.

- The uniform prior (in some particular range) serves to express our maximal ignorance concerning a variable that occupies that range. When positivity constraints apply to that variable, the uniform prior may no longer be justifiable and the Jeffreys prior should be considered.

- Eq. (4.80) shows clearly that the influence of the prior decreases with sample size. Lupton (1993) puts it nicely; the greater amount of information in the sample drowns out the information in the prior.

- The role and the meaning of the likelihood is paramount. Certainly, the likelihood $L(\mathbf{d}|\mathbf{m})$ has a meaning and use quite apart from the function it plays in Bayes' theorem. We wish to explore this issue here.

.5.2 Likelihood and Things

'he likelihood 'explains' the data, the fixed set of observations, $\mathbf{d}$, actually obtained. n $L(\mathbf{d}|\mathbf{m})$, $\mathbf{m}$ represents a vector of parameters that could have given rise to the bserved data. If $\mathbf{m}$ is considered to be a random variable, it is naturally associated rith a prior pdf. If it is not considered to be random, $p(\mathbf{m})$ plays no part and, aving only $L(\mathbf{m})$ at our disposal we obtain the well known maximum likelihood, ML, estimates. What we are really saying is that if we adopt the view that the model ector is not random, Bayes' theorem simply does not apply. There is danger in using nything indiscriminately. This certainly applies to the use of the Bayesian approach.

The best vision of the likelihood function comes from its relationship to the ubiquious method of least squares. The observed data $\mathbf{d}$ are modelled as the superposition f 'true' signal $\mathbf{s}$ corrupted by additive noise $\mathbf{n}$, i.e., $\mathbf{d} = \mathbf{s} + \mathbf{n}$. $\mathbf{s}$ is related to $\mathbf{m}$ by $(\mathbf{m}) = \mathbf{s}$, where g represents some functional relationship. The likelihood function s constructed by taking the difference between observed data and the signal. This ifference is the noise to which we can assign the most noninformative pdf that is onsistent with the available information. The common practice, and a well supported one, is to assign the Gaussian pdf. As we have discussed in Section 3.3.5.1, his assignment follows from the PME (Jaynes, 1982) (as well as from the Central imit Theorem, Section 6.7.3.6).

Supposing that the noise values $[n_1, n_2, \ldots, n_N]$ are independent

$$\begin{aligned} L(\mathbf{d}|\mathbf{m}) &= \prod_i \left[\frac{1}{\sqrt{2\pi\sigma^2}} \exp\left\{ -\left(\frac{n_i^2}{2\sigma^2} \right) \right\} \right] \\ &= \sigma^{-N} \exp\left(\frac{1}{2\pi\sigma^2} \sum_{i=1}^{N} [d_i - g(m_i)]^2 \right) \end{aligned} \quad (4.81)$$

Finding the model parameters that minimize the sum in the exponent of Eq. (4.81) is the method of least squares. Of course as shown in Section 4.2, in the under determined case, we must use the singular value decomposition that allows the construction of the smallest model (an excellent treatment of related issues can be found in Farquharson and Oldenburg, 1998). The procedure that, in the case of the Gaussian pdf is equivalent, is to maximize $L(\mathbf{d}|\mathbf{m})$, and represents the method of maximum likelihood. We see immediately that the latter method is much more flexible since it does not depend on the assignment of a Gaussian pdf.

4.5.3 Non Random Model Vector

Let us look in a little more detail at the fixed model vector approach. A logical thing to do might be to modify the Bayes procedure in order to eliminate the expectation (or average) over $p(\mathbf{m})$ since $\mathbf{m}$ is now a constant. For a start, we consider a mean squared error (MSE) criterion, written as

$$\text{MSE}(\mathbf{m}) = \int (\hat{\mathbf{m}} - \mathbf{m})^2 p(\mathbf{d}|\mathbf{m}) d\mathbf{d} \quad (4.82)$$

where we have taken expectations over the only random variable in the problem, $\mathbf{d}$. We see immediately that a minimum of MSE($\mathbf{m}$) is obtained when $\hat{\mathbf{m}} = \mathbf{m}$. True but hardly useful. What we want is to obtain an unbiased estimate, $E[\hat{\mathbf{m}}] = \mathbf{m}$, that at the same time has minimum variance. The way that most people do that is to maximize the likelihood $L(\mathbf{d}|\mathbf{m})$ (or, more often, the natural logarithm of $L(\mathbf{d}|\mathbf{m})$ that is called $\mathcal{L}(\mathbf{d}|\mathbf{m})$) with respect to $\mathbf{m}$ to obtain the ML estimate that we have mentioned briefly previously. The literature abounds with details of the properties of the ML estimator, its bias and variance for various probability distributions and, in particular, its relationship to the method of least squares. The important conclusion to be drawn in the present discussion stems from the relationship between the MAP and ML estimators. Taking logarithms in Eq. (4.75), differentiating and setting the result to zero, we obtain

$$\frac{\partial}{\partial \mathbf{m}} \log p(\mathbf{m}|\mathbf{d}) \bigg|_{\mathbf{m}=\hat{\mathbf{m}}} = \frac{\partial}{\partial \mathbf{m}} \mathcal{L}(\mathbf{d}, \mathbf{m}) \bigg|_{\mathbf{m}=\hat{\mathbf{m}}} + \frac{\partial}{\partial \mathbf{m}} \log p(\mathbf{m}) \bigg|_{\mathbf{m}=\hat{\mathbf{m}}} = 0$$

Clearly, the ML estimator is equivalent to the MAP estimator when $\log p(\mathbf{m}) = 0$. This implies that, in practice, a random hypothesis for the model parameters coupled with a uniform prior density is equivalent to the fixed vector hypothesis.

4.5.4 The Controversy

The controversy that exists between the Bayesian and Classical approaches arises because, as so eloquently stated by Frieden (2001), these are fundamentally different. The Classical approach is not wrong. (J.R. Oppenheimer, as paraphrased by Frieden (2001), stated that "The word 'Classical' means only one thing in science: it's wrong!"). Both views are important and both give answers to the problems that concern us.

The main aspect to the controversy lies in the use of prior information. The classical, or frequentist view, is that prior information makes no sense since probabilities can only be measured and not assigned. In this regard, the result obtained in Eq. (4.80) may, indeed, be gratifying to the frequentists. Here is an estimate of variance that is different to the accepted ML estimate. But is it worse? To answer this question, let us look at the MSE($\hat{\sigma}^2$) associated with the Jeffreys prior, where MSE($\cdot$) is defined in Eq. (4.82). We consider N normally distributed variates with known mean μ that, for convenience, is taken to be $\mu = 0$. Note that, in this example, we are concerned with the estimation of the variance for known mean, not the estimation of the sample variance. The estimator, $\hat{\sigma}^2$ is computed as

$$\hat{\sigma}^2 = \frac{1}{K}\sum_{i=1}^{N} x_i^2 \tag{4.83}$$

where $K = N$ for the ML estimator and $K = N + 1$ for the JB estimator. The simplest procedure to compute MSE($\hat{\sigma}^2$) is via the characteristic function, $\phi(t)$, that is defined as

$$\begin{aligned}\phi(t) &= E[e^{itx}] \\ &= \int e^{itx} p(x)dx\end{aligned}$$

If x is a $N(0,1)$ variate, the characteristic function of x^2 is (Lupton, 1993)

$$\begin{aligned}\phi_{x^2}(t) &= E[e^{itx^2}] \\ &= \frac{1}{\sqrt{2\pi}}\int e^{itx^2 - x^2/2}dx \\ &= \frac{1}{(1-2it)^{1/2}}\end{aligned}$$

The characteristic function of the estimator in Eq. (4.83), when x is distributed as $N(0,\sigma^2)$ is, consequently

$$\phi_{\hat{\sigma}^2}(t) = \frac{1}{(1-2it\sigma^2/K)^{N/2}} \tag{4.84}$$

To find the mean and variance of the corresponding distribution, we use the fact that the moments about the origin, μ_r', are obtained from

$$\mu'_r = \left.\frac{d^r\phi}{d(it)^r}\right|_{t=0}$$

Using Eq. (4.84) and performing the necessary algebraic steps we obtain

$$E[\hat{\sigma}^2] = \mu'_1 = \frac{N}{K}\sigma^2$$

and

$$\text{Var}[\hat{\sigma}^2] = \mu'_2 - \mu^2 = 2\sigma^4\frac{N}{K^2}$$

Using the above relations, we compute MSE($\hat{\sigma}^2$) as

$$\text{MSE}(\hat{\sigma}^2) = 2\sigma^4\frac{n}{k^2} + (\frac{n}{k}\sigma^2 - \sigma^2)^2 \tag{4.85}$$

Minimizing Eq. (4.85) with respect to K obtains $K = N+2$ and the minimum mean square estimator for this Gaussian example with known mean is

$$\hat{\sigma}^2 = \frac{1}{N+2}\sum_{i=1}^{N} x_i^2 \tag{4.86}$$

So, the ML estimator is unbiased, the estimator of Eq. (4.86) is minimum mean squared error and the JB estimate is somewhere in between. Not a bad choice, perhaps.

Our principal reason for using the Bayesian philosophy is because it provides a particularly flexible approach to the solution of our underdetermined inverse problem. Bayes allows us to construct objective functions that provide particularly desirable flavors to the solutions that we seek. Before describing our approach, we will lead the reader through the steps required to achieve a Bayesian solution to an inverse problem. We will omit the details but, where necessary, we point out various features that may not, at first hand, be obvious.

4.5.5 Inversion via Bayes

We use our previous notation to state the inverse problem. We have

$$g(\mathbf{m}) = \mathbf{d} = \mathbf{s} + \mathbf{n}$$

where $g(\mathbf{m})$ could represent a linear or nonlinear function. The likelihood, in light of the above discussion, is assumed Gaussian. We also assume some knowledge of the data covariance matrix, $\mathbf{C}_d$, in terms of the data variances

$$\mathbf{C}_d = \begin{bmatrix} \sigma_{d1}^2 & 0 & \dots & 0 \\ 0 & \sigma_{d2}^2 & \dots & 0 \\ \vdots & \vdots & \ddots & \vdots \\ 0 & 0 & \dots & \sigma_{dN}^2 \end{bmatrix}$$

Denoting the determinant of $\mathbf{C}_d$ by $|\mathbf{C}_d|$, we obtain the likelihood as

$$p(\mathbf{d}|\mathbf{m}) = \frac{1}{((2\pi)^N |\mathbf{C}_d|)^{1/2})} \exp - \left[\frac{1}{2}(\mathbf{d} - f(\mathbf{m}))^T \mathbf{C}_d^{-1} (\mathbf{d} - f(\mathbf{m}))\right] \tag{4.87}$$

Now comes the question of assigning the prior probability of our model. We suppose, in this example, that we know a priori that the model is smooth in a first derivative sense. Clearly, this would not be an appropriate supposition for a problem concerning the earth's reflectivity. It might be, however, if the problem is to find a source wavelet. For the smooth prior, we define $p(\mathbf{m})$ by

$$p(\mathbf{m}) = \left(\frac{\eta}{2\pi}\right)^{(N-1)/2} \exp\left[-\frac{\eta}{2}\mathbf{m}^T \mathbf{D}^T \mathbf{D} \mathbf{m}\right] \tag{4.88}$$

where $\mathbf{D}$ is the derivative matrix and is represented by

$$\mathbf{D} = \begin{bmatrix} 0 & 0 & 0 & \dots & 0 \\ -1 & 1 & 0 & \dots & \vdots \\ 0 & -1 & 1 & \dots & \vdots \\ \vdots & \vdots & \vdots & \ddots & \vdots \\ 0 & \dots & \dots & -1 & 1 \end{bmatrix}$$

The parameter η, known as a hyperparameter, characterizes the distribution in that $1/\eta$ corresponds to the variance of the prior distribution. We now call upon Bayes

$$p(\mathbf{m}|\mathbf{d}) = \frac{p(\mathbf{d}|\mathbf{m})p(\mathbf{m})}{\int p(\mathbf{d}|\mathbf{m})p(\mathbf{m})d\mathbf{m}} \tag{4.89}$$

Although, as we have already mentioned, the denominator of Eq. (4.89) is nothing more than a normalization factor, it may, and in this case it does, play a very important role in the inversion. Specifically, we define $\Lambda(\mathbf{d}|\eta)$ as

$$\Lambda(\mathbf{d}|\eta) = \int p(\mathbf{d}|\mathbf{m})p(\mathbf{m}|\eta)d\mathbf{m} \tag{4.90}$$

where we have rewritten $p(\mathbf{m})$ as $p(\mathbf{m}|\eta)$ to indicate its dependence on η. In general, the inverse problem will be cast in terms of a vector $\boldsymbol{\eta}$ of hyperparameters. $\Lambda(\mathbf{d}|\eta)$ is known as the Bayesian likelihood and we will see in a moment how it is used in the inversion. For the time being, we compute the following form for the posterior distribution from Eqs. (4.87), (4.88) and (4.89)

$$p(\mathbf{m}|\mathbf{d}) = \frac{1}{((2\pi)^N|\mathbf{C}_d|)^{1/2}}(\frac{\eta}{2\pi})^{(N-1)/2}\exp - \left[\frac{1}{2}\Phi(\mathbf{m},\eta)\right]$$

where

$$\Phi(\mathbf{m},\eta) = \parallel \mathbf{C}^{-1}(\mathbf{d} - g(\mathbf{m})) \parallel^2 + \eta \parallel \mathbf{Dm} \parallel^2 \qquad (4.91)$$

and represents the objective, or cost, function for the problem. We note here that, in as much as the mean squared error, the MSE that we have dealt with above, is a trade-off between bias and variance, so our cost function is a trade-off between resolution and smoothness - a trade-off that is noise dependent.

We have built our first Bayesian objective function. We will see in the next section more examples of objective function building, an approach that we have found extremely useful in diverse applications. As is clear from Eq. (4.91), the hyperparameter η plays a central role. As its function would suggest, determining it is not an easy task. If η were known, the likeliest model parameters represented by $\mathbf{m}_{MAP}$ may be found by minimizing $\Phi(\mathbf{m},\eta)$. Since the problem of minimizing $\Phi(\mathbf{m},\eta)$ is nonlinear, it is linearized around a starting model and the minimization proceeds iteratively. We do not consider details here. We will, however, spend a little time looking into the determination of the hyperparameter η.

4.5.6 Determining the Hyperparameters

The approach that we will describe is based on the work of Hirotugu Akaike (Akaike, 1980) whose contributions have so influenced the field of probability and statistics and is based on the ABIC criterion. The ABIC is, in turn, based on Akaike's information criterion, the AIC, which we met in Section 3.5. Defining the AIC here, again, for convenience and clarity,

$$\begin{aligned} \text{AIC} = & -2\log[p(\mathbf{d}|\hat{\mathbf{m}})]_{max} \\ & + 2(\text{number of parameters}) \end{aligned}$$

Reiterating, the first term in the above expression is related to the sample variance and decreases with the number of parameters. The second is related to the fact that the error of fitting the parameters increases with their number. The minimum of the AIC allows the computation of the appropriate number of parameters, a quest particularly arduous in problems such as fitting time series models. The ABIC is similar to the AIC in form and is computed in terms of the Bayesian likelihood defined in Eq. (4.90)

$$\begin{aligned} \text{ABIC} = & -2\log[\Lambda(\mathbf{d}|\eta)] && (4.92) \\ & + 2(\text{number of hyperparameters}) && (4.93) \end{aligned}$$

The correct hyperparameters are evaluated at the minimum value of the ABIC. The full Bayesian inversion proceeds as follows

- [1] Evaluate the Jacobian matrix for the starting model vector, $\mathbf{m}_k$.
- [2] Minimize $\Phi(\mathbf{m}|\eta)$ for a given value of η.
- [3] Evaluate the ABIC.
- [4] Repeat steps [2] and [3] until the optimum value of η that minimizes the ABIC is reached.
- [5] If convergence properties are satisfied, stop. Else, return to step [1] with the updated model vector $\mathbf{m}_{k+1}$.

The convergence properties in step [5] are related to data uncertainties. An approach is to consider the RMS misfit that is given by

$$\text{RMS misfit} = \sqrt{\| \mathbf{C}^{-1}(\mathbf{d} - g(\mathbf{m}) \|^2 / N}$$

4.5.7 Parameter Errors: Confidence and Credibility Intervals

It is always important to say something about the confidence with which the model parameters that follow from our inverse computation are reported. Two issues need to be explored. The first entails the probability associated with the particular model that we are seeking given a perfectly noiseless set of observations. The second entails the confidence region for the parameters that results from the noise in the observations. Regarding the first issue, there are an infinity of possible solutions to our underdetermined inverse problem. Are all possible solutions equally probable? Certainly not. We have already, in Section 3.3.4.2, examined the probabilities associated with various possible solutions for the die problem and have reached this conclusion.

But what does a die have to do with real earth inverse problems? It is true that we can repeat the die experiment 1000 times. We certainly do not have that luxury in dealing with the real earth. We do, however, have the luxury of imagination. We can, if we imagine the earth from a random rather than fixed view point, think of the outcome of a measurement on the earth as the outcome of a random experiment. We measure our world and deduce the probability associated with the parameters that gave rise to our observations. It is the beauty of such an approach that we can then deduce all manner of model characteristics. It is our contention that a probabilistic approach is many times more flexible than the fixed vector approach. In our view, we can and should say something about the likelihood of possible solutions.

4.5.7.1 A Bit More About Prior Information

We have addressed the probability of a particular outcome for an underdetermined problem. In obtaining the MRE solution, we used a uniform prior in an attempt to express our prior ignorance concerning the unknown distribution. We could have assumed some different prior distribution. Our point is that, having chosen a prior, there is no point in talking about the probability of the prior probability. If we are not certain about the prior, we use the most noninformative, innocuous one we can (more on this later). A prior should express knowledge that is not in doubt. Examples, density is positive, compressional wave velocity is $< 10,000$ m/s, the Sun is bigger than Mars. If we are looking for a buried pipe, it seems appropriate to constrain the gravity inversion with a prior that expresses the pipe geometry. If we wish to regularize our inversion by computing the truncated SVD solution we so should, providing that we have a good a priori reason for choosing the smallest model as the appropriate solution.

4.5.8 Parameter Uncertainties

We now come to the issue involved with uncertainties of the inverse solution. All observations have errors. These errors propagate through the inversion and attack our parameters. We examine the behavior of such errors both in the fixed and random views that we have introduced.

Since, at this stage, it is our belief that we have chosen an approach to the inversion that gives us what we believe to be the most probably correct result, we consider the treatment of data errors only. In the case of the fixed vector approach, the customary statistic that is used is ξ^2. It is well known and we will not review it here. We do, however, examine the concept of confidence regions. Once again, we delve into Lupton (1993).

The frequentist approach to confidence intervals, for example for determining the intervals associated with an estimate of μ when the sample is drawn from a $N(\sigma^2, \mu)$ population, is to compute the estimated mean, $\bar{\text{x}}$, and to assign probability bounds like $|\bar{\text{x}} - \mu| < 1.96\sigma/\sqrt{N}$ at the 95% confidence level. In this statement μ is fixed and $\bar{\text{x}}$ is the random variable. Bayesians take a different point of view. Here, μ is the random variable and Eq. (4.75), given a uniform prior for μ, allows us to compute $p(\mu|\mathbf{x})$ and hence the associated confidence regions. In the Gaussian case, we get the same result as we do by means of the conventional approach but, we stress again, the view is different.

We now complicate the situation a little by considering the case when σ^2 is unknown. Again, using Eq. (4.74), we compute the posterior pdf $p(\mu, \sigma|\mathbf{x})$ assuming a Gaussian likelihood, a uniform prior for μ, the Jeffreys prior for σ and independence of μ and σ. To simplify matters, let s^2 indicate the sample variance computed as

$$s^2 = \frac{1}{N}\sum_i (x_i - \bar{x})^2$$

where $\bar{x}$ is the ML estimate of the mean. By direct substitution, we can verify that

$$s^2 + (\bar{x} - \mu)^2 = \frac{1}{N}\sum_i (x_i - \mu)^2$$

and hence, substituting into Eq. (4.75) we obtain

$$p(\mu, \sigma|\mathbf{x}) \propto \sigma^{-(N+1)} \exp\left\{-\left[\frac{N(s^2 + (\bar{x} - \mu)^2)}{2\sigma^2}\right]\right\}$$

We are interested in obtaining the distribution for $p(\mu|x)$. To do this, we integrate over σ. This is called marginalization. Integrating, we obtain

$$\begin{aligned} p(\mu|\mathbf{x}) &= \int_0^\infty p(\mu,\sigma|\mathbf{x})d\sigma \\ &\propto \int_0^\infty e^{-N(s^2+(\bar{x}-\mu)^2)/2\sigma^2} d\sigma \end{aligned}$$

Following the procedure outlined in Lupton (1993)

$$p(\mu|\mathbf{x}) \propto \left[1+\left(\frac{t^2}{\nu}\right)\right]^{-(\nu+1)/2}$$

where $\nu = N-1$ and $t^2 = (N-1)[(\bar{x}-\mu)^2]/s^2$. This is the well known t-distribution. We are now free to evaluate any type of confidence regions for μ that we wish. The procedure of obtaining the marginal density involves integration over the parameter that we are not interested in. Such parameters are called *nuisance* parameters and are dealt in the next Section.

The title of this subsection contained the word credibility. We credit Press (1989) with this description and adopt it with a sigh of relief. Credibility intervals are to Bayesian estimation what confidence intervals are to likelihood estimation. We need this differentiation because, as we have seen above, what is considered to be fixed and random in the two approaches differs completely. From now on, we consider parameter credibility regions.

4.5.8.1 A Little About Marginals

Remembering Bayes' theorem in the form expressed by Eq. (4.52), we imagine a hypothesis of the form

$$H \equiv f(x) = \sum_{i=1}^{M} a_i\phi_i(x,\{k_x\})$$

where $f(x)$ is some function of the spatial parameter x that is modelled in terms of the decomposition into M components consisting of M amplitudes, a_i, and basis functions ϕ_i, $i=1,2,\ldots,M$. Each ϕ_i, in turn depends on a set of parameters, wavenumbers for example, that we have designated as $\{k_x\}$. Imagine that, in our problem, what is of consequence to uncover is this set of wavenumbers. The amplitudes are of less relevance. The Bayesian approach allows the formulation of the posterior probability so that it is independent of the a_i, which we term nuisance parameters. To see how this is done, we begin with a very simple problem of two parameters: k_x is desired, a is the nuisance. In other words, we wish the posterior density $p(k_x|D,I)$. We first compute the joint posterior pdf of both k_x and a using Bayes' theorem

$$p(k_x,a|D,I) = \frac{p(D|k_x,a,I)p(k_x,a|I)}{P(D|I)}$$

We now integrate a out of the equation to obtain the *marginal* posterior pdf for k_x

$$p(k_x|D,I) = \int p(k_x,a|I)da$$

which expresses the information contained about k_x in the data and the prior information, regardless of the value of a. This apparently somewhat miraculous result is easily explained. Consider the joint pdf $p(x,y)$ of two random variables $\mathbf{x}$ and $\mathbf{y}$. $p(x)\Delta x$ is the probability of observing $\mathbf{x}$ in the interval $(x, x+\Delta x)$. The value of $\mathbf{y}$ is immaterial and may lie anywhere in the interval $(-\infty, +\infty)$. Hence

$$\begin{aligned}
p(x)\Delta x &= \lim_{\Delta x\to 0} \text{probability}(x < \mathbf{x} \le x+\Delta x), \quad (-\infty < \mathbf{y} \le +\infty) \\
&= \lim_{\Delta x\to 0} \int_x^{x+\Delta x} \int p(x,y)dxdy \\
&= \lim_{\Delta x\to 0} \int_x^{x+\Delta x} \left[\int p(x,y)dy\right] dx
\end{aligned}$$

The integral inside the brackets is a function of x and is constant over the interval $(x, x+\Delta x)$ since $\Delta x \to 0$. Hence

$$\begin{aligned}
p(x)\Delta x &= \lim_{\Delta x\to 0} \left[\int p(x,y)dy\right] \left[\lim_{\Delta x\to 0} \int_x^{x+\Delta x} dx\right] \\
&= \Delta x \int p(x,y)dy \qquad (4.94)
\end{aligned}$$

and therefore

$$p(x) = \int p(x,y)dy$$

We have illustrated the computation of a marginal distribution from a joint distribution of two variables. The strength of this approach is that it is quite general.

4.5.8.2 Parameter Credibility Intervals

We have dealt with marginal distributions, nuisance parameters and credibility intervals. It is time to put all this together for a more general treatment. We assume that, following the outline presented above, we have computed the a posteriori pdf for our model. We now wish to present credibility intervals associated with each parameter of the model.

The Bayesian approach has led us to an a posteriori pdf that, given the prior distribution, has allowed us to determine a complete statistical description of our model. To obtain the credibility intervals associated with individual model parameters, we obtain the marginal pdf's as follows. For a given parameter m_i

$$p(m_i) = \int_{m_1} \cdots \int_{m_{i-1}} \int_{m_{i+1}} \cdots \int_{m_M} p(\mathbf{m}|\mathbf{d})dm_1 \ldots\ dm_{i-1}dm_{i+1} \ldots dm_M$$

We now obtain the means and variances of each marginal to characterize the statistics of the individual model parameters. Of course, as pointed out by Tarits et al. (1994), if the marginal pdf is not unimodal, the interpretation of the moments of the pdf is not simple.

4.5.9 Computational Tractability and Minimum Relative Entropy

As pointed out by Tarits et al. (1994), for a large number of model parameters, the full Bayesian solution may not be tractable and the common practice is to settle for maximum likelihood or asymptotic solutions. In Section 4.4, we outlined another approach that is 'Bayesian like' that is the MRE method. The gist of the this approach is to use first moment constraints. By the same token, we treat the prior model parameters as expectations over the prior pdf that is determined by adopting the PME. The MRE approach is in our opinion very flexible and computationally much less demanding than the full Bayesian approach, yet it allows the incorporation of probabilistic constraints. Woodbury and Ulrych (1998) and Jacobs and van der Geest (1991) compare the Bayesian and MRE approaches. The basic difference is that, as expressed by the latter authors, the MRE posterior pdf is obtained from the minimization of the entropy functional constrained not by the full equation, $\mathbf{d} = \mathbf{Gm}$, but by the first moments, $E[\mathbf{d} = \mathbf{Gm}]$, a weakened form of the constraints. In fact, the MRE solution coincides with the Bayesian solution when $I(q, \pi)$ in Eq. (4.57) is minimized over the M moments of the data.

4.5.10 More About Priors

We have seen the pivotal role played by the prior in the Bayesian inverse approach. It is therefore of importance to examine the effect of our choice on the answer. In particular, following Scales and Tenorio (2001), we examine the issue of the noninformative prior. To do this we must deal briefly with the risk associated with an estimator. Our definitions follow from the work of Efron and Morris (1973). First, we define loss and risk. For simplicity of discussion we consider M samples, $x_i, i = 1, 2, \ldots, M$ of the random variable $\mathbf{x}$ distributed as $\mathbf{x}|\mu \overset{\text{ind}}{\sim} N(\mu, 1)$, and the associated ML estimator of μ

$$\boldsymbol{\delta}^0(\mathbf{x}) = \frac{1}{M} \sum_i x_i$$

The loss, L, may be defined in the usual way as

$$L(\boldsymbol{\mu}, \hat{\boldsymbol{\mu}}) = \sum_{i=1}^{k} (\hat{\mu}_i - \mu_i)^2$$

where $\hat{\boldsymbol{\mu}}$ is the estimate of $\boldsymbol{\mu}$. The risk, $\mathbf{R}(\cdot\,,\cdot)$, in turn is formally defined for the ML estimator as

$$\mathbf{R}(\boldsymbol{\mu}, \boldsymbol{\delta}^0) = E_\mu \sum_{i=1}^{k} (\boldsymbol{\delta}^0(x_i) - \mu_i)^2$$

where E_μ is the expectation over the distribution. $\boldsymbol{\delta}^0$ is a very special estimator indeed. Statisticians have shown that, given $\mathbf{x}|\mu \overset{\text{ind}}{\sim} N(\mu, 1)$, $\boldsymbol{\delta}^0$ has lowest risk of any linear or non-linear unbiased estimator. We have previously seen that $\boldsymbol{\delta}^0$ is the ML estimator as well as the Bayesian estimator obtained with a uniform prior.

We have already met the risk in the guise of the MSE. Specifically, we dealt with the MSE associated with the Bayesian estimation of variance using the Jeffreys noninformative prior. Using the above notation, MSE($\hat{\sigma}^2$) is equal to $\mathbf{R}(\boldsymbol{\sigma}^2, \hat{\boldsymbol{\sigma}}^2)$. Bayes with the Jeffreys prior leads to an estimator that, at least for the simple case considered here, constitutes a trade-off between bias and variance.

Scales and Tenorio (2001) consider the following problem. Given a datum, d, from $N(\mu, 1)$ and the prior information that μ is bounded by $(-b, b)$, what are the risks associated with a Bayesian and a frequentist approach to the estimation of μ where the Bayesian prior is taken to be uniform (maximally noninformative?) in $(-b, b)$. The answer is that, the risk of Bayesian estimator is, in a certain range, lower than the lower bound risk of the minimax estimator that incorporates the constraint as a hard bound. The conclusion drawn from this result, since the minimax risk is not based on any distribution, is that the uniform pdf has introduced information beyond that of just the bounds. Putting it another way, whereas the frequentist approach, at least in this case, is truly non committal with respect to what we do not know, the Bayesian method of introducing a uniform prior is not - a very interesting observation.

4.5.11 Bayes, MaxEnt and Priors

As we have seen, the assignment of prior probabilities is crucial in the Bayesian methodology. In this section we work through an example that couples the PME with the Bayesian formalism. Specifically, we consider the problem of estimating material properties from finite data, a problem, as we will see, that lends itself particularly well to such issues as imposing positivity constraints, determining least biased probability distributions and estimating confidence intervals for estimated attributes. The development here is based on the work of Jowitt (1979), a paper that we consider to be a classic.

4.5.11.1 The MaxEnt pdf

The problem is simply stated as follows:
We have some test data, $\mathbf{d} = [m_1, m_2, \ldots, m_N]^T$, concerning an earth parameter of interest, m, a random variable that we know, a priori, can assume only positive values[7]. The task is to determine some inferences concerning m. Our approach is first, to estimate a pdf that describes m, secondly to assimilate the effect of a finite sample size describing m and, thirdly, to establish confidence intervals for our inference. We view the problem via the Jaynes approach, which is MaxEnt + Bayes. We remark that, in general of course, our earth parameter is not observed directly, but is related to the test data by means of some linear or nonlinear transformation. In this example, however, we wish not to obscure the central task by means of the required Jacobians, and we treat the simpler problem. In traditional manner, we determine the sample statistics

[7]There is a reason for this apparently peculiar notation, that soon comes to light.

$$\bar{m} = \frac{1}{N}\sum_i m_i \tag{4.95}$$

and

$$s_m^2 = \frac{1}{N-1}\sum_i (m_i - \bar{m})^2 \tag{4.96}$$

For the moment, assuming that these statistics are equal to the respective expectations, we may estimate the MaxEnt pdf in the manner outlined in Section (3.3.5) where, because of the positivity constraint, all integrals are evaluated on $(0, \infty)$. We have already seen (Section (3.3.5.1)) that, given the constraint on the variance, the resulting MaxEnt pdf is Gaussian. The case here is somewhat different owing to the positivity constraint. Following some elementary algebra, we obtain the following pdf.

$$\mathbf{p}(m) = \exp[-\lambda_0 - \lambda_1 m - \lambda_2 m^2] \tag{4.97}$$

where the λ's are the Lagrange multipliers that are determined from Eqs. (4.95) and (4.96) together with the unimodular constraint. Eq. (4.97) defines a truncated Gaussian distribution. The actual form of this pdf depends strongly on a measure which we define as $\beta = s_m/\bar{m}$. As emphasized by Jowitt (1979) and illustrated in Fig (4.11) when β is sufficiently small, ($\beta < 0.5$, say), $p(m)$ is approximately Gaussian. For much larger values of β, $p(m)$ can no longer be approximated by a normal distribution. We will make the simplifying assumption that β, for the pdf describing our parameter, is sufficiently small that the Gaussian approximation holds.

4.5.11.2 Incorporating Sample Size via Bayes

We now consider the important aspect of how our inference concerning m is affected by sample size. In the previous discussion, we tentatively assumed that the sample statistics were equivalent to the respective expectations. We now drop this assumption and we account for parameter uncertainty in a rigorous, Bayesian manner.

As discussed above, the pdf that describes m is Gaussian and we write it as

$$p(m|\mu, \sigma_m) = \frac{1}{\sqrt{2\pi}\sigma_m}\exp\left[-\frac{1}{2}\left(\frac{m-\mu}{\sigma_m}\right)^2\right] \tag{4.98}$$

where the pdf is conditional on μ and σ_m, the actual mean and standard deviation of m. These two parameters are unknown, however, and we need to determine the pdf that is conditional on the measured data. i.e., we require $p(m|\mathbf{d})$ which may be written as

$$p(m|\mathbf{d}) = \int_\mu\int_{\sigma_m} p(m, \mu, \sigma_m|\mathbf{d})\,d\mu\,d\sigma_m \tag{4.99}$$

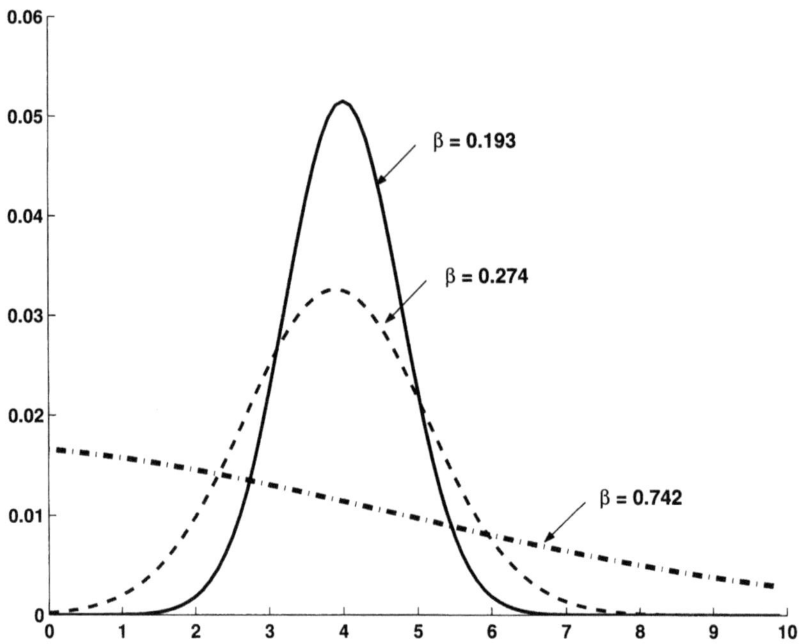

Figure 4.11: Truncated Gaussian distributions as a function of β.

a marginalization of the joint pdf $p(m, \mu, \sigma_m|\mathbf{d})$. Since, according to the product rule

$$p(m, \mu, \sigma_m|\mathbf{d}) = p(m|\mu, \sigma_m, \mathbf{d})p(\mu, \sigma_m|\mathbf{d}) \tag{4.100}$$

and since μ and σ_m determine the pdf of m completely

$$p(m, \mu, \sigma_m|\mathbf{d}) = p(m|\mu, \sigma_m)p(\mu, \sigma_m|\mathbf{d}) \tag{4.101}$$

The first term in Eq. (4.101) is given by Eq. (4.98) and the second term is determined by means of Bayes' theorem

$$p(\mu, \sigma_m|\mathbf{d}) = \frac{p(\mathbf{d}|\mu, \sigma_m)p(\mu, \sigma_m)}{\int_\mu \int_{\sigma_m} p(\mathbf{d}|\mu, \sigma_m)p(\mu, \sigma_m)d\mu d\sigma_m} \tag{4.102}$$

A priori, we know that $-\infty < \mu < \infty$ and $0 \leq \sigma_m < \infty$. The range on μ may seem peculiar given that m is positive but, since we have assumed β to be small leading to a Gaussian distribution for m, μ must conform and be unrestricted.

The prior joint probability density, $p(\mu, \sigma_m)$, in Eq. (4.102) can now be determined if we remember the discussion concerning the Jeffrey's prior (Section (4.5.1)). Specifically, since μ and σ_m are independent random variables, assigning the Jeffrey's prior to σ_m and a uniform prior to μ, we can write

$$p(\mu, \sigma_m) \quad \propto \quad \frac{1}{\sigma_m} \tag{4.103}$$

and inserting Eq. (4.103) into Eq. (4.102)

$$p(\mu, \sigma_m|\mathbf{d}) \quad = \quad \frac{p(\mathbf{d}|\mu, \sigma_m)\frac{1}{\sigma_m}}{\int_\mu \int_{\sigma_m} p(\mathbf{d}|\mu, \sigma_m)\frac{1}{\sigma_m} d\mu d\sigma_m} \tag{4.104}$$

But

$$p(\mathbf{d}|\mu, \sigma_m) \quad = \quad \prod_i p(d_i|\mu, \sigma_m) \tag{4.105}$$

and consequently

$$p(\mathbf{d}|\mu, \sigma_m)\frac{1}{\sigma_m} \quad = \quad \frac{1}{\sqrt{2\pi}\sigma_m^N} \exp\left[-\frac{1}{2\sigma_m^2}\sum_i (m_i - \mu)^2\right] \tag{4.106}$$

We now need some useful algebra that will allow us to transform the exponential term in the above equation. Expanding as below

$$\sum_i (m_i - \mu)^2 \quad = \quad \sum_i m_i^2 - 2\mu \sum_i m_i + N\mu^2$$

Since $\sum_i m_i = N\bar{m}$, we obtain

$$\sum_i (m_i - \mu)^2 \quad = \quad \sum m_i^2 - 2N\mu\bar{m} + N\mu^2 \tag{4.107}$$

Eq. (4.96) can be written as

$$s_m^2 \quad = \quad \frac{1}{N-1}(\sum_i m_i^2 - N\bar{m}^2) \tag{4.108}$$

which leads to

$$\sum_i m_i^2 \quad = \quad (N-1)s_m^2 + N\bar{m}^2 \tag{4.109}$$

Substituting Eq. (4.109) into Eq. (4.107) gives

$$\sum_i (m_i - \mu)^2 \quad = \quad (N-1)s_m^2 + N\bar{m}^2 - 2N\mu\bar{m} + N\mu^2$$

$$= N\left[\frac{(N-1)}{N}s_m^2 + (\bar{m}-\mu)^2\right] \tag{4.110}$$

Finally, using Eqs. (4.106) and (4.110), we obtain the required pdf, $p(\mathbf{d}|\mu, \sigma_m)$, which we write in terms of a function $g(\mu, \sigma_m)$

$$p(\mathbf{d}|\mu, \sigma_m) = \frac{g(\mu, \sigma_m)}{\int_\mu \int_{\sigma_m} g(\mu, \sigma_m) d\mu d\sigma_m} \tag{4.111}$$

where

$$g(\mu, \sigma_m) = \sigma_m^{-(N+1)} \exp\left\{-\frac{N}{2\sigma_m^2}\left[\frac{(N-1)}{N}s_m^2 + (\bar{m}-\mu)^2\right]\right\}$$

The denominator of Eq. (4.111) can be evaluated in terms of gamma functions as follows (Jowitt, 1979)

$$\int_\mu \int_{\sigma_m} g(\mu, \sigma_m) d\mu d\sigma_m = \frac{1}{2}\sqrt{\frac{2\pi}{N}}\left[\frac{(N-1)}{2}s_m^2\right]^{-(n-1)/2} \Gamma[\tfrac{(N-1)}{2}] \tag{4.112}$$

where the gamma function is defined as

$$\Gamma[\alpha] = \int_0^\infty e^{-x} x^{\alpha-1} dx \quad |\alpha| > 0$$

The marginal distributions, $p(\mu|\mathbf{d})$ and $p(\sigma_m|\mathbf{d})$, can be found quite simply. Carrying out the integration for $p(\mu|\mathbf{d})$ obtains a pdf that has some interesting properties. Specifically, (see Jowitt, 1979)

$$p(\mu|\mathbf{d}) = \frac{\Gamma[\frac{N}{2}]}{\Gamma[\frac{(N-1)}{2}]} \frac{\left[1 + \frac{N}{(N-1)}\left(\frac{\mu-\bar{m}}{s_m}\right)^2\right]^{-N/2}}{s_m\sqrt{\pi\frac{(N-1)}{N}}} \tag{4.113}$$

Letting

$$t = \sqrt{N}\left(\frac{\mu-\bar{m}}{s_m}\right)$$

and

$$\nu = N-1$$

we obtain a new distribution

$$p(t|\mathbf{d}) = \frac{\Gamma[\frac{(\nu+1)}{2}]}{\Gamma[\frac{\nu}{2}]} \frac{\left[1+\frac{t^2}{\nu}\right]^{-(\nu+1)/2}}{\sqrt{\pi\nu}} \tag{4.114}$$

Not surprisingly, in view of the discussion in Section (4.5.8), we have obtained a Student's t-distribution with ν degrees of freedom. We emphasize, however, that Eq. (4.114) is not a sampling distribution that would be obtained from a frequentist point of view.

In order to illustrate the relationship of the $\Gamma[\cdot]$ function to factorials, we rewrite this pdf as

$$p(t|\mathbf{d}) = \frac{1}{\sqrt{\pi\nu}} \frac{(\nu/2-1/2)!}{(\nu/2-1)!} \left[1+\frac{t^2}{\nu}\right]^{-(\nu+1)/2} \tag{4.115}$$

where we have used the relationship $\Gamma[k+1] = k!$ It can be shown that $p(t|\mathbf{d})$ has zero mean and variance given by $\sigma^2 = (\nu+1)/(\nu-1)$. Further, we also have that

$$p(t|\mathbf{d}) \rightarrow N(0,1) \quad \text{as} \quad N \rightarrow \infty$$

and, using the facts that $\Gamma[\alpha+1] = \alpha\Gamma[\alpha]$, $\Gamma[1] = 1$ and $\Gamma[\frac{1}{2}] = \sqrt{\pi}$

$$p(t|\mathbf{d}) = \frac{1}{\pi(1+t^2)} \quad \text{for} \quad \nu = 1$$

which is the Cauchy distribution. The above discussion is illustrated in Fig (4.12).

Our final objective, that of computing $p(m|\mathbf{d})$, is achieved by combining Eqs. (4.99), (4.101), (4.98), (4.111) and (4.104) and performing the integrations as outlined above. The final result is equivalent to that expressed by Eq. (4.114) but with t defined by

$$t = \sqrt{\frac{N}{N+1}} \left(\frac{m-\bar{m}}{s_m}\right)$$

Again, we emphasize that the resulting pdf is not a sampling distribution derived from a frequentist point of view. Rather, it is a Bayesian distribution that incorporates objective priors.

4.5.11.3 Summary

We have derived a prior pdf for a parameter that is constrained to take positive values. The importance of this distribution is that it contains only information that is known to us, i.e., the sample mean and standard deviation. It has been derived taking the sample size into account and therefore lends itself to the determination of confidence intervals for our inference concerning m. The particular form derived does contain an assumption on the value of the parameter β but the validity of this assumption

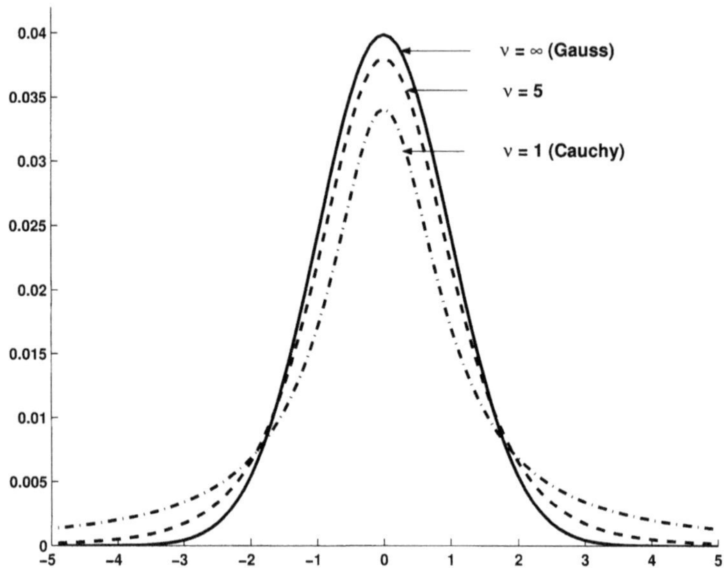

Figure 4.12: t-distributions as a function of the number of degrees of freedom, ν.

can clearly be evaluated from the available data. An additional benefit inherent in the MaxEnt/Bayesian approach is that other marginal pdf's, such as the distribution for the mean, which is sometimes of importance, can be readily derived.

4.5.12 Bayesian Objective Functions

The two previous sections have drawn particular attention to the information in the prior. In this section, the prior model is used in a very different sense. From the infinity of possible models we will choose a set that is constrained to be of a certain quality by the prior distribution. We do not, in this approach, pay any heed whatsoever to the 'truth' of the prior. We choose it with an aim in mind, an example of what John Scales would, perhaps, call 'a post prior'. We illustrate the approach with the work of Sacchi et al. (1998) who consider the objective of obtaining an 'aperture-free' Fourier transform.

A common approach to signal analysis and decomposition is based on mapping the data into a new domain where the support of each signal is considerably reduced and, consequently, decomposition can be easily attained. This is applicable to the discrete Fourier transform. Since we are always concerned with a finite amount of data the correct decomposition of events is complicated by sidelobe artifacts. Often, our problem may be posed as a linear inverse where the correct regularization is crucial to the resolution of signals when the aperture is below the resolution limit. The regularization is obtained by incorporating into the problem a long-tailed prior using Bayes' rule.

For simplicity we start with the 1-D DFT since extensions to higher dimensions are straightforward. Consider a N-sample time or spatial series $x_1, x_2, \ldots, x_N$. The

DFT of the discrete series is given by

$$X_k = \sum_n x_n e^{-i2\pi nk/N} \qquad k = 1, 2, \ldots, N$$

Similarly, the inverse DFT is given by

$$x_n = \frac{1}{N} \sum_k X_k e^{i2\pi nk/N} \qquad n = 1, 2, \ldots, N \tag{4.116}$$

We wish to estimate M spectral samples where $M > N$. A standard approach to solving this problem is by means of zero padding. Defining a new time series consisting of the original series plus a zero extension for $n = N+1, N+2, \ldots, M$, we can estimate M spectral samples using the DFT. This procedure helps to remove ambiguities due to discretization of the Fourier transform but, as is well know, it does not reduce the sidelobes created by the temporal/spatial window or improve the resolution. Let us therefore consider the estimation of M spectral samples without zero padding. In other words we want to estimate the DFT using only the available information. The underlying philosophy is similar to Burg's maximum entropy method (Section 3.4.1) except that in the MaxEnt method the target is a power spectral estimate, a phaseless function.

To avoid biasing our results by the discretization we also impose the condition $M >> N$. Rewriting Eq. (4.116) as

$$x_n = \frac{1}{M} \sum_k X_k e^{i2\pi nk/M}$$

gives rise to a linear system of equations

$$\mathbf{x} = \mathbf{FX}$$

where the vectors $\mathbf{x} \in \mathbf{R^N}$ and $\mathbf{X} \in \mathbf{C^M}$ denote the available information and the unknown DFT, respectively, and $\mathbf{F}$, contains the exponential terms $F_{n,k} = \frac{1}{M} \sum_k e^{i2\pi nk/M}$. We now have an underdetermined problem that must be suitably regularized to impose uniqueness. We examine two regularization strategies.

4.5.12.1 Zero Order Quadratic Regularization

In standard fashion, we assume data contaminated with noise which is distributed as $N(0, \sigma_n^2)$. The first regularization scheme used by Sacchi et al. (1998) imposed a Gaussian prior. After combining the likelihood with the prior probability of the model by means of Bayes' rule, the MAP solution is computed by minimizing

$$J_{GG} = \lambda ||\mathbf{x}||_2^2 + ||\mathbf{x} - \mathbf{Fx}||_2^2 \tag{4.117}$$

where the subscript GG stands for the Gauss-Gauss model (Gaussian model and Gaussian errors). The scalar is the ratio of variances, $\lambda = \sigma_n^2/\sigma_x^2$. Eq. (4.117) is the objective function of the problem. The first term represents the model norm while the second term is the misfit function. The hyperparameter λ enables us to move our estimate along the trade-off curve. Taking derivatives and equating to zero yields the estimator

$$\begin{aligned} \mathbf{x} &= (\mathbf{F}^H\mathbf{F} + \lambda\mathbf{I}_M)^{-1}\mathbf{F}^H\mathbf{x} \\ &= \mathbf{F}^H(\mathbf{F}\mathbf{F}^H + \lambda\mathbf{I}_N)^{-1}\mathbf{x} \end{aligned} \quad (4.118)$$

where $\mathbf{I}_M$ and $\mathbf{I}_N$ are the $(M \times M)$ and $(N \times N)$ identity matrices, respectively. Since $\mathbf{F}^H\mathbf{F} = (1/M)\mathbf{I}_N$, Eq. (4.118) may be rewritten as

$$\mathbf{x} = (\frac{1}{M} + \lambda)^{-1}\mathbf{F}^H\mathbf{x} \quad (4.119)$$

The result is simply the DFT of x_n modified by a scale factor. The solution expressed by Eq. (4.119) becomes

$$X_k = \frac{1}{1+\lambda M}\sum_n x_n e^{-i2\pi nk/(M-1)} \quad (4.120)$$

Eq. (4.120) represents the DFT of the windowed time series and is equivalent to padding with zeros in the range $n = N+1, N+2, \ldots, M$. It is clear that the zero order regularization yields a scaled version of the conventional DFT. The associated periodogram

$$P_k = X_k {X^*}_k \quad (4.121)$$

will show the classical sidelobe artifacts due to truncation effects. We do not wish to minimize the problem associated with choosing the appropriate value of λ. Estimating hyperparameters in a sensible and robust manner remains a central problem in inversion. We mention a non-optimal approach below.

4.5.12.2 Regularization by the Cauchy-Gauss Model

To obviate the sidelobe structure, Sacchi et al. (1998) proposed the use of a regularization derived from a pdf that mimics a sparse distribution of spectral amplitudes. A heavy-tailed distribution, like the Cauchy pdf, will induce a sparse model consisting of only a few elements different from zero. If the data consist of a few number of harmonics (1-D case) or a limited number of plane waves (2-D case) a sparse solution will help to attenuate sidelobes artifacts. The Cauchy pdf of the complex variable, X_k, is given by

$$p(X_k|\sigma_c) \propto \frac{1}{(1+\frac{X_k X_k^*}{2\sigma_c^2})}$$

where σ_c is a scale parameter. When we combine the Cauchy prior with the data likelihood the cost function becomes

$$J_{CG} = S(\mathbf{X}) + \frac{1}{2\sigma_n^2}(\mathbf{x}-\mathbf{F}\mathbf{X})^H(\mathbf{x}-\mathbf{F}\mathbf{X})$$

where the subscript CG stands for the Cauchy-Gauss model. The function, $S(\mathbf{X})$, which is expressed by

$$S(\mathbf{X}) = \sum_k \log(1+\frac{X_k X_k^*}{2\sigma_c^2})$$

is the regularizer imposed by the Cauchy distribution and is a measure of the sparseness of P_k, the vector of spectral powers defined in Eq. (4.121). The constant σ_c controls the amount of sparseness that can be attained by the inversion. Taking derivatives of $J_{CG}(\mathbf{x})$ and equating to zero yields the following result

$$\mathbf{X} = (\lambda\mathbf{Q}^{-1}+\mathbf{F}^H\mathbf{F})^{-1}\mathbf{F}^H\mathbf{x}$$

where $\lambda = \sigma_n^2/\sigma_c^2$ and $\mathbf{Q}$ is a $M\times M$ diagonal matrix with elements given by

$$Q_{ii} = 1+\frac{X_i X_i^*}{2\sigma_c^2}, \quad i=1,2\ldots,M$$

Sacchi et al. (1998) show that the solution may also be written as

$$\mathbf{x} = \mathbf{Q}\mathbf{F}^H(\lambda\mathbf{I}_N+\mathbf{F}\mathbf{Q}\mathbf{F}^H)^{-1}\mathbf{y}. \quad (4.122)$$

The hyperparameters of the problem are fitted by Sacchi et al. (1998) by means of a χ^2 criterion. This approach is not optimum, as we have pointed out above, but certainly requires a much less intensive computational effort. Eq. (4.122) is iteratively solved with the initial model being the DFT of the truncated signal. An excellent view of objective functions without appeal to Bayes is the work of Farquharson and Oldenburg (1998).

An example of the approach that we have outlined above is illustrated in Figs. 4.13-4.16. In the first example we attempt to determine a highly resolved power spectral estimate from irregularly sampled and noisy data. Fig. 4.13(a) shows the continuous data without noise (solid line) and the noisy samples (squares) where the added noise was Gaussian with a standard deviation of 0.4. The sampled data clearly demonstrate a gapped appearance.

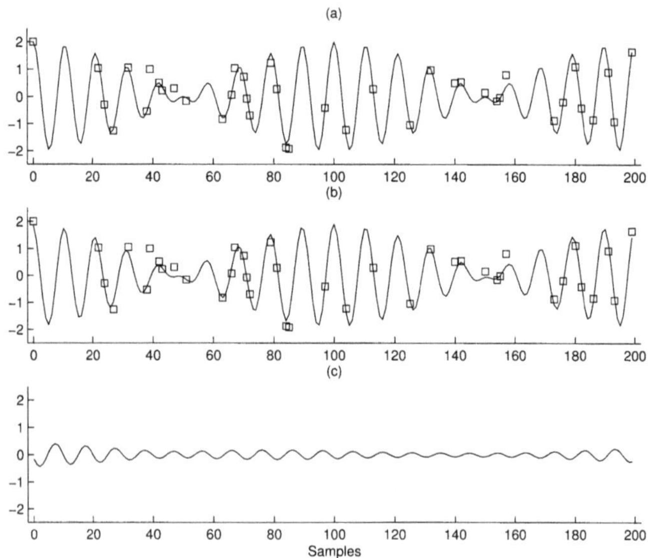

Figure 4.13: (a) Input data. (b) Reconstructed data. (c) Error.

We have used Eq. (4.122) to obtain the results illustrated in Figs. 4.13(b), 4.13(c) and Fig. 4.14(a). The former shows the reconstituted data (solid line) using the noisy, irregularly sampled input repeated as squares in this figure for convenience. Fig. 4.13(c) shows the error panel. The computed high resolution spectrum is shown in Fig. 4.14(a) and is to be compared with Fig. 4.14(b) that shows the periodogram of the complete time series (the solid curve in Fig. 4.13(a)). Figs. 4.15 and 4.16 illustrate the results of our Bayesian approach to a 2D example. Specifically, we are interested in computing the 2D spectrum of the vertical seismic profile shown in Fig. 4.15. Fig. 4.16(a) illustrates the 2D periodogram, whereas Fig. 4.16(b) demonstrates the resulting power spectrum of the discrete Fourier transform computed via the Cauchy-Gauss method. The enhancement in the resolution that has been obtained is very clear.

4.5.13 Summary and Discussion

We have attempted to summarize some of the concepts that are central in the Bayesian approach to inverse problems. Of course, the center piece is the prior distribution in function as well as in controversy. We follow the conclusion of Scales and Sneider (1997) who posed the question "To Bayes or not to Bayes?" and answered it with a careful yes. We would, perhaps, be somewhat more affirmative and say yes, by way of an honest appraisal of what it is that we really do know.

The Bayesian approach may be applied either 'hierarchically' (pure Bayes) or 'empirically' (less pure Bayes). In the former, we are much concerned about the information content in the prior and its effect upon our estimate. Thus, issues such as, how informative is a noninformative prior, the risk of a Bayesian estimate, etc.,

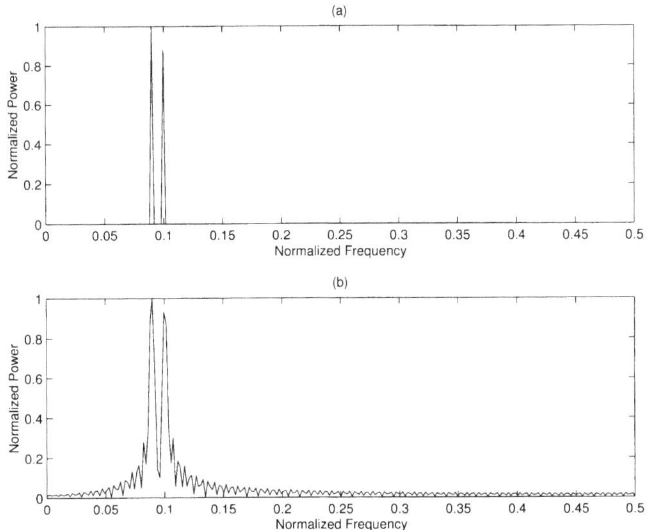

Figure 4.14: (a) High resolution spectrum. (b) Periodogram.

are indeed uppermost. In the empirical approach, these questions are less imperative.

4.5.13.1 Hierarchical Issues

Let us return, briefly, to the Jeffreys prior. Certainly $1/\sigma$ can hardly be construed as a pdf. It is, in fact, known as an improper prior since it is not normalizable. But how important is this fact really? In practice, σ is always bounded, it is certainly greater than zero and less than infinity and can therefore be normalized. In any case, since the prior gets multiplied by the likelihood which certainly goes to zero as σ approaches zero or infinity, the bounds are not really significant.

Scales and Tenorio (2001) discuss two interesting issues. The first deals with the information contained in a seemingly noninformative prior which indicates that, from the point of view of the risk associated with the minimax and Bayes estimators for a particular problem, the Bayesian estimator is less non committal than the frequentist estimator. Certainly, this is worth pondering about. It is interesting that a uniform prior adds more information than a hard bound, particularly since incorporating the constraint via a distribution is referred to as 'softening' the constraint. From a practical point of view, given all other uncertainties involved in a real data inversion problem, the differences incurred in the solution as a result of information possibly added due to the above effect is, in all probability, insignificant.

The second issue discussed by Scales and Tenorio (2001) is that of added difficulties when hard constraints are replaced by probability constraints that occur in higher dimensions. This topic is most certainly worthy of further investigation.

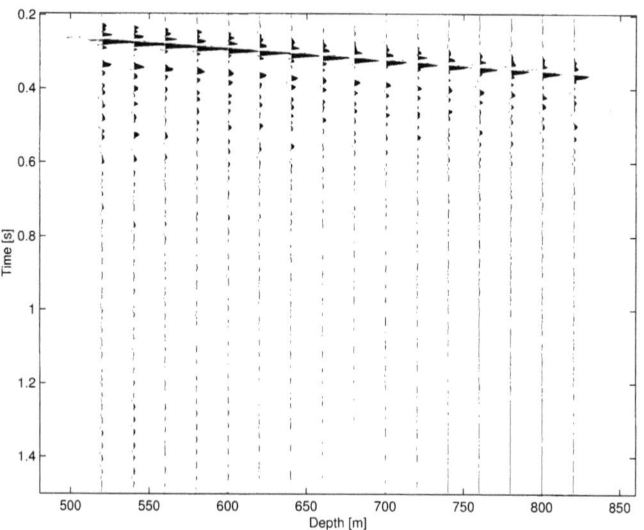

Figure 4.15: Vertical seismic profile.

4.5.13.2 Empirical Issues

The idea behind the empirical Bayes approach is that the prior is based on information contained in the input data. The methodology for constructing objective functions from prior pdf's and Bayes rule is not exactly empirical but it certainly is not hierarchical. In this case, the 'truth' of the prior is irrelevant. We use it to constrain the solution to a form that is a priori known to us to be the desired one. We have illustrated this by the computation of a high resolution DFT (Sacchi et al., 1998). Another application, with which we have had considerable success, is the compensation of aperture effects in computing the Radon transform (Sacchi and Ulrych, 1995b; Trad et al., 2003), a subject that is explored at some length in Chapter 5. We simply look for that transform that has a limited support in the Radon domain. The Bayesian approach here is, in some sense, empirical, in that the data tell us how sparse the model should be.

We conclude by saying that, in our opinion, Bayes is to likelihood what likelihood is to least squares (or SVD in this case). No one would argue that likelihood is not that much more a flexible approach than least squares. So we argue that Bayes, when used diligently, is that much more flexible than plain old likelihood.

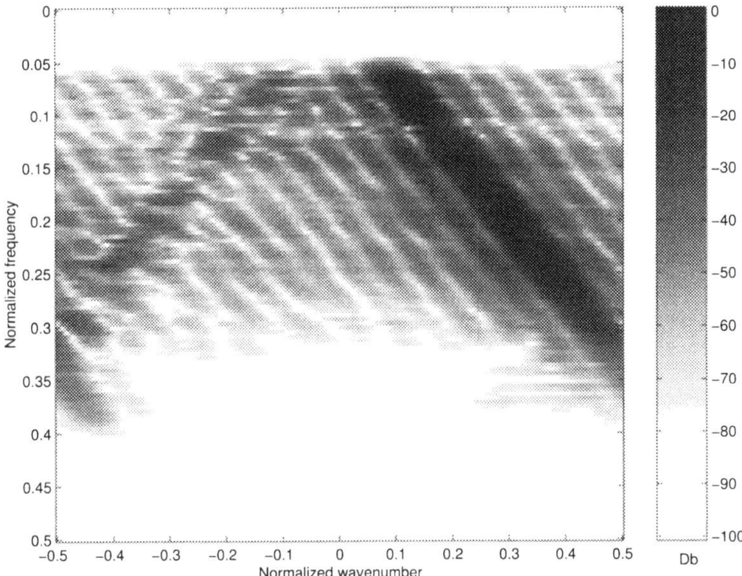

Figure 4.16: (a) 2D periodogram.

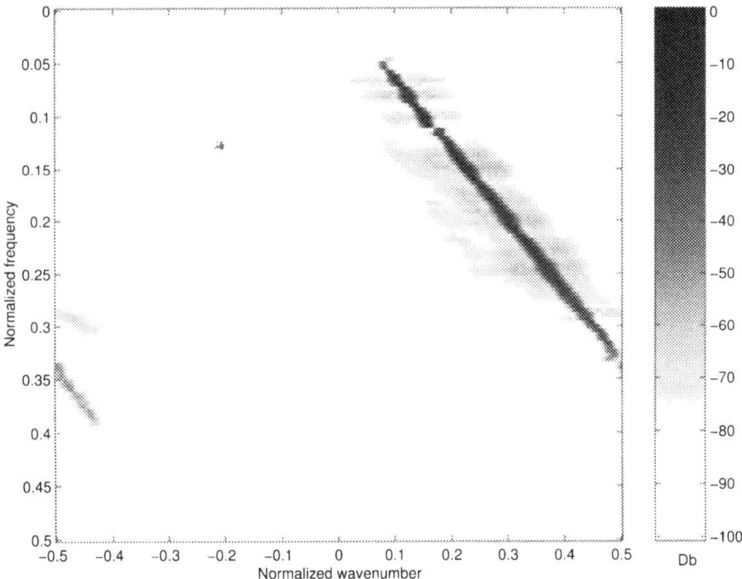

Figure 4.16: (b) 2D high resolution spectrum

APPENDIX

A.1 Singular Value Decomposition, SVD

We present a very brief and heuristic derivation of the singular value decomposition of an arbitrary $(M \times N)$ matrix, $\mathbf{A}$. Details may be found in Lanczos (1956), whose invention this is (we think), and Golub and Van Loan (1996).

Consider a linear system of equations

$$\mathbf{A}\mathbf{x} = \sigma\mathbf{u} \tag{A.1}$$

where, as will become clear, the vectors $\mathbf{x}$ and $\mathbf{u}$ are rather special, and σ is a positive constant ($\neq 0$). Multiplying through by $\mathbf{A}^T$, we obtain

$$\mathbf{A}^T\mathbf{A}\mathbf{x} = \sigma\mathbf{A}^T\mathbf{u} \tag{A.2}$$

$\mathbf{A}^T\mathbf{A}$ is a square, symmetric matrix and admits the eigen decomposition

$$\mathbf{A}^T\mathbf{A}\mathbf{v} = \lambda\mathbf{v} \tag{A.3}$$

where $\mathbf{v}$ is an eigenvector corresponding to the eigenvalue λ. We now delineate the 'special' nature of the equations in Eq. (A.1) by insisting that the vectors $\mathbf{x}$ in Eq. (A.1) and $\mathbf{v}$ in Eq. (A.3) are equivalent, thus defining $\mathbf{u}$. Eq. (A.1) now becomes

$$\mathbf{A}\mathbf{v} = \sigma\mathbf{u} \tag{A.4}$$

Premultiplying by $\mathbf{A}^T$ and given Eq. (A.3), we have

$$\mathbf{A}^T\mathbf{A}\mathbf{v} = \sigma\mathbf{A}^T\mathbf{u} = \lambda\mathbf{v}$$

Consequently

$$\mathbf{A}^T\mathbf{u} = \frac{\lambda}{\sigma}\mathbf{v} \tag{A.5}$$

Premultiplying Eq. (A.5) by $\mathbf{A}$ and substituting Eq. (A.4), gives

$$\mathbf{A}\mathbf{A}^T\mathbf{u} = \frac{\lambda}{\sigma}\mathbf{A}\mathbf{v} = \lambda\mathbf{u} \tag{A.6}$$

We have, therefore, two eigenvector equations, Eqs. (A.3) and (A.6).

Writing the equations in terms of eigenvector and eigenvalue matrices rather than vectors and scalars, Eq. (A.4) becomes

$$\mathbf{A}\mathbf{V} = \mathbf{U}\mathbf{\Sigma}$$

Left multiplying by $\mathbf{V}^T$ and remembering the eigenvector orthogonality condition

$$\mathbf{V}\mathbf{V}^T = \mathbf{I}$$

results in

$$\mathbf{A} \;=\; \mathbf{U}\boldsymbol{\Sigma}\mathbf{V}^T \tag{A.7}$$

the famous SVD of $\mathbf{A}$, where $\mathbf{U}$ is the left eigenvector matrix, $\mathbf{V}$ is the right eigenvector matrix and $\boldsymbol{\Sigma}$ is the diagonal matrix of singular values. The dimensions of these matrices are important and we sketch the structure for $M = 2$ and $N = 3$ for illustrative purposes.

$$\mathbf{A} = \begin{bmatrix} u_{11} & u_{12} \\ u_{21} & u_{22} \end{bmatrix} \begin{bmatrix} \sigma_1 & 0 & 0 \\ 0 & \sigma_2 & 0 \end{bmatrix} \begin{bmatrix} v_{11} & v_{21} & v_{31} \\ v_{12} & v_{22} & v_{32} \\ v_{13} & v_{23} & v_{33} \end{bmatrix}$$

Since $\mathbf{A}$ is (2×3) (and $\mathbf{A}\mathbf{A}^T$ is of rank 2), there are only two non zero σ's, as reflected in $\boldsymbol{\Sigma}$. Naturally, $\mathbf{U}$ is (2×2) and $\mathbf{V}$ is (3×3). The relationship between the λ's and σ's can be easily derived. Using Eq. (A.7) and the orthogonality conditions, we can write

$$\mathbf{A}\mathbf{A}^T \;=\; \mathbf{U}\boldsymbol{\Sigma}\mathbf{V}^T\mathbf{V}\boldsymbol{\Sigma}\mathbf{U}^T = \mathbf{U}\boldsymbol{\Sigma}^2\mathbf{U}^T \tag{A.8}$$

and Eq. (A.6) can be written as

$$\mathbf{A}\mathbf{A}^T \;=\; \mathbf{U}\boldsymbol{\Lambda}\mathbf{U}^T \tag{A.9}$$

Comparing Eqs. (A.8) and (A.9), gives

$$\boldsymbol{\Sigma} \;=\; \boldsymbol{\Lambda}^{\frac{1}{2}} \tag{A.10}$$

Both the inner and outer product matrices have the same eigenvalues, λ_k, and the singular values, σ_k, are given by $\sigma_k = \sqrt{\lambda_k}$. The SVD may, consequently, be computed using Eqs. (A.3), (A.6) and (A.10).

Chapter 5

Signal to Noise Enhancement

5.1 Introduction

The problem of signal and noise separation is ubiquitous in all disciplines. It is important, therefore, to attempt to delineate exactly what one implies by signal and noise for specific scenarios. Ours is seismic and we define our problem accordingly.

Let **d** represent the record of our seismic experiment in whatever form. **d** contains all the information that is at our disposal. Our task is to chisel away at **d** to unearth **s**, the signal therein contained. We define signal as that energy that is coherent from trace to trace. Noise, $\mathbf{n}_r$, on the other hand, is that energy that is incoherent from trace to trace. Unfortunately, the most expensive noise is also spatially coherent and we must modify the above definition. We define signal as that energy that is **most** coherent. This definition immediately introduces one of the central themes of this chapter, resolution. In order to extract the signal from the background of noise we must be able to resolve the difference, often very subtle and highly dependent on the acquisition, between the coherent signal and coherent noise. We now write the model

$$\mathbf{d} = \mathbf{s} + \mathbf{n}_c + \mathbf{n}_r \tag{5.1}$$

where $\mathbf{n}_c$ and $\mathbf{n}_r$ represent the coherent and incoherent noise components, respectively and $\mathbf{n}_c + \mathbf{n}_r = \mathbf{n}$.

In the task of the identification and suppression of $\mathbf{n}$, two processes share the honors for most significant quantum leaps forward. Variable area plotting (according to Mike Graul) and stacking. The problem with stacking is, as we know, that it is not always desirable, since it may destroy important signal attributes. We have chosen three techniques, $f-x$ filtering, eigenimage and KL decomposition and Radon transformation which can lead to signal and noise separation without, we hope, signal distortion. This chapter deals with these three approaches.

5.2 $f - x$ Filters

SNR enhancement in the $f - x$ domain, introduced by Canales (1984) and described and further developed by various authors (Spitz, 1983; Soubaras, 1994; Sacchi and

Kuehl, 2001) is widely accepted and used in the oil and gas industry. The method is very effective in attenuating random noise, is easy to implement and very efficient computationally. The method depends on signal predictability which has been extensively studied in the context of AR filters and harmonic retrieval via ARMA models (Ulrych and Clayton, 1976b).

The idea is quite simple, and can be summarized as follows. Events that are linear or quasilinear in the $t - x$ domain, manifest themselves as a superposition of harmonics in the $f - x$ domain. If noise is taken into account, an optimal model to predict a superposition of harmonics is an ARMA model. However, given the fact that ARMA models might not be very stable (they involve the solution of an eigenvalue problem), the ARMA model is often approximated by a long AR filter. In this case, the predictability is not optimal, but the problem can be easily solved using predictor error filters of the type we have already analyzed in the context of deconvolution in Chapter 2. We first illustrate the concept of predictability by considering a very simple model composed of a sum of harmonics. Later, we discuss a novel approach to noise attenuation proposed by Soubaras (1994), based on the concept of quasi-predictability.

5.2.1 The Signal Model

The signal model is based on the assumption that seismic data can be represented as a superposition of events with linear moveout. This model can be made valid in cases where the events are not linear by dividing the section into overlapping windows. Consider, first, a seismic section, $s(t,x)$, that consists of a single waveform. The frequency domain representation of $s(t,x)$ is given by

$$S(f,x) = A(f)\, e^{i2\pi f\theta x}$$

where $A(f)$ indicates the source spectrum, f the temporal frequency, x the spatial variable or offset and θ the apparent slowness along x. We assume that x is regularly sampled, $x = (k-1)\Delta x \;\; k = 1, 2, \ldots, N$. For any f, we can write

$$S_n = Ae^{i\alpha n} \qquad n = 1, 2, \ldots N$$

where $\alpha = 2\pi f\theta\Delta x$ and we have written A for $A(f)$. The following recursion is obtained by combining S_n and S_{n-1}

$$S_n = a_1 S_{n-1}$$

where $a_1 = \exp(i\alpha)$. The last equation is a first order difference equation that allows us to recursively predict the signal along the spatial variable x. Similarly, it can be shown that the superposition of p complex harmonics (p linear events in $x - t$) can be represented by a difference equation of order p

$$S_n = a_1 S_{n-1} + a_2 S_{n-2} + \ldots a_p S_{n-p} \tag{5.2}$$

We have already met these ideas in Section 2.5.1 in the time domain. Here, the problem is cast in the $f - x$ domain and is, consequently, in complex form. The last equation can be written in prediction error form as

$$\sum_{k=0}^{p} g_k S_{n-k} = 0 \tag{5.3}$$

where the coefficients of the prediction error filter, PEO, are related to the coefficients a_k in Eq. (5.2) by

$$g_0 = 1, \quad g_k = -a_k, \quad k = 1, 2, \ldots, p$$

(please see Section 2.4).

So far we have been able to define a recursive expression to predict a noise-free superposition of complex harmonics. In real applications, however, additive noise will corrupt the data

$$Y_n = S_n + Z_n$$

where Z_n represents a white noise sequence. Substituting $S_{n-k} = Y_{n-k} - Z_{n-k}$ into Eq. (5.3) leads to the following system of equations that defines the signal model in terms of the PEO

$$\begin{aligned} \sum_{k=0}^{p} g_k Y_{n-k} &= \sum_{k=0}^{p} g_k Z_{n-k} \\ &= E_n \end{aligned} \tag{5.4}$$

The latter is an ARMA(p,p) process in which the AR and MA components are identical. The signal E_n in Eq. (5.4) designates the nonwhite innovation sequence $\sum_{k=0}^{p} g_k W_{n-k}$. We have met a real version of this special model before in Section 2.5.1, where we related it to the Pisarenko spectral estimator.

5.2.2 AR $f - x$ Filters

Rather than trying to solve the ARMA equations one can replace the ARMA model by a long AR model

$$Y_n + a_1 Y_{n-1} + a_2 Y_{n-2} \ldots + a_p Y_{n-p} = Q_n \tag{5.5}$$

where this time, the a_k are the coefficients of the AR(p) process (we use the same notation to avoid clutter). The parameter p is the order of the model, which in this case should be large enough to adequately represent the original noisy harmonics. This kind of model has been extensively studied in the context of spectral and regression analysis. The determination of p is not an easy task, and demands the application of information theoretical criteria like the AIC (Section 3.5). It is important to realize, that Z_n and Q_n are quite different sequences. Both are white, but Z_n is uncorrelated with the harmonic signal, whereas Q_n is the innovation sequence that is correlated with Y_n. Writing Eq. (5.5) for $p = 3$, obtains

$$\begin{pmatrix} Y_1 & 0 & 0 \\ Y_2 & Y_1 & 0 \\ Y_3 & Y_2 & Y_1 \\ Y_4 & Y_3 & Y_2 \\ 0 & Y_4 & Y_3 \\ 0 & 0 & Y_4 \end{pmatrix} \begin{pmatrix} a_1 \\ a_2 \\ a_3 \end{pmatrix} - \begin{pmatrix} Y_2 \\ Y_3 \\ Y_4 \\ 0 \\ 0 \end{pmatrix} = \begin{pmatrix} q_2 \\ q_3 \\ q_4 \\ 0 \\ 0 \end{pmatrix}$$

The matrix in this equation is by now familiar to us as a convolution matrix. The last equation can be written in matrix form as

$$\mathbf{Ya} - \mathbf{d} = \mathbf{q}$$

where $\mathbf{d} = [Y_2, Y_3, Y_4, 0, 0]^T$. The least squares estimate of the filter $\mathbf{a}$ is computed by minimizing the power of the innovation, $\mathbf{q}$. The cost function is given by

$$J = ||\mathbf{Ya} - \mathbf{d}||^2$$

and minimizing in traditional manner

$$\mathbf{Y}^H\mathbf{Ya} = \mathbf{Y}^H\mathbf{d}$$

$\mathbf{Y}^H\mathbf{Y}$ is Toeplitz and $\mathbf{a}$ can be efficiently determined using Levinson's recursion. The solution is

$$\hat{\mathbf{a}} = (\mathbf{Y}^H\mathbf{Y})^{-1}\mathbf{Y}^H\mathbf{d}$$

The 'clean' data vector $\hat{\mathbf{d}}$ is obtained as

$$\hat{\mathbf{d}} = \mathbf{Y}\hat{\mathbf{a}}$$

$\hat{\mathbf{d}}$ constitutes the predicted data. The predicted noise sequence is

$$\hat{\mathbf{q}} = \hat{\mathbf{d}} - \mathbf{d}$$

In general, regularization (e.g., via ridge regression) is required to obtain a stable filter

5.2.3 The Convolution Matrix

We have adopted a very simple convolution matrix in the design of our filter via Levinson's recursion. Bear in mind, however, that other data matrices can also be used. The Canales (1984) original formulation uses the following model

$$\begin{pmatrix} Y_1 & 0 & 0 \\ Y_2 & Y_1 & 0 \\ Y_3 & Y_2 & Y_1 \\ Y_4 & Y_3 & Y_2 \\ 0 & Y_4 & Y_3 \\ 0 & 0 & Y_4 \end{pmatrix} \begin{pmatrix} a_1 \\ a_2 \\ a_3 \end{pmatrix} = \begin{pmatrix} Y_2 \\ Y_3 \\ Y_4 \\ 0 \\ 0 \end{pmatrix}$$

Ulrych and Clayton (1976b) proposed the transient-free convolution matrix, where zero extension is avoided. In the simple $p = 3$ example, the transient-free matrix formulation is given by

$$\begin{pmatrix} Y_3 & Y_2 & Y_1 \\ Y_4 & Y_3 & Y_2 \\ Y_5 & Y_4 & Y_3 \\ Y_6 & Y_5 & Y_4 \end{pmatrix} \begin{pmatrix} a_1 \\ a_2 \\ a_3 \end{pmatrix} = \begin{pmatrix} Y_4 \\ Y_5 \\ Y_6 \\ Y_7 \end{pmatrix}$$

The solution of the above system gives a filter essentially free of truncation artifacts. However, $\mathbf{Y}^H\mathbf{Y}$ is no longer Toeplitz and Levinson's recursion cannot be used.

It is important to mention that the above analysis involved only forward prediction filters. In other words, we are trying to predict the future based only upon the past. A more sophisticated scheme, as discussed in Section 2.4.4, involves the simultaneous minimization of a forward and backward prediction error. In this case, one can show that the system of equations in the transient-free case has the following form (we assume a filter of length p and a signal composed of N samples)

$$\begin{pmatrix} Y_p & \dots & Y_1 \\ . & \dots & . \\ . & \dots & . \\ . & \dots & . \\ Y_{N-1} & \dots & Y_{N-p} \\ Y_2^* & \dots & Y_{p+1}^* \\ . & \dots & . \\ . & \dots & . \\ . & \dots & . \\ Y_{N-p+1}^* & \dots & Y_N^* \end{pmatrix} \begin{pmatrix} a_1 \\ a_2 \\ . \\ . \\ . \\ a_p \end{pmatrix} = \begin{pmatrix} Y_{p+1} \\ . \\ . \\ . \\ Y_N \\ Y_1^* \\ . \\ . \\ . \\ Y_{N-p}^* \end{pmatrix}$$

5.2.4 Some Examples

Figs. 5.1 and 5.2 illustrate two simulations where the predictability of a single harmonic component in the time or space domains is investigated. Two dimensional simulations are displayed in Figs. 5.3 and 5.4.

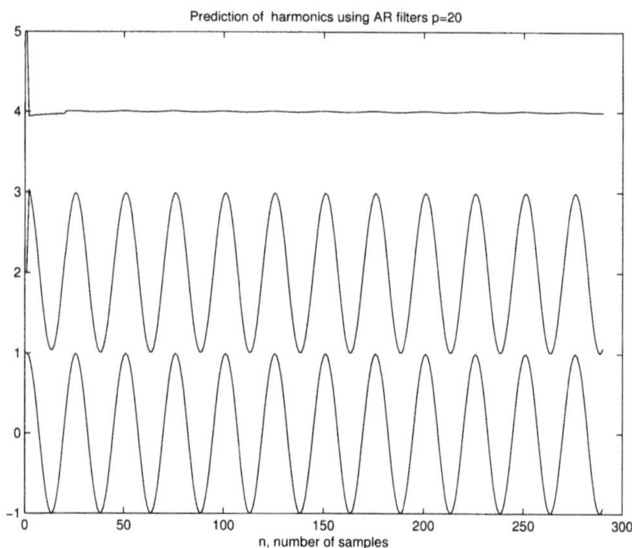

Figure 5.1: Prediction of a single harmonic (no noise) using an AR filter.

The $f-x$ noise attenuation algorithm can be summarized as follows:

1. Transform the data to the $f-x$ domain,
2. Estimate the AR prediction filter, $\mathbf{a}$, at each temporal frequency, f.
3. Convolve the filter with the transformed data (this is a complex convolution).
4. Transform back to the $t-x$ domain.

5.2.5 Nonlinear Events: Chirps in $f-x$

Linear events, in the transformed domain, are well modelled by an ARMA process. It is interesting to investigate the possibility of modelling nonlinear events which predominate seismic sections. We begin by considering a single event with parabolic moveout

$$s(t,x) = a(t-q\,x^2)$$

In the $f-x$ domain, we can write

$$S_n = A\exp(\beta n^2) \tag{5.6}$$

where the parameter $\beta = 2\pi f q \Delta x^2$ is the coefficient of the chirp. Although a simple recursion via a difference equation is not possible in this case, we can combine S_n, S_{n-1} and S_{n-2} to obtain the following recursion

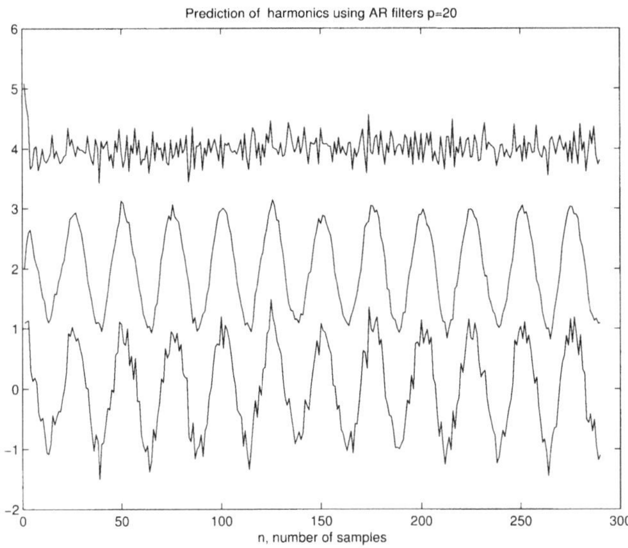

Figure 5.2: Prediction of a single harmonic ($\sigma_{noise} = 0.2$) using an AR filter.

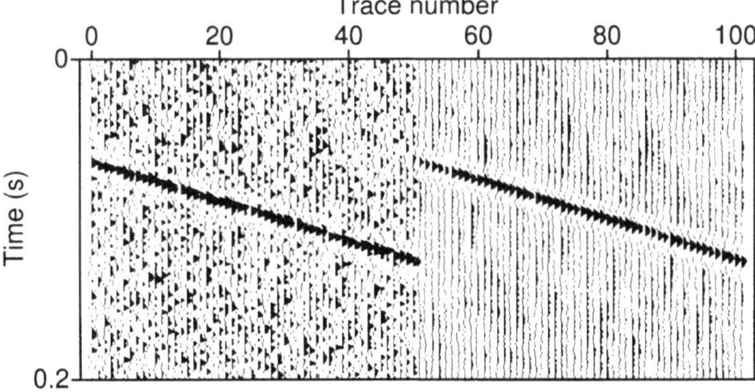

Figure 5.3: $f - x$ filtering of a single linear event immersed in noise.

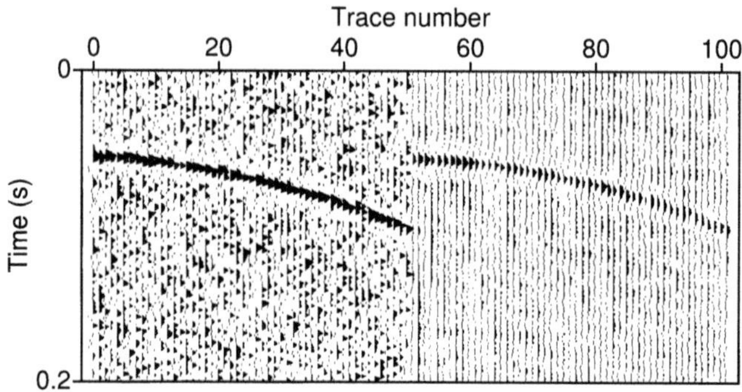

Figure 5.4: $f - x$ filtering of a single linear hyperbolic event immersed in noise.

$$S_n = \frac{bS_{n-1}^2}{S_{n-2}}$$

The nonlinearity is evident, implying that such signals cannot be optimally modelled via ARMA or AR models. So far, we have been unable to find a recursion (nonlinear) in the $f - x$ domain for a superposition of parabolic events immersed in white noise.

5.2.6 Gap Filling and Recovery of Near Offset Traces

The recovery of near offset traces is important in many procedures associated with seismic data processing. Prediction of missing traces in the frequency-offset domain depends on the assumption of linear moveout. As discussed in the previous section, when the signal exhibits nonlinear moveout, the $f - x$ domain is no longer composed of a finite set of harmonics. In fact, it is composed of a superposition of chirps, which makes conventional AR prediction a suboptimal solution. Specifically, for a single signal with parabolic moveout a non-linear recursion would be required. The latter is a very simple scenario, yet it provides an interesting framework to analyze the complexity of the problem.

We have developed a gap filling technique which, although based on linear prediction ideas, has shown considerable promise in the resolution of this difficult problem. This technique was first used by Wiggins and Miller (1972) for the prediction of glitches in earthquake records, and further developed by Fahlman and Ulrych (1982) for determining the power spectra of astronomical data. In this section we present a suboptimal solution to the problem of gap filling. First, we assume that the data in $f - x$ can be modelled by an AR model of order p and that the AR parameters have been estimated. In this case, we can transform the gap filling problem into an inverse problem. The prediction error for the AR model in terms of the AR coefficients is given by

$$E_n = g_0 Y_n + g_1 Y_{n-1} + g_2 Y_{n-2} + \ldots g_p Y_{n-p}$$

Since the gap comprises the points $Y_{l_1}, Y_{l_1+1}, \ldots Y_{l_2}$, we minimize $\sum_{k=l_1}^{l_2} E_k E_k{}^*$, the MSE in the gap. Taking derivatives with respect to the samples in the gap leads to the following system

$$\mathbf{R}_g [Y_{l_1}, Y_{l_1+1}, \ldots, Y_{l_2}]^T = \mathbf{b}$$

where $\mathbf{R}_g$ is the Toeplitz autocorrelation matrix of the PEO and $\mathbf{b}$ is a vector that depends on the data outside the gap and on the PEO, and is known.

So far we have assumed that the PEO is known. By means of various synthetic and field data experiments we have substantiated the strategy of Fahlman and Ulrych (1982) which is to estimate the PEO from consecutive traces to the left and to the right of the gap. The PEO that is utilized in the gap filling is the average PEO. A first pass can be used to fill the gap and a second pass, using the complete data, can be used to recalculate the PEO and the traces within the gap.

Figs. (5.6) and (5.7) show real data examples. In this case we have tested the algorithm with two land data shots. It is convenient to balance the energy of each trace to a common value before proceeding with the gap filling algorithm. In these examples traces $49, 50, \ldots, 56$ are missing. A 15 point PEO and a regularization parameter (1% of $\text{tr}[\mathbf{R}_g]$) were used. The stabilization will minimize the energy brought into the gap, and the recovered amplitudes within the gap may have to be balanced.

5.2.7 $f-x$ Projection Filters

Another approach to the problem of noise attenuation is by means of the concept of quasi-predictability, introduced by Soubaras (1994), and projection filters.

5.2.7.1 Wavenumber Domain Formulation

Quasi-predictability may be understood by considering a signal, $s(f,x)$ in $f-x$, defined as

$$s(f,x) = s_0(f,x) + n(f,x) \quad (5.7)$$

where $s_0(f,x)$ is the signal and $n(f,x)$ the additive noise. Consider the spatial prediction problem for a fixed frequency f. In the wavenumber domain, Eq. (5.7) can be written as

$$S(k) = S_0(k) + N(k) \quad (5.8)$$

where f has been omitted for simplicity. Now, assume that $S_0(k)$ can be predicted using a PEO, $A(k)$

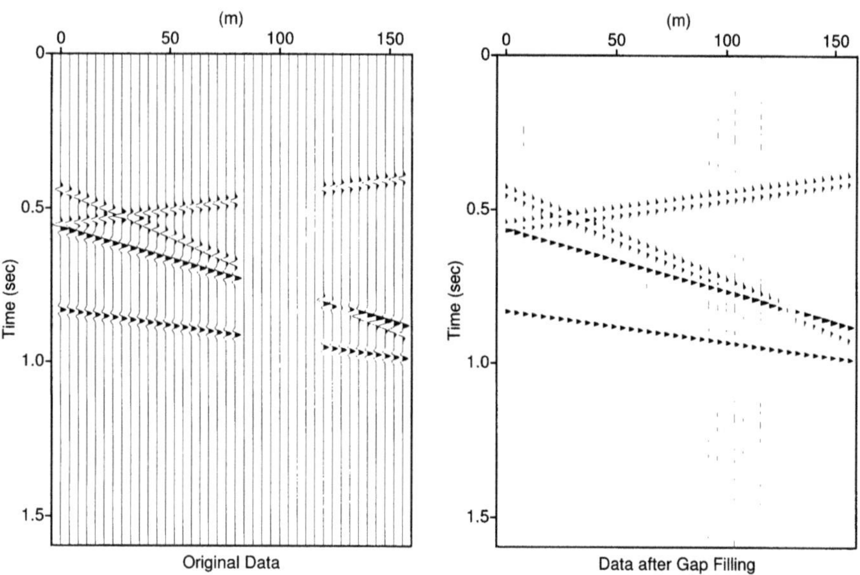

Figure 5.5: Predictability of linear events. The AR gap filling technique is used to estimate a PEO used to fill the gap.

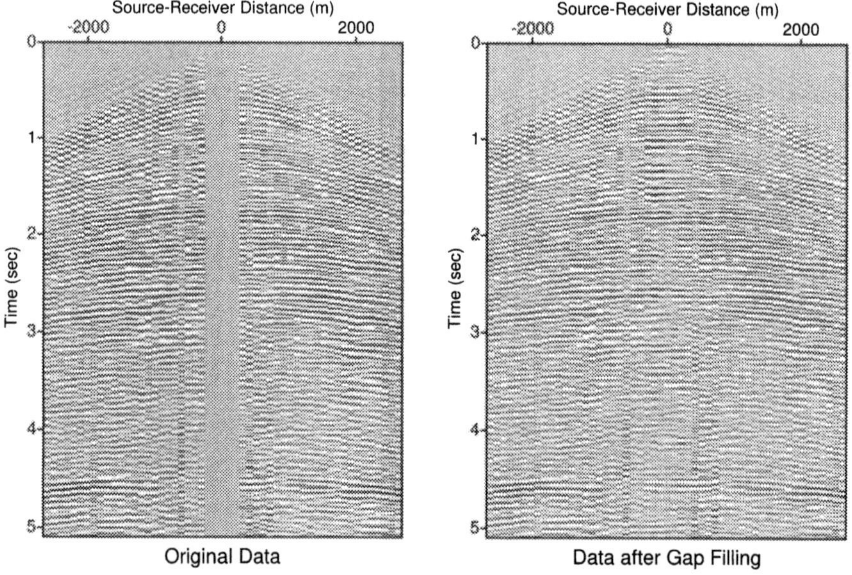

Figure 5.6: Original (left), after gap filling with a PEO(15).

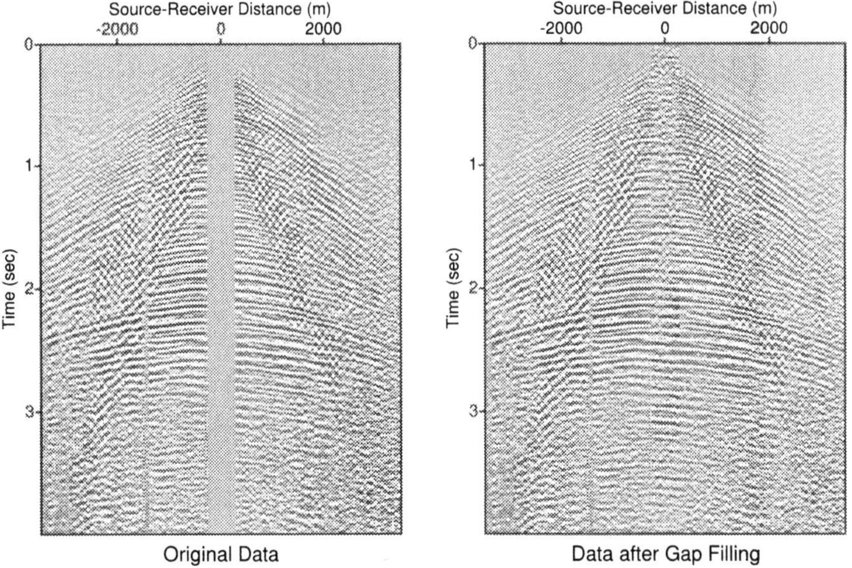

Figure 5.7: Original (left), after gap filling with a PEO(15).

$$A(k)S_0(k) \quad = \quad E(k) \tag{5.9}$$

where $E(k)$ is the prediction error or innovation. If $A(k)$ is known, we can construct a cost function, J, to find the noise sequence

$$J \quad = \quad |N(k)|^2 + \mu|E(k)|^2$$

J combines two desires, noise attenuation and predictability and may be rewritten, using Eqs. (5.8) and (5.9), as

$$J \quad = \quad |S(k) - S_0(k)|^2 + \mu|A(k)S_0(k)|^2$$

Note that, if $\mu = 0$, we obtain the trivial solution $N(k) = 0$, $S_0(k) = S(k)$ and noise attenuation is not realized. Minimizing J with respect to $S_0(k)$ obtains

$$S_0(k) \quad = \quad \frac{S(k)}{1 + \mu A(k)A(k)^*}$$

and the noise term becomes

$$N(k) \quad = \quad S(k) - S_0(k) = \frac{A(k)A(k)^*}{\mu^{-1} + A(k)A(k)^*}S(k) \tag{5.10}$$

Eq. (5.10) is Soubaras' (Soubaras, 1994) expression for the projection filter $M(k)$ where $N(k) = M(k)S(k)$. In other words

$$M(k) = \frac{A(k)A(k)^*}{\mu^{-1} + A(k)A(k)^*}$$

Summarizing, we have

1. when $\mu \to 0$
$$N(k) = 0, \quad S_0(k) = S(k), \quad E(k) = A(k)S(k)$$

2. when $\mu \to \infty$
$$N(k) = S(k), \quad S_0(k) = 0, \quad E(k) = 0.$$

3. All intermediate cases, $\mu \in [0, \infty)$, correspond to quasi-predictability (tradeoff between additive noise attenuation and predictability of signal).

5.2.7.2 Space Domain Formulation

Soubaras (1994) also derived a space domain expression for the projection filter

$$\mathbf{s} = \mathbf{s_0} + \mathbf{n}$$

$$\mathbf{A}\mathbf{s_0} = \mathbf{e}$$

In this case, $\mathbf{s}$ and the unknowns $\mathbf{s_0}$ and $\mathbf{n}$ are spatial vectors. $\mathbf{A}\mathbf{s_0}$ corresponds to a convolution with the PEO. The cost function for the problem becomes

$$J = (\mathbf{s} - \mathbf{s_0})^T(\mathbf{s} - \mathbf{s_0}) + \mu \mathbf{s_0}^T\mathbf{A}^T\mathbf{A}\mathbf{s_0} \tag{5.11}$$

and, taking derivatives, yields

$$\mathbf{s} = (\mathbf{I} + \mu\mathbf{A}^T\mathbf{A})\mathbf{s_0}$$

Since $\mathbf{s} = \mathbf{s_0} + \mathbf{n}$

$$\mathbf{s} = (\mathbf{I} + \mu\mathbf{A}^T\mathbf{A})(\mathbf{s} + \mathbf{n}) \tag{5.12}$$

and

$$\mathbf{n} = \mu(\mathbf{I} + \mu\mathbf{A}^T\mathbf{A})^{-1}\mathbf{A}^T\mathbf{A}\mathbf{s} \tag{5.13}$$

where, after using the identity

$$\mathbf{A}^T(\mathbf{I}+\mu\mathbf{A}\mathbf{A}^T)^{-1} = (\mathbf{I}+\mu\mathbf{A}\mathbf{A}^T)^{-1}\mathbf{A}^T$$

one obtains

$$\mathbf{n} = \mathbf{A}^T(\mu^{-1}\mathbf{I}+\mathbf{A}\mathbf{A}^T)^{-1}\mathbf{A}\mathbf{s} \tag{5.14}$$

The projection filter in fact deconvolves the PEO. To prove this, suppose that we know $\mathbf{A}$ and apply it to $\mathbf{s}$

$$\mathbf{A}\mathbf{s} = \mathbf{A}(\mathbf{s_0}+\mathbf{n}) = \mathbf{A}\mathbf{s_0}+\mathbf{A}\mathbf{n}$$

We see that the filter output has two components, the innovation, $\mathbf{A}\mathbf{s_0}$, which is zero if the noiseless signal is perfectly predictable, and the filtered noise sequence, $\mathbf{An}$. A careful look at Eq. (5.14) shows that the projection filter deconvolves the PEO signature from the noise sequence

$$\mathbf{n} = \underbrace{\mathbf{A}^T(\mu^{-1}\mathbf{I}+\mathbf{A}\mathbf{A}^T)^{-1}}_{2}\underbrace{(\mathbf{A}\mathbf{s})}_{1} \tag{5.15}$$

Operation 1 does the filtering of the signal and of the noise, operation 2 removes the color that the filter has introduced into the noise (deconvolution of the PEO). The term denoted 2 in Eq. (5.15) is also the pseudo-inverse of the PEO matrix, $\mathbf{A}$, defined here as $\mathbf{A}^+$.

5.2.7.3 A Wrong Formulation of the Problem

It is instructive to try another, apparently legal, formulation of the problem. Let us suppose, once again, that the PEO is known and is applied to $\mathbf{s}$

$$\mathbf{A}\mathbf{s} = \mathbf{A}(\mathbf{s_0}+\mathbf{n}) = \underbrace{\mathbf{A}\mathbf{s_0}}_{1}+\underbrace{\mathbf{A}\mathbf{n}}_{2} \tag{5.16}$$

It is evident that the desideratum is that term 1 be small (predictability) and term 2 also be small (noise attenuation). We may try to satisfy both wishes by defining the following objective function

$$J = \mathbf{s_0}\mathbf{A}^T\mathbf{A}\mathbf{s} + \mu(\mathbf{s}-\mathbf{s_0})^T\mathbf{A}^T\mathbf{A}(\mathbf{s}-\mathbf{s_0}) \tag{5.17}$$

Minimizing with respect to $\mathbf{s_0}$ yields

$$\mathbf{s_0} = \frac{\mu}{1+\mu}\mathbf{s} \tag{5.18}$$

Something is clearly wrong with Eq. (5.18). This is a consequence of having defined an incorrect objective function. In fact, what we want to minimize is the term denoted 1 in Eq. (5.16) plus the actual noise term, $\mathbf{n} = \mathbf{s}-\mathbf{s_0}$, and not the filtered noise,

An. The proper objective function for the problem is given by Eq. (5.11) and not by Eq. (5.17). Soubaras' projection filter is an attempt to estimate an optimum noise sequence from a given PEO. In other words, the projection filter is the solution of an inverse problem where an unfiltered noise sequence is the unknown.

5.2.8 ARMA Formulation of Projection Filters

We have seen that the underlying signal model for linear events in $t - x$ is an ARMA model in $f - x$. Thus far, we have done prediction and noise removal using two concepts: 1- Converting the ARMA problem into an AR problem and performing conventional linear prediction (Canales' method, Canales, 1984) and 2- Invoking the concept of quasi-predictability and solving a problem where both the PEO and additive noise are unknowns (Soubaras' method, Soubaras, 1994). One of the problems of AR filtering, as we have mentioned, is that the noise enters into the problem as an innovation rather than as additive noise. Soubaras solved this problem by introducing the theory of projection filters. In this section we show that projection filters can be computed by solving the original ARMA problem without introducing the concept of quasi-predictability. Let us return to the original system of equations for the ARMA formulation, Eq. (5.4), that we repeat here for convenience

$$
\begin{aligned}
\sum_{k=0}^{p} g_k Y_{n-k} &= \sum_{k=0}^{p} g_k Z_{n-k} \\
&= E_n
\end{aligned}
$$

an equation that describes the special ARMA(p,p) process with identical AR and MA components, where E_n is the non-white innovation sequence. The problem can be summarized as follows. Given the ARMA representation of the noise signal

1. How do we estimate the PEO, g_k?

2. How do we use g_k to estimate the additive noise sequence Z_k?

5.2.8.1 Estimation of the ARMA Prediction Error Filter

Writing Eq. (5.4) in matrix form

$$
\begin{aligned}
\mathbf{Yg} &= \mathbf{Zg} \\
&= \mathbf{e}
\end{aligned}
$$

where $\mathbf{Y}$ is the convolution matrix of the signal constructed to properly represent discrete convolution. For a signal of length $N = 4$ and a PEO of length 3 ($p = 2$), $\mathbf{Y}$ takes the form

$$\mathbf{Y} = \begin{pmatrix} Y_0 & 0 & 0 \\ Y_1 & Y_0 & 0 \\ Y_2 & Y_1 & Y_0 \\ Y_3 & Y_2 & Y_1 \\ 0 & Y_3 & Y_2 \\ 0 & 0 & Y_3 \end{pmatrix}$$

It is important to stress that the transient-free convolution matrix proposed by Ulrych and Clayton (1976b) can also be adopted (Harris and White, 1997). $\mathbf{Z}$ in Eq. (5.19) is the convolution matrix of the noise constructed in similar fashion.

Assuming that the noise is spatially uncorrelated and stationary, $\mathbf{g}$ can be estimated by transforming Eq. (5.19) into an eigenvalue problem as we did in Section 2.5.1. Since noise and signal are spatially uncorrelated, we obtain

$$\mathbf{R_{yy}g} = P_{\mathbf{z}}\mathbf{g} \tag{5.19}$$

where $\mathbf{R_{yy}} = \mathbf{Y}^H\mathbf{Y}$ is the Toeplitz covariance matrix of the noisy data. $P_{\mathbf{z}}$ represents the noise power. $\mathbf{R_{yy}}$ is positive definite with $p+1$ eigenvalues and eigenvectors (Marple, 1987) and Eq. (5.19) admits $p+1$ solutions. Each solution corresponds to an eigenvector and its associated eigenvalue (see Section 2.5.1).

Assuming that the i-th eigenvector is the solution to our problem, i.e., $\mathbf{g} = \mathbf{v}_i$, the energy in $\mathbf{e}$ is

$$\begin{aligned} \mathbf{e}^H\mathbf{e} &= \mathbf{v}_i^H\mathbf{R_{yy}}\mathbf{v}_i \\ &= P_{\mathbf{z}}\mathbf{v}_i^H\mathbf{v}_i \\ &= \lambda_i \end{aligned}$$

Since the eigenvectors are normalized, $\lambda_i = P_{\mathbf{z}}$. In other words, the eigenvector corresponding to the minimum eigenvalue minimizes the power of the additive white noise[1]. If the eigenvalues are sorted in descending order, the estimator of the PEO is $\mathbf{g} = \mathbf{v}_{p+1}$ and the minimum eigenvalue, λ_{p+1}, gives an estimate of the noise power. The other eigenvectors provide a solution that corresponds to local minima of the quadratic form, $\mathbf{e}^H\mathbf{e}$.

5.2.8.2 Noise Estimation

After estimating $\mathbf{g}$, the remaining problem is to estimate the noise sequence $\hat{Z}_k$ that is used to estimate the "clean" signal, $\hat{S}_k = Y_k - \hat{Z}_k$. The noise is estimated by

[1]As previously discussed, the eigenvector solution arises as the result of the often encountered problem, stated as follows: Find the vector $\mathbf{x}$ that maximizes or minimizes the quadratic form $\mathbf{x}^H\mathbf{A}\mathbf{x}$. Of course, since $\pm\infty$ are not interesting answers, a constraint in the form of $\mathbf{x}^H\mathbf{x} =$ constant must be added via a Lagrange multiplier. The solution is $\mathbf{A}\mathbf{x} = b\mathbf{x}$ and shows that $\mathbf{x}$ is the eigenvector corresponding to b, the eigenvalue. Maximization leads to the maximum eigenvalue, minimization to the minimum.

deconvolving the PEO from the non-white innovation in Eq. (5.19). In order to facilitate the algebra, we rewrite Eq. (5.19) by commuting the sequences involved in the convolution

$$\mathbf{G}\mathbf{y} = \mathbf{G}\mathbf{z} \tag{5.20}$$

where $\mathbf{G}$ is the convolution matrix of the PEO, and $\mathbf{y}$ and $\mathbf{z}$ are vectors containing the observations and the white noise sequence, respectively. We stress that, after computing the PEO, the RHS term in Eq. (5.20) is known. A trivial solution of this equation is $\hat{\mathbf{z}} = \mathbf{y}$. This solution implies that $\mathbf{s} = 0$ a shortcoming that is avoided by solving a constrained minimization problem

$$J = [\mathbf{G}(\mathbf{y} - \mathbf{z})]^H [\mathbf{G}(\mathbf{y} - \mathbf{z})] + \mu(\mathbf{z}^H \mathbf{z} - P_Z) \tag{5.21}$$

J is minimized when

$$\hat{\mathbf{z}} = (\mathbf{G}^H \mathbf{G} + \mu \mathbf{I})^{-1} \mathbf{G}^H \mathbf{G} \mathbf{y} \tag{5.22}$$

and the 'clean signal', $\hat{\mathbf{s}} = \mathbf{y} - \hat{\mathbf{z}}$, is estimated as

$$(5.23)$$

$$\hat{\mathbf{s}} = [\mathbf{I} - (\mathbf{G}^H \mathbf{G} + \mu \mathbf{I})^{-1} \mathbf{G}^H \mathbf{G}] \mathbf{y} \tag{5.24}$$

Some comments are in order. First, we note that when $\mu = 0$, $\hat{\mathbf{z}} = \mathbf{y}$. In other words, the signal has been annihilated ($\hat{\mathbf{s}} = 0$). If μ is too large, $\hat{\mathbf{z}} = \mathbf{0}$. In this case, there is no noise attenuation ($\hat{\mathbf{s}} = \mathbf{y}$), and represents the case of perfect predictability in the absence of noise, where Eq. (5.20) becomes $\mathbf{G}\mathbf{s} = \mathbf{0}$. In general, a line search procedure could be used to determine the value of μ that yields a noise sequence with MSE that agrees with the estimated noise power obtained after solving Eq. (5.19). In practice, μ is determined by assessing the quality of the final results in the same way as is done in seismic deconvolution.

5.2.9 ARMA and Projection Filters

Our algorithm is composed of two stages:

1. Determination of the prediction error filter or eigenfilter.

2. Noise estimation, that is equivalent to the projection filtering technique proposed by Soubaras (1994).

In Soubaras' technique the noise is estimated by minimizing an objective function that establishes a tradeoff between predictability (or pseudo-predictability) and random noise attenuation. Soubaras first assumes that the PEO is known and then uses Eq. (5.22) to determine the noise sequence. The first term is associated with signal predictability, the second with random noise attenuation. In the projection filter approach the PEO is bootstrapped from the data in an iterative manner. The

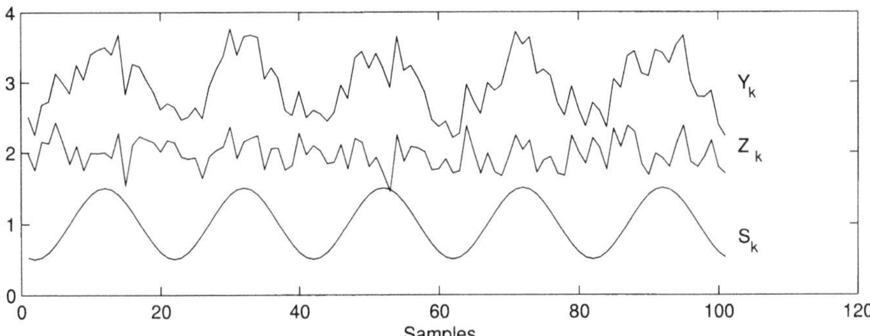

Figure 5.8: 1D synthetic example. A sinusoid of normalized wavenumber $k_0 = 0.05$ is used to test the ARMA filtering method. The data are contaminated with white noise, $\sigma = 0.5$.

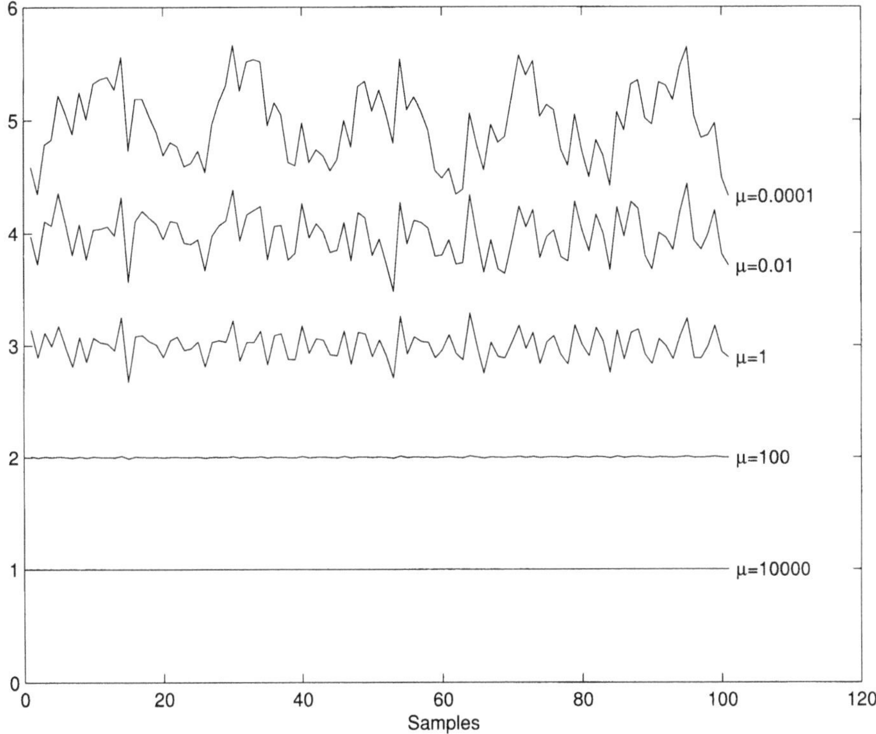

Figure 5.9: The eigenfilter, estimated from the noisy data, is used to estimate the noise. Figure portrays the estimated of the noise sequence vs. μ. Large values of μ completely annihilate the noise. Small values overestimate the noise (the signal leaks into the noise sequence.)

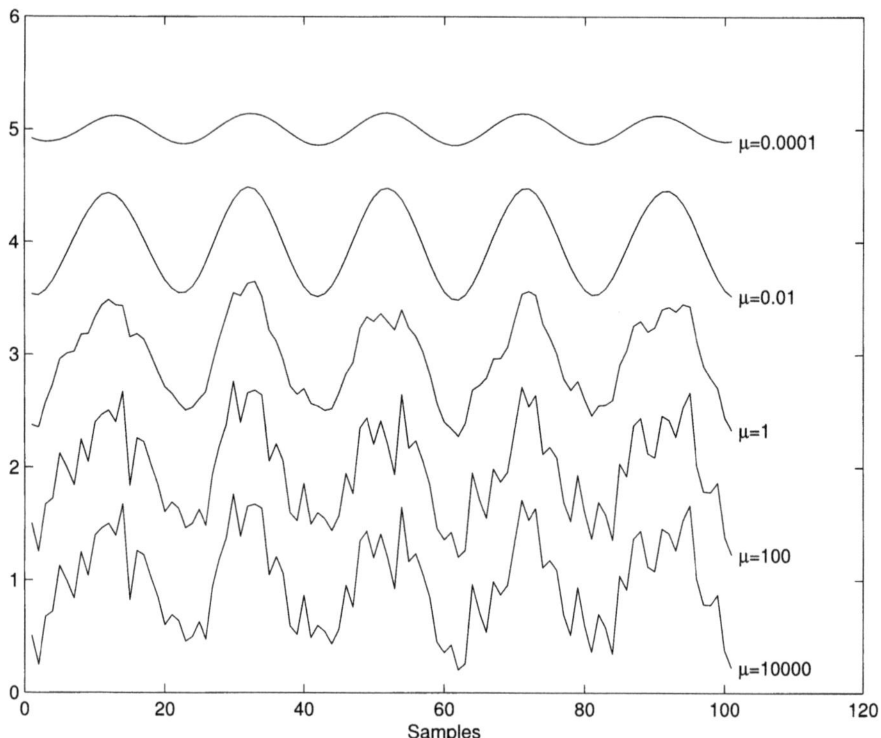

Figure 5.10: Estimates of the signal vs. μ. The signal is recovered for $\mu = 0.01$.

ARMA representation of the seismic signal in the $f-x$ domain, on the other hand, leads not only to a technique to estimate the PEO, but also to the projection filter estimator of the noise sequence.

In Fig. 5.8 we present a 1D synthetic example. The signal is composed of one real sinusoid plus white noise (with standard deviation $\sigma = 0.5$). The number of signals is $p = 2$ (1 complex sinusoid is represented by 2 real harmonics). We have estimated the noise sequence using different tradeoff parameters, illustrated in Fig. 5.9. Fig. 5.10 shows the estimated signal versus the tradeoff parameter μ. The optimum value is $\mu = 0.01$ and leads to a noise sequence with $\hat{\sigma} = 0.53$. The transfer function of the projection filter used in the example is shown in Fig. 5.11.

The wavenumber response of the projection filter is given by

$$A(k) = \frac{|G(k)|^2}{\mu + |G(k)|^2} \tag{5.25}$$

where $G(k)$ is the FT of the eigenfilter. In this example the eigenfilter has a root on the unit circle at $k_0 = 0.05$ and, therefore, Eq. (5.25) represents a notch filter that eliminates that particular spectral component. When the proper value of μ is chosen, the transfer function, $A(k)$, selectively attenuates the signal letting only the

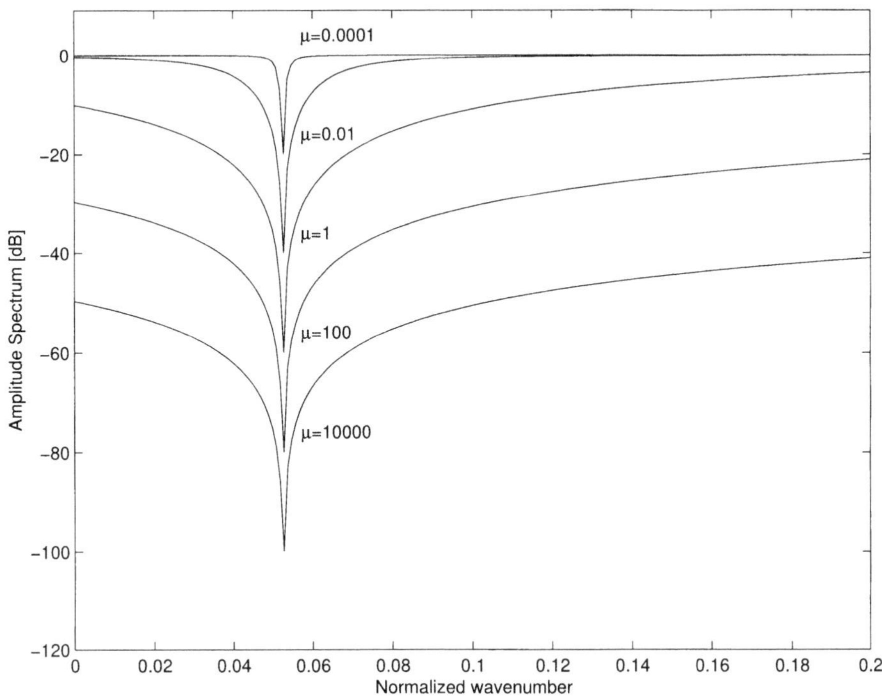

Figure 5.11: Amplitude response of the projection filter used to estimate e the noise in Fig. 5.9. Large values of μ attenuate both signal and noise. Small value of μ do not properly attenuate the noise.

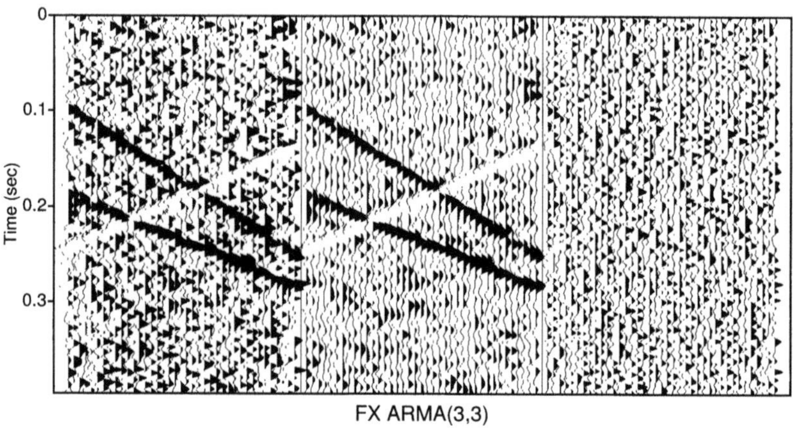

Figure 5.12: Left: The data consisting of 3 waveforms immersed in spatially uncorrelated noise. Center: Filtered data using the ARMA representation. Right: Estimate of the noise.

noise pass. It is clear that when μ is increased the transfer function will attenuate both signal and noise.

A comparison of the Canales' $f-x$ technique, using a 15 point PEO, with the $f-x$ ARMA approach using a 4 point eigenfilter ($p=3$), is shown in Figs. 5.12 and 5.13. Although the results look quite similar, one can observe that, in the AR formulation, a portion of the signal leaks into the noise space. Fig. 5.14 provides a field data example. We have used both the AR and ARMA methods to process a common offset section. The fact that the two methods produce similar results [2] leads us to conclude that, in real data applications, both models do not properly represent the true underlying signal.

5.2.10 Discussion

There is undeniable elegance in the $f-x$ approach to SNR enhancement. The realization by Canales (1984), that linearity in $t-x$ implies a sum of harmonics at each frequency and hence allows AR modelling in $f-x$, is quite beautiful. The problem arises, of course, when events are nonlinear. We have explored the possibility of finding appropriate models for this nonlinear case, but, thus far, none have appeared. The best manner to proceed, in our opinion, is to model the nonlinear events as approximately linear in suitable windows and derive the special ARMA models that incorporate the presence of additive random noise.

[2]For the ARMA method, $p=2$ and $\mu=0.1$. The AR approach used a forward/backward PEO with $p=15$.

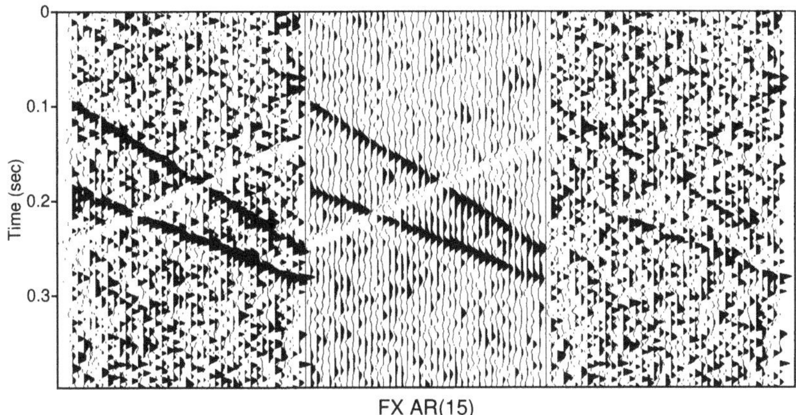

Figure 5.13: Left: The data consisting of 3 waveforms immersed in spatially uncorrelated noise. Center: Filtered data using the AR representation (Conventional $f-x$ random noise attenuation). Right: Estimate of the noise.

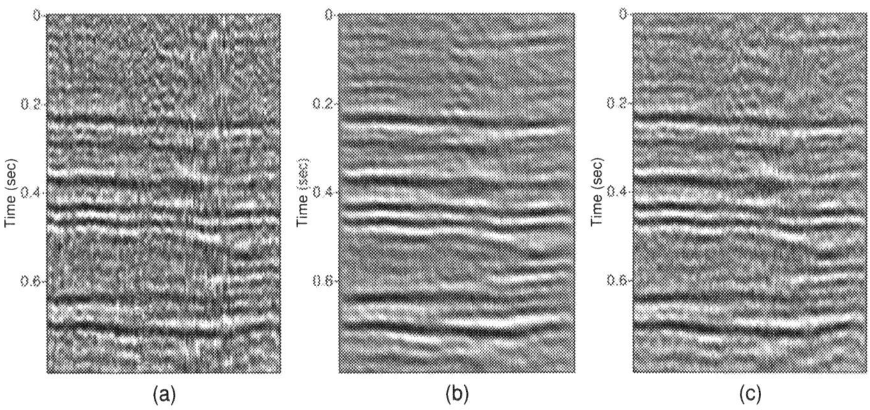

Figure 5.14: (a) A window from a common offset gather. (b) The filtered section using the ARMA representation with a 3 points eigenfilter ($p=2$). (c) Conventional $f-x$ random noise attenuation ($p=15$).

5.3 Principal Components, Eigenimages and the KL Transform

5.3.1 Introduction

This section describes an application of a particularly well known statistical technique, the decomposition into principle components, to data processing. Although we will be mostly concerned with examples in seismology, the technique which, basically, attempts to maximize the information content in recorded data, has a wide range of applicability.

We begin with a little history. The first publication which introduced and applied aspects of principle component analysis, PCA, to seismic data was the paper by Hemon and Mace (1978). These authors investigated the application of a particular linear transformation known as the Karhunen-Loève, KL , transformation. The KL transformation is also known as the principal component, PC, transformation, the eigenvector transformation, or the Hotelling transformation (Anderson, 1971; Loéve, 1977). It has been used by various authors for the purpose of one and two dimensional data compression and as a means of feature selection for pattern recognition. Of particular relevance to the ensuing discussion is the excellent paper by Ready and Wintz (1973) which deals with information extraction and SNR improvement in multispectral imagery. In 1983, the work of Hemon and Mace was extended by a group at the University of British Columbia which culminated in the work of Jones and Levy (1987).

Freire and Ulrych (1988) applied the KL transformation in a somewhat different manner to the processing of vertical seismic profiling data. The actual approach which was adopted in this work was by means of singular value decomposition, which is another way of viewing the KL transformation (the relationship between the KL and SVD transformations is discussed in Section 5.3.3). In later works, (Ulrych et al., 1999, please see), the SVD approach was applied to various other problems, including the attenuation of multiple reflections, and the nomenclature of eigenimage decomposition was adopted for this method of data processing. Eigenimages were first introduced into the literature by Andrews and Hunt (1977) in the context of image processing and, in our opinion, this description is one which is the most succinct for the purpose at hand.

PCA has a long history in other fields as well. To mention a couple only, multi spectral imaging (Richards, 1993) has seen a plethora of applications where the main focus of PCA is data compression. A recent application of much importance, particularly in light of the events of Sept 11th, is to the problem of facial recognition (Pentland and Turk, 1991). In fact, it is this field that motivated us into the extension of 2D eigenimages to 3D eigensections described in Section 5.3.7.

5.3.1.1 PCA and a Probabilistic Formulation

PCA has been mainly used in the statistical literature for dimension reduction. Other applications are, for example, in the fields of pattern recognition and image processing. The common method of deriving the PCA formalism is in terms of the maximization of the variance of a standardized linear projection.

Consider a set of N observed data vectors, $\mathbf{x}_j$, $j = 1, 2, \ldots, N$, of length L. The aim of PCA is to obtain the M, $M \leq N$, orthogonal principal axes, $\mathbf{w}_j, j = 1, 2, \ldots, M$ onto which the retained variance under projection is maximal (under the constraint that $\mathbf{w}_j^T \mathbf{w}_j = 1$). It may be shown (see, for example Cooley and Lohnes, 1971) that the $\mathbf{w}_j$ are given by the M eigenvectors corresponding to the largest eigenvalues (called the dominant eigenvectors) of the sample covariance matrix, $\mathbf{C_{xx}}$, such that

$$\mathbf{C_{xx}} \mathbf{w}_j = \lambda_j \mathbf{w}_j$$

Because of the imposed orthogonality, the M principal components, PCs, are given by

$$\mathbf{y}_j = \mathbf{W}^T (\mathbf{x}_j - \bar{\mathbf{x}})$$

where $\mathbf{W}^T = [\mathbf{w}_1, \mathbf{w}_2, \ldots, \mathbf{w}_M]$ and $\bar{\mathbf{x}}$ is the mean vector computed from the set of N vectors $\mathbf{x}_j$. The PCs have the property that they are decorrelated, in the sense that $E[\mathbf{y}\mathbf{y}^T]$ is a diagonal matrix with elements λ_j.

A complementary and important derivation of the PCs (Ahmed and Rao, 1975) is to look for that linear and orthogonal projection that minimizes the reconstruction error, E_{MSE}, given by

$$E_{MSE} = \sum_{k=1}^{K} ||\mathbf{x}_k - \hat{\mathbf{x}}_k||^2, \quad K < N$$

where $\hat{\mathbf{x}}_k = \mathbf{W}_K \mathbf{y}_k + \bar{\mathbf{x}}$ and $\mathbf{W}_K = [\mathbf{w}_1, \mathbf{w}_2, \ldots, \mathbf{w}_K]$.

Hidden in both derivations is the fact that PCs obey a probabilistic interpretation (Tipping and Bishop, 1999). We will not repeat here the derivation of Tipping and Bishop (1999) which shows how PCA may be viewed as a maximum likelihood procedure based on a pdf model of the observed data. Briefly, however, we follow the above authors by considering a general latent variable model (which is of use when one considers nonlinear PCs) of the form

$$\mathbf{x} = \mathbf{g}(\mathbf{y}, \boldsymbol{\theta}) + \boldsymbol{\epsilon} \tag{5.26}$$

where $\mathbf{g}(\mathbf{y}, \boldsymbol{\theta})$ is a function of the latent variable vector, $\mathbf{y}$, and parameters denoted by $\boldsymbol{\theta}$. $\boldsymbol{\epsilon}$ is a noise process that is independent of $\mathbf{y}$. We have omitted subscripts for the sake of notational simplicity. Assuming that the mapping is linear, yields

$$\mathbf{x} = \mathbf{W}\mathbf{y} + \boldsymbol{\epsilon} \tag{5.27}$$

Assuming further, in the spirit of MaxEnt, that $\boldsymbol{\epsilon}$ is isotropically Gaussian, i.e., $p(\boldsymbol{\epsilon}) \sim N(\mathbf{0}, \sigma_{\boldsymbol{\epsilon}}^2 \mathbf{I})$, we can write the likelihood as

$$L(\mathbf{x}|\mathbf{y}) \;\sim\; \exp\left(-\frac{1}{2\sigma_{\boldsymbol{\epsilon}}^2} ||\mathbf{x} - \mathbf{W}\mathbf{y} - \bar{\mathbf{x}}||^2\right) \tag{5.28}$$

(5.29)

With a Gaussian prior over $\mathbf{y}$ and by the use of Bayes theorem, Tipping and Bishop (1999) obtain the Normal posterior pdf, $p(\mathbf{y}|\mathbf{x})$, given by

$$p(\mathbf{y}|\mathbf{x}) \sim N(\mathbf{M}^{-1}\mathbf{W}^T(\mathbf{x}-\bar{\mathbf{x}}), \sigma_\epsilon^2\mathbf{M}^{-1}) \tag{5.30}$$

where $\mathbf{M}$ is the posterior covariance matrix, given by

$$\mathbf{M} = \mathbf{W}^T\mathbf{W} + \sigma_\epsilon^2\mathbf{I} \tag{5.31}$$

The Gaussian form of Eq. (5.30) is expected, of course, and is in accord with the fact that both of the alternative derivations of PCA follow the principle of least squares.

5.3.2 Eigenimages

We now consider eigenimages which describe the decomposition of 2D images in the form of data matrices. A seismic section which consists of M traces with N points per trace may be viewed as a data matrix $\mathbf{X}$ where each element x_{ij} represents the ith point of the jth trace. A singular value decomposition decomposes $\mathbf{X}$ into a weighted sum of orthogonal rank one matrices which have been designated by Andrews and Hunt (1977) as eigenimages of $\mathbf{X}$, i.e.,

$$\mathbf{E}_i = \mathbf{u}_i\mathbf{v}_i^T. \tag{5.32}$$

A particularly useful aspect of the eigenimage decomposition is its application in the complex form. In this instance, if each trace is transformed into the analytic form, then the eigenimage processing of the complex data matrix allows both time and phase shifts to be considered, certainly of relevance in applications such of the correction of residual statics. We first develop the required theoretical principles, illustrate with some applications, compare the eigenimage technique to the well known frequency-wavenumber, $f-k$, method (Treitel et al., 1967) and discuss important differences which arise especially with respect to spacial aliasing and the separation of signal and noise.

As discussed in Appendix A.1 of Chapter 4, we write the SVD of the $M \times N$ data matrix $\mathbf{X}$, where we assume without loss of generality that $M < N$ and the M traces form the rows of $\mathbf{X}$, as

$$\mathbf{X} = \mathbf{U}\boldsymbol{\Sigma}\mathbf{V}^T \tag{5.33}$$

$$= \sum_{i=1}^{r} \sigma_i\mathbf{u}_i\mathbf{v}_i^T \tag{5.34}$$

where r is the rank of $\mathbf{X}$, $\mathbf{u}_i$ is the ith eigenvector of $\mathbf{X}\mathbf{X}^T$, $\mathbf{v}_i$ is the ith eigenvector of $\mathbf{X}^T\mathbf{X}$ and σ_i is the ith singular value of $\mathbf{X}$. Andrews and Hunt (1977) designate the outer dot product $\mathbf{u}_i\mathbf{v}_i^T$ as the ith eigenimage of the matrix $\mathbf{X}$. Owing to the orthonormality of the eigenvectors, the eigenimages form an orthonormal basis which

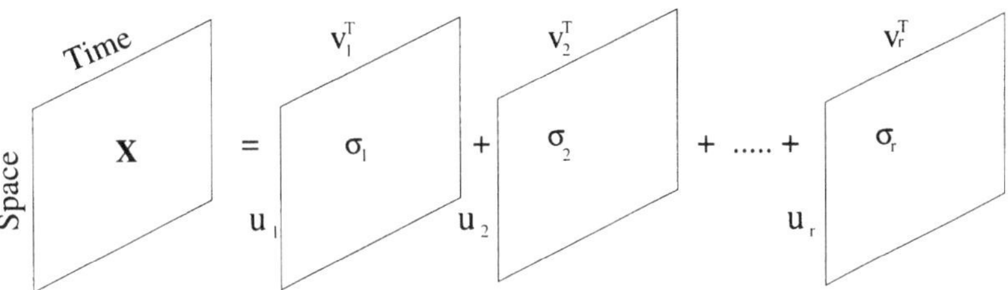

Figure 5.15: Eigendecomposition of **X** into the sum of weighted eigenimages.

may be used to reconstruct **X** according to Eq. (5.34). This concept is illustrated diagrammatically in Fig. 5.15

It is clear from Eq. (5.34) that the contribution of a particular eigenimage in the reconstruction of **X** is proportional to the magnitude of the associated singular value. Since in the SVD, the singular values are always ordered with decreasing magnitude, it is possible, depending of course on the data, to reconstruct the matrix **X** using only the first few eigenimages. Suppose, for example, that **X** represents a seismic section and that all M traces are linearly independent. In this case **X** is of full rank M, all the σ_i are different from zero and a perfect reconstruction of **X** requires all eigenimages. On the other hand, in the case where all M traces are equal to within a scale factor, all traces are linearly dependent, **X** is of rank one and may be perfectly reconstructed by the first eigenimage $\sigma_1 \mathbf{u}_1 \mathbf{v}_1^T$. In the general case, depending on the linear dependence which exists among the traces, **X** may be reconstructed from only the first few eigenimages. In this case, the data may be considered to be composed of traces which show a high degree of trace-to-trace correlation. Indeed, $\mathbf{X}\mathbf{X}^T$ is, of course, a weighted estimate of the zero lag covariance matrix of the data **X** and the structure of this covariance matrix, particularly the distribution of the magnitudes of the corresponding eigenvalues, indicates the parsimony or otherwise of the eigenimage decomposition. If only $p, p < r$, eigenimages are used to approximate **X**, a reconstruction error ϵ is given by

$$\epsilon = \sum_{k=p+1}^{r} \sigma_k^2 \,.$$

Freire and Ulrych (1988) defined band-pass $\mathbf{X}_{BP}$, low-pass $\mathbf{X}_{LP}$ and high-pass $\mathbf{X}_{HP}$ eigenimages in terms of the ranges of singular values used. The band-pass image is reconstructed by rejecting highly correlated as well as highly uncorrelated traces and is given by

$$\mathbf{X}_{BP} = \sum_{i=p}^{q} \sigma_i \mathbf{u}_i \mathbf{v}_i^T \qquad 1 < p \leq q < r \tag{5.35}$$

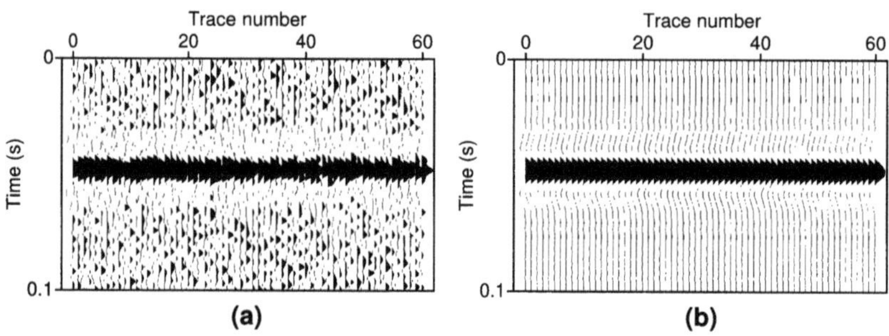

Figure 5.16: A flat event immersed in noise and the reconstruction using only the first eigenimage

The summation for $\mathbf{X}_{LP}$ is from $i = 1$ to $p - 1$ and for $\mathbf{X}_{HP}$ from $i = q + 1$ to r. It may be simply shown that the percentage of the energy which is contained in a reconstructed image $\mathbf{X}_{BP}$ is given by E, where

$$E = \frac{\sum_{i=p}^{q} \sigma_i^2}{\sum_{i=1}^{r} \sigma_i^2} . \tag{5.36}$$

The choice of p and q depend on the relative magnitudes of the singular values, which are a function of the input data. These parameters may, in general, be estimated from a plot of the eigenvalues $\lambda_i = \sigma_i^2$ as a function of the index i. This is reasonable given the form of Eq. (5.36). In certain cases, an abrupt change in the eigenvalues is easily recognized. In other cases, the change in eigenvalue magnitude is more gradual and care must be exercised in the choice of the appropriate index values.

Figs. 5.16 and 5.17 illustrate the reconstruction of a flat event immersed in noise using the first eigenimage of the data. In this example only the most energetic singular value was retained. When the data exhibit some type of moveout, one eigenimage is not sufficient to properly reconstruct the data. This can be observed in Figs. 5.18 and 5.19.

5.3.3 Eigenimages and the KL Transformation

As we have seen, decomposition of an image $\mathbf{X}$ into eigenimages is performed by means of the SVD of $\mathbf{X}$. Many authors also refer to this decomposition as the Karhunen-Loève or KL transformation. We believe however, that the SVD and KL approaches are not equivalent theoretically for image processing and, in order to avoid confusion, we suggest the adoption of the term eigenimage processing. Some clarification is in order.

A wide sense stationary process $\xi(t)$ allows the expansion

$$\hat{\xi}(t) = \sum_{n=1}^{\infty} c_n \psi_n(t) \quad 0 < t < T \tag{5.37}$$

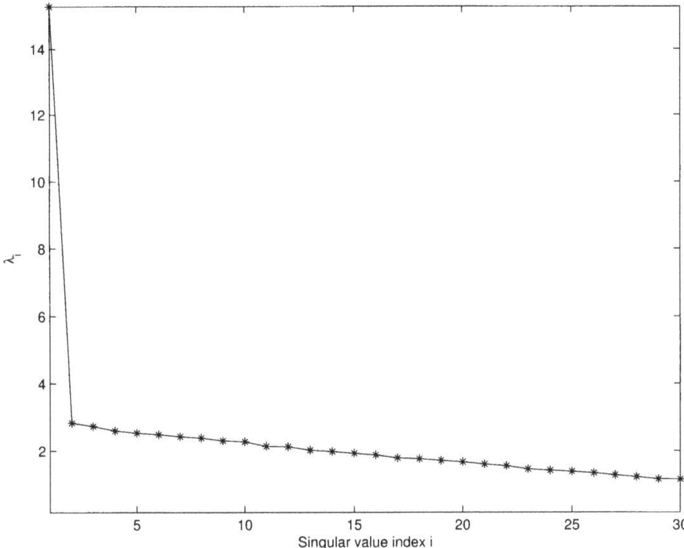

Figure 5.17: Singular value spectrum for the data in Fig. 5.16.

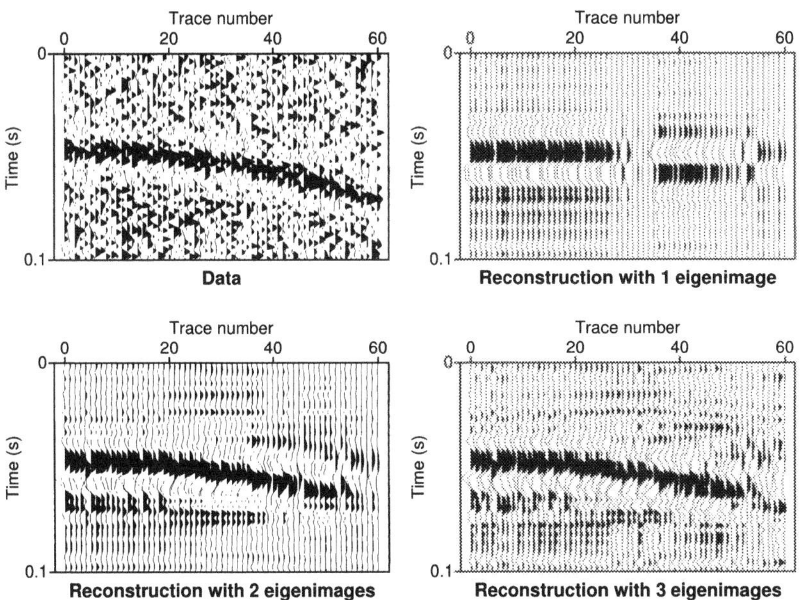

Figure 5.18: A parabolic event immersed in noise and the reconstruction using the first 3 eigenimages

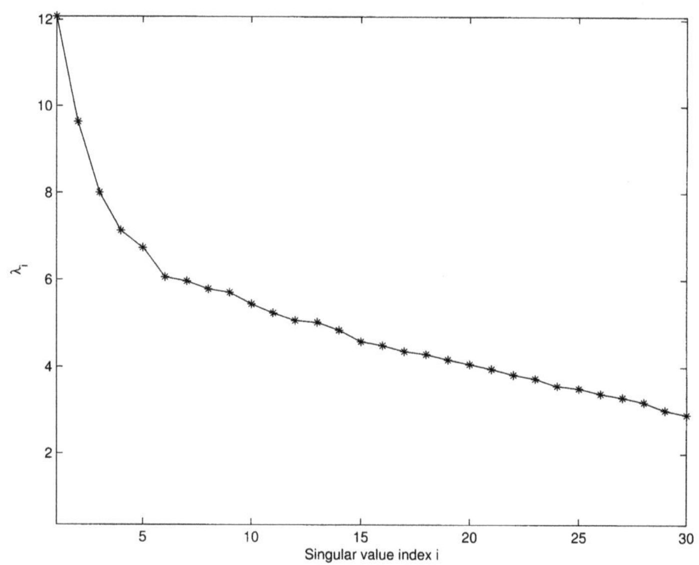

Figure 5.19: Singular value spectrum for the data in Fig. 5.18.

where $\psi_n(t)$ is a set of orthonormal functions in the interval $(0, T)$ and the coefficients c_n are random variables. The Fourier series is a special case of the expansion given by Eq. (5.37) and it can be shown that, in this case, $\xi(t) = \hat{\xi}(t)$ for every t and the coefficients c_n are uncorrelated only when $\xi(t)$ is mean squared periodic. Otherwise, $\xi(t) = \hat{\xi}(t)$ only for $|t| < T/2$ and the coefficients c_n are no longer uncorrelated. In order to guarantee that the c_n are uncorrelated and that $\xi(t) = \hat{\xi}(t)$ for every t without the requirement of mean squared periodicity, it turns out that the $\psi_n(t)$ must be determined from the solution of the integral equation

$$\int_0^T R(t_1, t_2)\psi(t_2)dt_2 = \lambda\psi(t_1) \quad 0 < t_1 < T \tag{5.38}$$

where $R(t_1, t_2)$ is the autocovariance of the process $\xi(t)$. Substituting the eigenvectors, which are the solutions of Eq. (5.38), into Eq. (5.37) gives the KL expansion of $\xi(t)$. An infinite number of basis functions is required to form a complete set. For a $N \times 1$ random vector, $\mathbf{x}$, we may write Eq. (5.37) in terms of a linear combination of orthonormal basis vectors, $\mathbf{w}_i = [w_{i1}, w_{i2}, \ldots, w_{iN}]^T$ as

$$x_k = \sum_i y_i w_{ik} \quad k = 1, 2, \ldots, N \tag{5.39}$$

which is equivalent to

$$\mathbf{x} = \mathbf{W}\mathbf{y} \tag{5.40}$$

where $\mathbf{W} = [\mathbf{w}_1, \mathbf{w}_2, \ldots, \mathbf{w}_N]$. Now only N basis vectors are required for completeness. The KL transformation or, as it is also often called, the KL transformation to principal components, is obtained as

$$\mathbf{y} = \mathbf{W}^T\mathbf{x} \tag{5.41}$$

where $\mathbf{W}$ is determined from the covariance matrix $\mathbf{C_{xx}}$ of the process

$$\mathbf{C_{xx}} = \mathbf{W\Lambda W}^T \tag{5.42}$$

where $\mathbf{\Lambda}$ is the diagonal matrix of eigenvalues.

5.3.3.1 Eigenimages and Entropy

A particularly interesting property of the transformation described by Eq.(5.41) is its relationship to the entropy of a stochastic process. With the discussion concerning entropy in Chapter 3 in mind, we reiterate here the definition of the entropy of a process, $\mathbf{x}$, assuming jointly Gaussian variables with zero mean (please also see the discussion in Section 3.2.1). Defining $p(\mathbf{x})$ to be the Gaussian density function, it follows that the entropy $H(\mathbf{x})$ is given by

$$\begin{aligned} H(\mathbf{x}) &= -\int p(\mathbf{x}) \log p(\mathbf{x}) d\mathbf{x} \\ &= \int p(\mathbf{x}) \left[\frac{N}{2} \log 2\pi + \frac{1}{2} \log |\mathbf{C_{xx}}| + \frac{1}{2} \mathbf{x}^T \mathbf{C_{xx}^{-1}} \mathbf{x} \right] d\mathbf{x} \end{aligned}$$

where $|\mathbf{C_{xx}}|$ is the determinant of $\mathbf{C_{xx}}$. Taking expectations, obtains

$$E[\mathbf{x}\mathbf{C_{xx}^{-1}}\mathbf{x}^T] = E[\mathrm{tr}[\mathbf{C_{xx}^{-1}}]\mathbf{x}\mathbf{x}^T] = \mathrm{tr}[\mathbf{I}] = \frac{N}{2}$$

and allows us to write

$$H(\mathbf{x}) = \frac{1}{2} \log |\mathbf{C_{xx}}| + \frac{N}{2} \log 2\pi e$$

Since the determinant of a matrix is given by the product of its eigenvalues, the above equation becomes

$$H(\mathbf{x}) = \frac{1}{2} \sum_{i=1}^{N} \log \lambda_i + \frac{N}{2} \log 2\pi e$$

which defines the entropy of the Gaussian process $\mathbf{x}$. Using the above definition of the entropy, Young and Calvert (1974) show that, given a positive definite matrix $\mathbf{W}$, maximizing the entropy of $\mathbf{y}$ in Eq. (5.41) subject to $\mathbf{w}_i^T\mathbf{w}_i = 1, \quad i = 1, 2, \ldots, M$, results in $\mathbf{W} = \mathbf{U}$ where $\mathbf{U}$ is obtained from $\mathbf{X}\mathbf{X}^T = \mathbf{U}\boldsymbol{\Sigma}^2\mathbf{U}^T$. In other words, principal component transformation constrains the output vector to have maximum entropy.

5.3.3.2 KL Transformation in Multivariate Statistics

Let us now turn our attention to the issue of the KL transformation in multivariate statistical analysis. In this case we consider M vectors $\mathbf{x}_i, \quad i = 1, 2, \ldots, M$ arranged in a $M \times N$ data matrix, $\mathbf{X}$. The M rows of the data matrix are viewed as M realizations of the stochastic process, $\mathbf{x}$, and consequently the assumption is that all rows have the same row covariance matrix $\mathbf{C}_r$. The KL transform now becomes

$$\mathbf{Y} = \mathbf{W}^T\mathbf{X} \tag{5.43}$$

Assuming a zero mean process for convenience, an unbiased estimate of the row covariance matrix is given by

$$\hat{\mathbf{C}}_r = \frac{1}{M-1}\sum_i \mathbf{x}_i\mathbf{x}_i^T \tag{5.44}$$

Since the factor $M-1$ does not influence the eigenvectors, we can see from Eq. (5.44) and the definition of $\mathbf{U}$, that $\mathbf{W} = \mathbf{U}$. Consequently, we can rewrite Eq. (5.43) as

$$\mathbf{Y} = \mathbf{U}^T\mathbf{X} \tag{5.45}$$

Substituting Eq. (5.33) into Eq. (5.44), we obtain

$$\begin{aligned}\mathbf{Y} &= \mathbf{U}^T\mathbf{U}\boldsymbol{\Sigma}\mathbf{V}^T \\ &= \boldsymbol{\Sigma}\mathbf{V}^T\end{aligned} \tag{5.46}$$

The principal components contained in the matrix $\mathbf{Y}$ may be viewed as the inner product of the eigenvectors of $\mathbf{X}\mathbf{X}^T$ with the data, or as the weighted eigenvectors of $\mathbf{X}^T\mathbf{X}$. Since $\mathbf{X}$ may be reconstructed from the principal component matrix, $\mathbf{Y}$, via the inverse KL transformation

$$\mathbf{X} = \mathbf{U}\mathbf{Y} \tag{5.47}$$

we may combine Eqs. (5.46) and (5.47) to obtain

$$\mathbf{X} = \mathbf{U}\boldsymbol{\Sigma}\mathbf{Y}^T \tag{5.48}$$

Eq. (5.48) is identical to Eq. (5.33), showing that, provided that we are considering a multivariate stochastic process, the SVD and the KL transformation are computationally equivalent.

(5.49)

5.3.3.3 KL and Image Processing

We turn now to the problem at hand, which is image processing. In this instance the situation encountered is essentially different. From a stochastic point of view, we now have one realization, $\mathbf{X}$, of a two dimensional random process. The KL transformation and the SVD of $\mathbf{X}$ will be the same only if we assume the separability of the covariance matrix of the process into a product of the covariance matrices of the rows and columns of the image, and if these covariance estimates are computed from the one realization which is our image (Gerbrands, 1981). It is clear that in this case the use of stochastic terminology, like the KL transformation, is not appropriate. On the other hand, eigenimage decomposition, which depends on a deterministic decomposition using the SVD, is perfectly descriptive. It should also be pointed out that in the case when the SVD and KL decompositions are equivalent, as in the situation described above, the KL transformation is generally referred to as the zero-lag KL transformation since it is computed from only the zero-lag covariance matrix.

5.3.4 Eigenimages and the Fourier Transform

An interesting aspect concerning eigenimages which is also relevant in the comparison of eigenimage and velocity filtering in both the $t-x$ and the $f-k$ domains, is the behavior of eigenimages under Fourier transformation. To this end, we take the Fourier transform, FT, of Eq. (5.34) to obtain

$$\mathcal{F}_2[\mathbf{X}] = \sum_{i=1}^{r} \sigma_i \mathcal{F}_2[\mathbf{u}_i \mathbf{v}_i^T] \tag{5.50}$$

where $\mathcal{F}_2[\cdot]$ represents the 2D FT. It is clear from Eq. (5.50) that the $f-k$ representation of $\mathbf{X}$ may also be viewed in terms of eigenimages in that domain. Further, since the M rows of the ith eigenimage, $(u_{ki}\mathbf{v}_i^T,\ \ k=1,2,\ldots,M)$, are equal to within a scale factor, we may write

$$\mathcal{F}_2[\mathbf{u}_i \mathbf{v}_i^T] = \mathcal{F}_1[\mathbf{u}_i]\mathcal{F}_1[\mathbf{v}_i]^T$$

where $\mathcal{F}_1[\cdot]$ represents the 1D FT. Consequently we may write Eq. (5.50) as

$$\mathcal{F}_2[\mathbf{X}] = \sum_{i=1}^{r} \sigma_i \mathcal{F}_1[\mathbf{u}_i]\mathcal{F}_1[\mathbf{v}_i]^T \tag{5.51}$$

This expression demonstrates that the 2D FT of any matrix $\mathbf{X}$ may be obtained as the weighted sum of 1D FTs of the eigenvectors associated with $\mathbf{X}\mathbf{X}^T$ and $\mathbf{X}^T\mathbf{X}$. In the case when $\mathbf{X}$ may be sufficiently well reconstructed from only the first few eigenimages, its 2D FT may be efficiently computed using only a small number of 1D FTs.

5.3.5 Computing the Filtered Image

Even with the availability of efficient algorithms the computation of the full SVD decomposition may be a time consuming task. This is particularly true in the seismic case where the dimension N may be large. Fortunately, in our case, the dimension M is often considerably less than N and we are also concerned with the reconstruction of $\mathbf{X}$ from only a few eigenimages. We can consequently reconstruct the filtered section, $\mathbf{X}_{LP}$ say, rapidly by computing only those eigenvectors of the $(M \times M)$ matrix $\mathbf{X}\mathbf{X}^T$ which enter into the summation in Eq. (5.33). In order to make the derivation quite general, we will concern ourselves with the construction of a general $\mathbf{X}_{BP}$ using the singular values $\sigma_p \geq \sigma_{p+1} \geq \ldots \geq \sigma_q$ where $p > 1$, $q < r$ and r is the rank of the matrix. We wish to compute $\mathbf{X}_{BP}$ without the necessity of computing the complete SVD of the data matrix $\mathbf{X}$. Using Eq. (5.34), the band-pass matrix $\mathbf{X}_{BP}$ is given by

$$\mathbf{X}_{BP} = \mathbf{U}_{BP}\mathbf{\Sigma}_{BP}\mathbf{V}_{BP}^T \tag{5.52}$$

where $\mathbf{U}_{BP}$, $\mathbf{V}_{BP}$ and $\mathbf{\Sigma}_{BP}$ are equal to $\mathbf{U}$, $\mathbf{V}$ and $\mathbf{\Sigma}$ respectively with the exception that the first $p-1$ and the last $r-q$ columns of each matrix are zeroed. Without loss of generality we consider the case where M, the number of traces, is less than N, the number of time samples per trace, and we compute the covariance matrix $\mathbf{X}\mathbf{X}^T$ of smaller dimension. Thus

$$\mathbf{X}\mathbf{X}^T = \mathbf{U}\mathbf{\Sigma}^2\mathbf{U}^T$$

With the availability of $\mathbf{U}$ we compute $\mathbf{U}_{BP}^T\mathbf{X}$ as

$$\mathbf{U}_{BP}^T\mathbf{X} = \mathbf{U}_{BP}^T\mathbf{U}\mathbf{\Sigma}\mathbf{V}^T \tag{5.53}$$

Owing to the orthonormality of the eigenvectors it follows that

$$\mathbf{U}_{BP}^T\mathbf{U}\mathbf{\Sigma} = \mathbf{\Sigma}_{BP} \tag{5.54}$$

and substituting Eq. (5.54) into Eq. (5.53) we obtain

$$\mathbf{U}_{BP}^T\mathbf{X} = \mathbf{\Sigma}_{BP}\mathbf{V}^T \tag{5.55}$$

Owing to the zero values along the diagonal of $\mathbf{\Sigma}_{BP}$ it is clear that

$$\mathbf{\Sigma}_{BP}\mathbf{V}^T = \mathbf{\Sigma}_{BP}\mathbf{V}_{BP}^T$$

Using the above expression in Eq. (5.55),

$$\mathbf{U}_{BP}^T\mathbf{X} \;=\; \mathbf{\Sigma}_{BP}\mathbf{V}_{BP}^T \tag{5.56}$$

Finally, substituting Eq. (5.56) into Eq. (5.33) we obtain

$$\mathbf{X}_{BP} \;=\; \mathbf{U}_{BP}\mathbf{U}_{BP}^T\mathbf{X} \tag{5.57}$$

It is interesting to note from Eq. (5.57) that in the case when $p = 1$ and $q = r < M$

$$\mathbf{X}_{LP} \;=\; \mathbf{U}_{LP}\mathbf{U}_{LP}^T\mathbf{X} = \mathbf{X} \tag{5.58}$$

It should be realized however that the product $\mathbf{U}_{LP}\mathbf{U}_{LP}^T$ is not in general equal to the identity matrix. $\mathbf{U}_{LP}\mathbf{U}_{LP}^T = \mathbf{I}$ only in the case when $q = M$. The above development is immediately extended to the complex case by replacing ^{T}, the matrix transpose, with ^{H}, the complex conjugate transpose.

5.3.6 Applications

In this section we illustrate, using synthetic and real data examples, some of the applications of eigenimage processing to seismic sections. In particular we are interested in discussing such issues as signal to noise enhancement, velocity analysis and residual static correction.

5.3.6.1 Signal to Noise Enhancement

In order to gain deeper insight into the eigenimage filter for SNR enhancement, we consider very simple examples. Applications to seismic problems will be presented in Section 5.3.6.2

Let $\mathbf{S}$ and $\mathbf{N}$ represent the noiseless data matrix, which is composed of M identical traces, and the noise matrix which is uncorrelated with the data, respectively. The input matrix is $\mathbf{X} = \mathbf{S} + \mathbf{N}$. Since $\mathbf{S}$ is of rank one, it may be reconstructed from the first eigenimage. Using the reconstruction of Eq. (5.58) we may write

$$\mathbf{S} \;=\; \mathbf{u}_1\mathbf{u}_1^T\mathbf{S}$$

where, in this particular case, the matrix $\mathbf{u}_1\mathbf{u}_1^T$ is composed of elements each equal to $1/M$. Since N is composed of traces of uncorrelated random noise, the data covariance matrix $\mathbf{C_{xx}}$ is a sum of the signal and noise covariance matrices. i.e.,

$$\begin{aligned}\mathbf{C_{xx}} &= \mathbf{C_{ss}} + \mathbf{C_{nn}}\\ &= \mathbf{C_{ss}} + \sigma_{\mathbf{n}}^2\mathbf{I}\end{aligned}$$

where $\mathbf{I}$ is the $M \times M$ identity matrix and $\sigma_{\mathbf{n}}^2$ is the noise variance. Consequently

$$\begin{aligned} \mathbf{C_{xx}u} &= \lambda \mathbf{u} \\ (\mathbf{C_{ss}} + \mathbf{C_{nn}})\,\mathbf{u} &= \lambda \mathbf{u} \end{aligned} \tag{5.59}$$

and

$$\mathbf{C_{ss}u} = \left(\lambda - \sigma_{\mathbf{n}}^2\right) \tag{5.60}$$

From Eqs. (5.59) and (5.60) it is evident that the eigenvectors of $\mathbf{XX}^T$ are insensitive to white noise, while the eigenvalues are increased by $\sigma_{\mathbf{n}}^2$. Consequently the first eigenimage of $\mathbf{X}$ is $\mathbf{u}_1\mathbf{u}_1^T$ and

$$\begin{aligned} \mathbf{u}_1\mathbf{u}_1^T\mathbf{X} &= \mathbf{u}_1\mathbf{u}_1^T\mathbf{S} + \mathbf{u}_1\mathbf{u}_1^T\mathbf{N} \\ &= \mathbf{S} + \mathbf{u}_1\mathbf{u}_1^T\mathbf{N} \end{aligned} \tag{5.61}$$

Eq. (5.61) explains the behavior of the SVD filter very clearly. The signal is recovered fully and the noise is suppressed in a manner equivalent to an average stack. In the more general case, when the signal may vary from trace to trace, the rank of $\mathbf{S}$ is greater than one and the noise will be suppressed by an optimally weighted average. This average is optimum in the sense that the weight vectors $\mathbf{u}_i$ are obtained as a result of a maximum variance criterion[3].

5.3.6.2 Eigenimage Analysis of Common Offset Sections

A particularly natural example of the use of eigenimage processing is in relationship to common offset sections. which can often be efficiently compressed using eigenimages. Eigenimage filtering of such sections, also serves as an example of SNR enhancement of pre-stack data.

As before, we consider the data matrix $\mathbf{X}$ to be composed of M traces with N data points per trace, the M traces forming the columns of $\mathbf{X}$. We write the SVD of $\mathbf{X}$ in the usual manner

$$\mathbf{X} = \sum_{i=1}^{r} \lambda_i \mathbf{u}_i \mathbf{v}_i^T \tag{5.62}$$

where r indicates the rank of $\mathbf{X}$. As we have discussed above, if all the M traces are linearly independent, $\mathbf{X}$ is of full rank and all the singular values are different from zero. A perfect reconstruction of $\mathbf{X}$ requires all eigenimages. On the other hand, in

[3]A useful manner of computing the SNR is by using the following expression

$$\text{SNR} = \frac{\text{tr}[\mathbf{SS}^T]}{\text{tr}[(\mathbf{Y} - \mathbf{S})(\mathbf{Y} - \mathbf{S})^T]}$$

where $\mathbf{Y}$ is the filtered matrix and $\text{tr}[\,\cdot\,]$ represents the trace of the matrix.

the case where all M traces are equal to within a scale factor, all traces are linearly dependent, $\mathbf{X}$ is of rank one and may be perfectly recovered from the first eigenimage, $\lambda_1 \mathbf{u}_1 \mathbf{v}_1^T$.

The eigenimage decomposition can be used to optimally extract laterally coherent waveforms. In general, common offset sections exhibit a good lateral coherence. Our approach is to first decompose the pre-stack data cube into common offset sections and then apply eigenimage analysis to compress each common offset section and improve the SNR ratio. Our strategy is summarized as follows:

1. The pre-stack data cube is decomposed into common offset sections. In the examples, 10 common offset sections are constructed containing traces with offsets indicated in Table 1.

2. Each common offset section is decomposed into eigenimages. and only the eigenvectors that correspond to the first p singular values are kept.

3. Eq. (5.62) is used to reconstruct the common offset section. If the misfit is acceptable, we save the vectors $u_i,\ v_i,\ \lambda_i,\ \ i=1\dots p$.

It is interesting to note that the amount of data compression that can be achieved using this procedure is remarkably high. Using the SVD we can represent each common offset section by N_2 floats:

$$N_2 = p \times N + p \times M + p$$

We define the compression ratio as follows

$$C = (N_1 - N_2)/N_2,$$

where $N_1 = M \times N$ is the total number of floats required to represent the common offset section, $\mathbf{X}$.

Table 1 summarizes the compression ratio for the ten common offset sections which make up the data cube. In this example p corresponds to the number of singular values that account for 30% of the total power encountered in the spectrum of singular values, shown in Fig. 5.20. We note that the eigen decomposition is in terms of a few energetic singular values that correspond to coherent events in the common offset domain.

Figs. 5.21a and 5.21b display the common offset section #2 after and before eigenimage filtering. Since the events are fairly flat, the information content of the section may be captured in a few eigenimages. Figs. 5.22a and 5.22b show a CDP after and before eigenimage processing in the common offset domain. It is clear that we cannot use eigenimages in the CDP domain but, after filtering in the common offset domain and sorting into CDPs, some high frequency noise at near and far offsets has been eliminated.

In summary, sorting data into common offset section allows the application of eigenimage filtering on individual common offset traces. The pre-stack volume is reconstructed with minimal distortion.

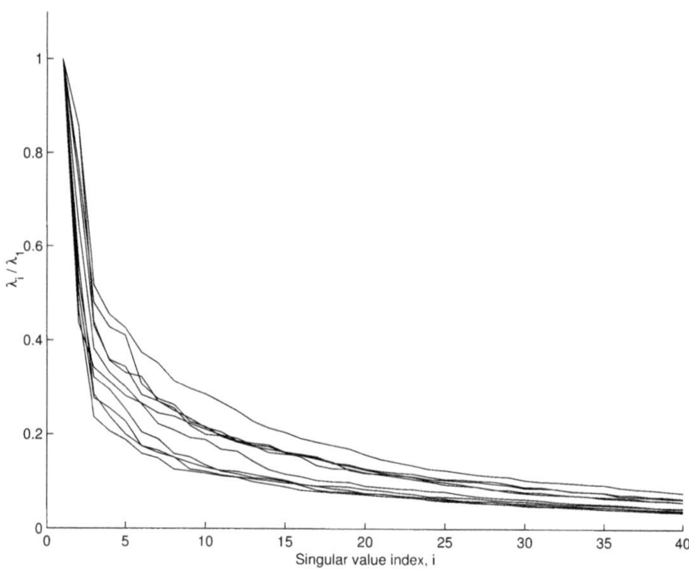

Figure 5.20: Spectra of singular values for the 10 common offset sections used to test the algorithm.

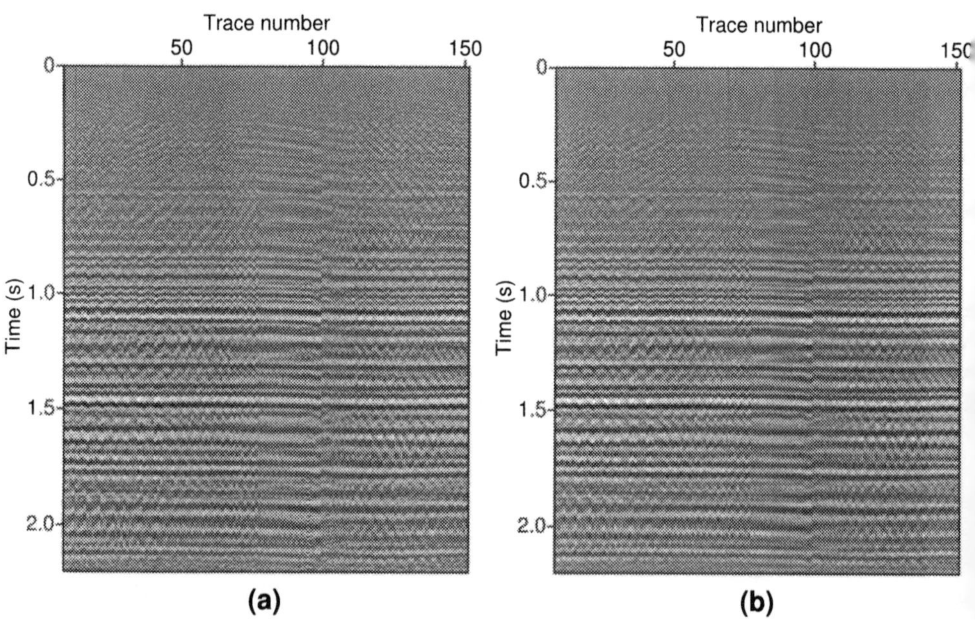

Figure 5.21: (a) Common offset section #2. (b) Common offset section after eigen-image filtering.

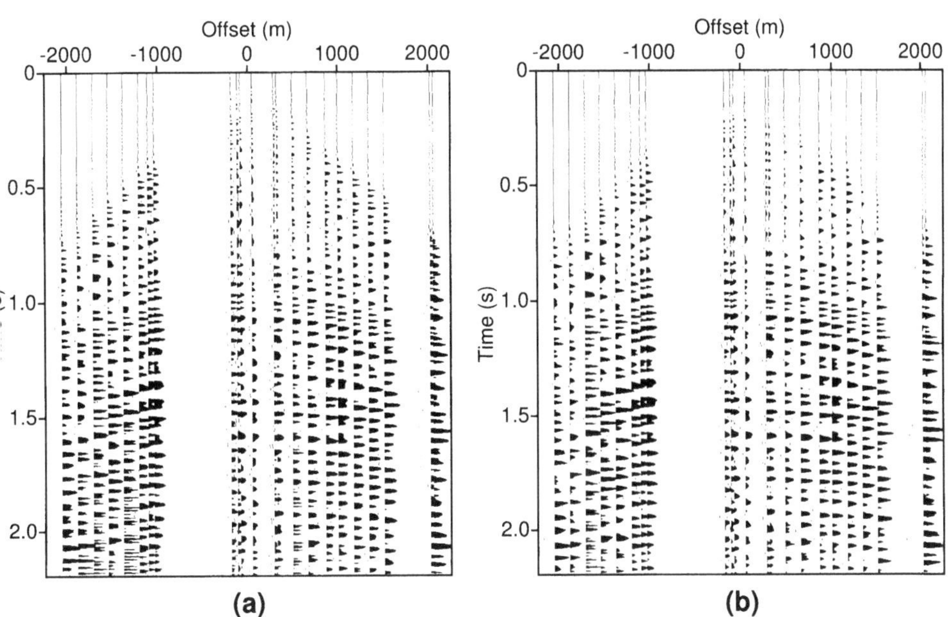

Figure 5.22: (a) Original CDP gahter. (b) CDP gather after eigenimage filtering in the common offset domain.

COS#	Offset [m]	p	$C = (n_1 - n_2) / n_2$
1	0-221	9	13.7
2	221-427	6	20.0
3	427-633	4	18.7
4	633-839	5	17.8
5	839-1045	4	23.7
6	1045-1250	5	21.2
7	1250-1456	6	18.2
8	1456-1662	6	14.4
9	1662-1868	6	15.2
10	1868-2780	7	13.0

Table 5.1: Compression ratios for 10 common offset sections. The variable p indicates the number of singular values used in the eigen decomposition.

5.3.6.3 Eigenimages and Velocity Analysis

Eigen decomposition of seismic data (hyperbolic windows in CMP gathers) can be used to design coherence measures for high resolution velocity analysis. The idea is to replace the semblance measure by a norm that is a function of the eigenvalues of the covariance matrix of the analysis windows. We present a very simple algorithm that can be used to compute high resolution coherence measures for velocity analysis. Techniques that exploit the eigenstructure of the covariance matrix have been borrowed from the field of array processing, (e.g., Bienvenu, 1883), and applied to velocity analysis by various researchers (e.g., Biondi and Kostov, 1989; Key and Smithson, 1990; Kirlin, 1992).

The seismic signal, in the presence of noise, at receiver i may be modelled by

$$x_i(t) \quad = \quad s(t - \tau_i) + n_i(t) \qquad\qquad i = 1, 2, \dots, N \tag{5.63}$$

where $\tau_i = (t_0^2 + d_i^2/V^2)^{1/2} - t_0$ is the delay of the signal between the i-th receiver and a receiver at a distance $d_0 = 0$ from the shot. For a waveform extracted along a hyperbolic path parameterized with velocity V, Eq. (5.63) may be rewritten as

$$x_i(t) \quad = \quad s(t) + n_i(t)$$

where, to avoid notational clutter, the same variable, $x_i(t)$, is used to designate the delayed and the corrected waveforms. The covariance matrix of the signal is defined as

$$R_{ij}(t) \quad = \quad E[x_i(t)x_j(t)] \qquad\qquad i, j = 1, 2, \dots, N \tag{5.64}$$

Assuming the noise and signal to be uncorrelated, the data covariance matrix becomes

$$R_{ij}(t) \quad = \quad R_{ij}^s(t) + \sigma_n^2(t)\delta_{ij}$$

where $R^s_{ij}(t)$ denotes the signal covariance matrix, and δ_{ij} is the Kronecker delta, as usual. Assuming stationarity, the dependence on t may be dropped. It is easy to verify that the eigenvalues of the covariance matrix become

$$\lambda_i = \lambda_i^s + \sigma_n^2 \tag{5.65}$$

where λ_i^s are the eigenvalues of the signal covariance matrix. Assuming that the signal is invariant across each trace, the signal covariance matrix is rank 1, and we can write

$$\lambda_1^s = NP_s \tag{5.66}$$
$$\lambda_i^s = 0, \quad i = 2, 3, \ldots, N \tag{5.67}$$

where $P_s = E[s(t)^2]$ denotes the signal power. Using Eq. (5.65), the eigenvalues of the data covariance matrix become

$$\lambda_1 = NP_s + \sigma_n^2 \tag{5.68}$$
$$\lambda_i = \sigma_n^2, \quad i = 2, 3, \ldots, N \tag{5.69}$$

For uncorrelated noise, the minimal $N-1$ eigenvalues of the data are equal to the variance of the noise. The largest eigenvalue is proportional to the power of energy of the coherent signal plus the variance of the noise.

In real situations, the eigenspectrum is retrieved from an estimate of the data covariance matrix. If the stationary random processes, $x_i(t)$ and $x_j(t)$, are ergodic, then, as we have previously discussed, ensemble averages defined in Eq. (5.64) can be replaced by time averages and the estimator of the covariance matrix becomes

$$\hat{R}_{ij} = \frac{1}{2M+1} \sum_{k=-M}^{M} x_i(k\Delta t) x_j(k\Delta t) \tag{5.70}$$

Using the results expressed by Eqs. (5.67) and (5.69), it is evident that an estimator of the noise variance is

$$\hat{\sigma}_n^2 = \frac{1}{N-1} \sum_{i=2}^{N} \hat{\lambda}_i \tag{5.71}$$

Similarly, an estimator of the signal energy is given by

$$\hat{P}_s = \frac{\hat{\lambda}_1 - \hat{\sigma}_n^2}{N} \tag{5.72}$$

and Eqs. (5.71) and (5.72) can be combined into a single measure of the SNR

$$\widehat{C} = \frac{1}{N} \frac{\hat{\lambda}_1 - \sum_i \hat{\lambda}_i/(N-1)}{\sum_i \hat{\lambda}_i/(N-1)}, \quad i = 2, 3, \ldots, N \tag{5.73}$$

The coherence measure, $\hat{C}$, was devised assuming the presence of a signal and that the proper velocity is used to extract the waveform. In general, $\hat{C}$ is computed for different gates and different trial velocities. It is convenient to explicitly emphasize the dependence of the coherence on these parameters by denoting $\hat{C}$ as $\hat{C}(t_0, V)$. When the analysis window contains only noise, $\hat{C}(t_0, V)$ tends to zero. When the trial velocity does not match the velocity of the reflection, the covariance matrix has a complete set of eigenvalues different from zero and it is not possible to recognize which part of the eigen-spectrum belongs to the noise and which belongs to the signal process.

Key and Smithson (1990) proposed another coherence measure based on a log-generalized likelihood ratio which tests the hypothesis of equality of eigenvalues,

$$\hat{W}_{ml} = M \log^N \left[\frac{(\sum_{i=1}^N \hat{\lambda}_i / N)^N}{\prod_{i=1}^N \hat{\lambda}_i} \right], \quad i = 1, 2, \ldots, N \tag{5.74}$$

In the absence of signal, $\hat{\lambda}_i = \hat{\sigma}_n^2$ for $i = 1, 2, \ldots, N$ and hence $\hat{W}_{ml} = 0$. In the presence of a single reflected signal, $\hat{\lambda}_1 \neq 0$, $\hat{\lambda}_i = 0$, $i = 2, 3, \ldots, N$ and $\hat{W}_{ml} \to \infty$. Therefore, $\hat{W}_{ml}$ is a strong discriminant between signal and noise. Key and Smithson (1990) combined Eqs. (5.73) and (5.74) into a single measure, K_{ml}, given by the product

$$\hat{K}_{ml} = \hat{W}_{ml} \hat{C} \tag{5.75}$$

Since the trace of a matrix is equal to the sum of its eigenvalues, (also the sum of its diagonal elements), it is clear and important, that only $\hat{\lambda}_1$ is required to compute $\hat{C}$. A comparison of semblance with a high resolution analysis of a CMP gather is shown in Fig. 5.23.

It is also of interest to mention that the velocity panel obtained via the SNR coherence measure can be further improved by adopting a bootstrap procedure (Sacchi, 1998). In this case, the seismic traces are randomly sampled to produce individual estimates of the coherence measure. From this information one can obtained an average coherence measure and a histogram (in fact a density kernel estimator) of the position of the peak that optimizes the coherence. The improved SNR coherence obtained with this techniques is illustrated in Fig. 5.24.

5.3.6.4 Residual Static Correction

A very interesting application of eigenimage decomposition occurs when the data matrix is mapped into the complex domain by forming the analytic data matrix. We only present some theoretical results here and refer the interested reader to Ulrych et al. (1999) for a fuller, illustrated, discussion. In this application of PCA, we are concerned not only with time shifts in the data but also with phase shifts which may occur in practice for various reasons as discussed by Levy and Oldenburg (1987). The philosophy behind this approach is the assumption that, under certain conditions,

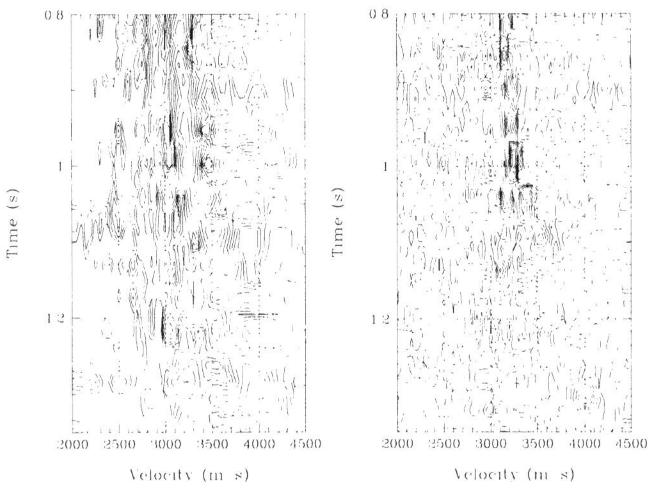

Figure 5.23: Left: Semblance of a CMP gather. Right: High resolution coherence analysis (SNR measure).

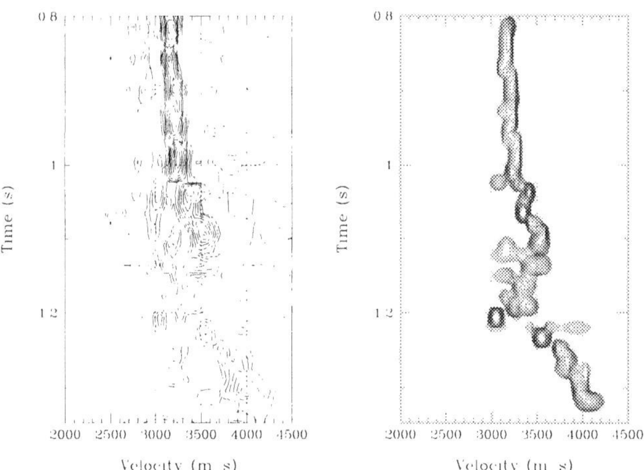

Figure 5.24: Left: Average SNR measure obtained via bootstrapping individual realizations. Right: Frequency distribution of the peak that maximizes the coherence after 50 bootstrap realizations.

frequency independent phase shifts in the seismic pulse may be viewed, approximately, as equivalent time shifts. We let

$$X(f) = \mathcal{F}_1[x(t)]$$

represent the FT of $x(t)$ as before. Then

$$\mathcal{F}_1[x(t-t_0)] = e^{i2\pi f t_0} X(f)$$

We now assume that, in the narrow seismic bandwidth

$$2\pi f t_0 \simeq 2\pi f_0 t_0 = \epsilon$$

This approximation has been used by Levy and Oldenburg (1987) very successfully in modelling residual wavelet phase variations. We can now write

$$\begin{aligned} \mathcal{F}_1[x(t-t_0)] &\simeq e^{-i\epsilon} X(f) \quad f > 0 \\ &\simeq e^{+i\epsilon} X(f) \quad f < 0 \end{aligned}$$

and hence

$$\mathcal{F}_1[x(t-t_0)] \simeq (\cos\epsilon - i\,\mathrm{sgn}[f \sin\epsilon])X(f) \quad \text{all } f$$

Transforming back to time

$$x(t-t_0) \simeq \Re[\hat{x}(t)e^{-i\epsilon}] \tag{5.76}$$

where $\hat{x}(t) = x(t) - i\mathcal{H}[x(t)]$ is the analytic signal and $\mathcal{H}[\,\cdot\,]$ represents the Hilbert transform. Eq. (5.76) shows that the frequency independent phase shift ϵ may, under the assumption of a narrow bandwidth, be approximated by the time shift t_0. From the point of view of residual phase shifts in the recorded seismic wavelet, $w(t)$, Eq. (5.76) may be written as

$$w(t) \simeq \Re[\,\hat{w}_0(t)e^{-i\epsilon}] \tag{5.77}$$

where $\hat{w}_o(t)$ is the analytic true wavelet.

The correction of residual statics when the seismic wavelet contains a degree of frequency independent phase variation, may be cast as a two-stage approach. In the first stage, the time shift associated with each individual trace which is determined by crosscorrelation with the first complex principal component of the section, is removed. This component is determined from the complex form of Eq. (5.41) as $\mathbf{y} = \mathbf{u}_1^H \mathbf{X}$, where $\mathbf{u}_1$ is the major eigenvector of $\mathbf{X}\mathbf{X}^H$. In the second stage, any frequency independent phase shifts are removed by rotating each trace with the phase determined from $\mathbf{u}_1$, which contains the information about ϵ. The final section is produced as a low-pass eigenimage reconstruction.

5.3.7 3D PCA - Eigensections

Eigenimages are defined as 2D images. Often, we are concerned with the analysis and processing of a volume of data and this is what this section is about. The extension of 2D eigenimages to 3D eigensections is based on a method used for the detection and recognition of human faces and is discussed by Kaplan and Ulrych (2002). This method, known as eigenface analysis, uses the SVD to project photographs of faces onto a vector basis, the components of which are called eigenfaces. From a collection of two dimensional seismic gathers, an analogous vector basis is computed, the components of which are called eigensections. These eigensections are used by Kaplan and Ulrych (2002) for two different purposes. The attenuation of incoherent noise in the prestack domain by exploiting section-to-section correlations in the signal, and the extraction of details from sections in the poststack domain by exploiting section-to-section global similarities and detail differences.

5.3.7.1 Introducing Eigensections

Eigensections form an orthogonal basis that is extracted from a volume of seismic data using SVD. The philosophy behind eigensections is similar to that which underlies analysis of multispectral satellite imagery (Richards, 1993) and is inspired by a technique used for the recognition and detection of human faces which uses SVD to extract, from a collection of photographs, an orthogonal basis, the components of which are called eigenfaces.

5.3.8 Eigenfaces

Because of the connection between eigensections and eigenfaces, we delve a little into the eigenface technique (Kirby and Sirovich, 1987; Pentland and Turk, 1991) which maps a set of photographs into a set of images, the eigenfaces. As an example, consider Figs. 5.25 and 5.26. Fig. 5.25 shows fifteen photographs which are mapped into the set of eigenfaces shown in Fig. 5.26.

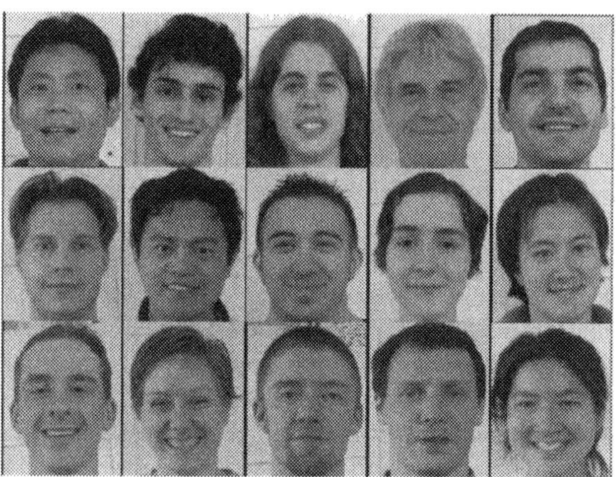

Figure 5.25: A tiny library of faces.

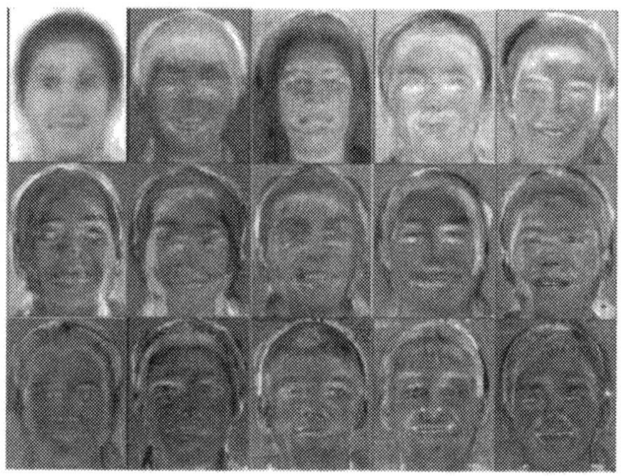

Figure 5.26: The computed eigenfaces.

The fifteen eigenfaces are mutually orthogonal and form a basis that spans the original photographs. As such, any one of the photographs can be reconstructed via a linear combination of the eigenfaces. Figs. 5.27a-f illustrate the projection of a face onto increasing numbers of eigenfaces. It can be seen that, in this particular choice of face, only three eigenfaces suffice to adequately recover the face.

Another example of a facial projection is shown in Fig. (5.28) where the same sequence of eigenfaces was used as that in Fig. (5.27). Unlike the previous example, in this case, a projection onto about fourteen eigenfaces is required for an adequate recovery. This difference in the required number illustrates the importance of having an adequate sample of representative faces in the library.

Although it is the coordinates of these projections that allow face recognition and detection algorithms, it is the images themselves that are of concern in the eigensection formulation. It is evident from Fig. 5.27 that some essential features of the

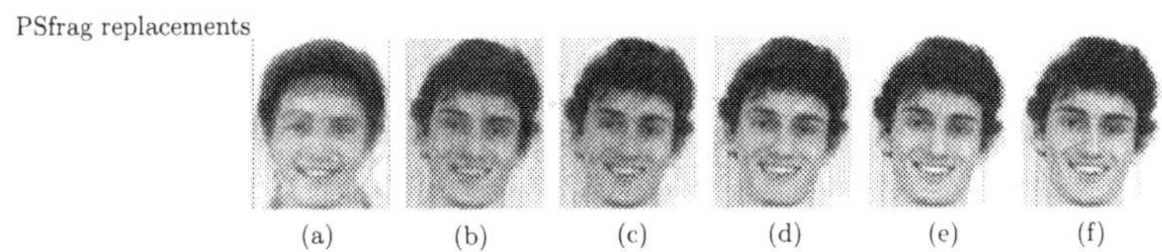

Figure 5.27: The projection of a photograph onto (a) the first three eigenfaces, (b) the first seven eigenfaces, (c) the first nine eigenfaces, (d) the first eleven eigenfaces, (e) the first fourteen eigenfaces, and (f) all fifteen eigenfaces.

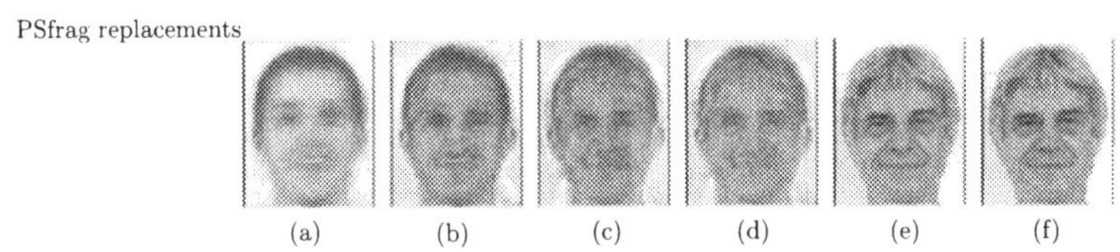

Figure 5.28: The projection of a photograph onto (a) the first three eigenfaces, (b) the first seven eigenfaces, (c) the first nine eigenfaces, (d) the first eleven eigenfaces, (e) the first fourteen eigenfaces, and (f) all fifteen eigenfaces.

photographed face are recovered in projections onto subsets of the eigenfaces. Of course, a perfect reconstruction is achieved only when all fifteen eigenfaces are used in the projection (Fig. 5.27f).

Just like the photographs in Fig. 5.25, seismic data in time and space are images. Application of the eigenface technique to seismic sections maps these sections to eigensections which are analogous to the eigenfaces shown in Fig. 5.26.

5.3.8.1 Computing the Eigensections

Returning to the definition of an eigenimage (Eq. (5.32)) which we repeat here for continuity

$$\mathbf{E}_i = \mathbf{u}_i \mathbf{v}_i^T \tag{5.78}$$

we note that the columns of $\mathbf{E}_i$ are parallel. In fact the columns of $\mathbf{E}_i$ are scalar multiples of $\mathbf{u}_i$, a fact which is easily seen by examining Eq. (5.78) in its matrix form

$$\mathbf{E}_i = \begin{bmatrix} u_{i1} \\ u_{i2} \\ \vdots \\ u_{im} \end{bmatrix} \begin{bmatrix} v_{i1} & v_{i2} & \cdots & v_{in} \end{bmatrix} \tag{5.79}$$

$$= \begin{bmatrix} v_{i1}\mathbf{u}_i & | & v_{i2}\mathbf{u}_i & | & \cdots & | & v_{in}\mathbf{u}_i \end{bmatrix}. \tag{5.80}$$

Therefore, the structural information of $\mathbf{E}_i$, and thus also of $\mathbf{X}$, the seismic section, can be expressed using only the vectors $\mathbf{u}_i$. This relationship will be used in the definition of the eigensections.

5.3.8.2 SVD in 3D

Performing SVD in 3D requires reordering the 3D data so that the 3D problem becomes a 2D problem and allows the use of SVD as in the computation of eigenimages. Each seismic section is mapped to a 1D vector via a lexicographic reordering. This transformation is achieved simply by taking each column of $\mathbf{X}$ and attaching it to the end of the previous column, thus stringing $\mathbf{X}$ out into a long one dimensional vector. This is done for p seismic sections giving p vectors $(\mathbf{x}_1, \mathbf{x}_2, \cdots, \mathbf{x}_p)$ each of length MN. Thus, each $\mathbf{x}_i$ is a lexicographically ordered vector. The new 2D matrix, $\mathbf{B}$, becomes

$$\mathbf{B} = \left[\begin{array}{c|c|c|c} \mathbf{x}_1 & \mathbf{x}_2 & \cdots & \mathbf{x}_p \end{array} \right]$$

and can be suitably decomposed. The above process is illustrated in Fig. (5.29).

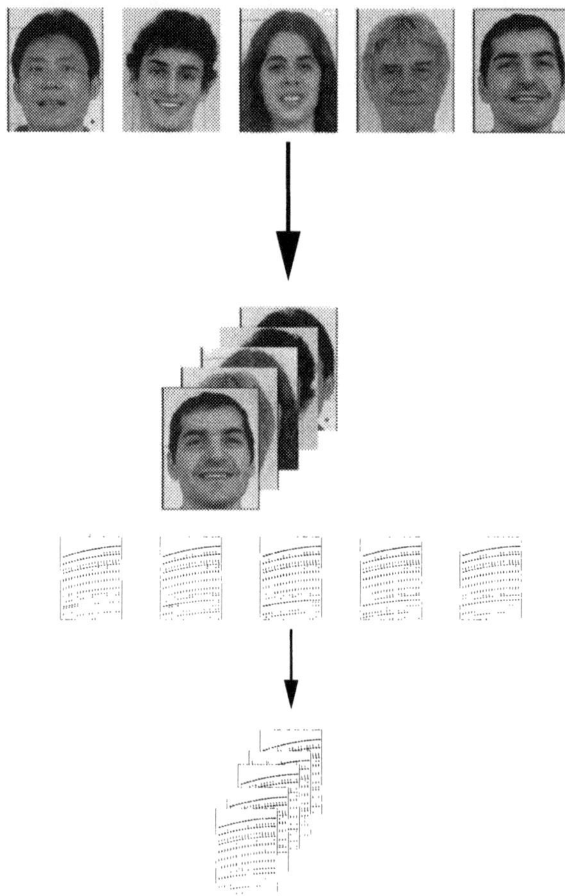

Figure 5.29: Ordering faces and sections.

The eigensections are computed by taking the SVD of $\mathbf{B}$,

$$\mathbf{B} = \mathbf{U\Sigma V}^T$$

where,

$$\mathbf{U} = \left[\begin{array}{c|c|c|c} \mathbf{u}_1 & \mathbf{u}_2 & \cdots & \mathbf{u}_p \end{array} \right]$$

The vectors $\mathbf{u}_i$ are the eigensections of the original three dimensional seismic data and contain all the structural information required for the recomposition of $\mathbf{B}$. We note some pertinent details.

(i) The vectors $\mathbf{u}_i$ are the eigenvectors of the symmetric $p \times p$ matrix, $\mathbf{B}\mathbf{B}^T$, and thus define an orthonormal basis that spans a p dimensional subspace. Therefore, the ordered eigensections also constitute an orthonormal basis in the subspace.

(ii) Projecting a column of $\mathbf{B}$, (a lexicographically ordered seismic section, $\mathbf{x}_i$), onto the subspace spanned by the eigensections maps this vector representation of the section to a new vector of dimension p via the relation

$$c_{ij} = \mathbf{u}_j^T \mathbf{x}_i . \tag{5.81}$$

A new set of coordinates, $\mathbf{c}_i$, is obtained for each seismic section, and the seismic sections can be reconstructed by applying these coordinates to the basis of eigensections from which the coordinates were computed.

Noise attenuation, optimal information extraction and data compression are achieved, just as in the 2D case, by projecting the sections onto the first $k < p$ eigensections where the projection is defined by $\mathbf{B}_k$. By eliminating the last $p - k$ eigensections from the projection, $\mathbf{B}_k$ is the approximation to $\mathbf{B}$ with the most incoherent information between the columns of $\mathbf{B}$ (the seismic sections) removed. Conversely, $\mathbf{B}_{p-k}$ is an approximation to $\mathbf{B}$ with the most coherent information between the columns of $\mathbf{B}$ removed. Applications of the eigensection technique to signal to noise enhancement of common midpoint sections are described in Kaplan and Ulrych (2002). An example of extracting detail information from seismic poststack data, also considered by Kaplan and Ulrych (2002), is presented here.

5.3.8.3 Detail Extraction

A very simple synthetic example, to illustrate the flavor of detail extraction, is modelled as ten poststack sections, nine of which are the same except for the addition of different realizations of random noise, and one which is slightly different. The goal is to recover the information (detail) that make the tenth section different from the other nine and, at the same time, attenuate the noise. This task may be achieved by eliminating the first few eigensections from the eigensection basis.
Fig. (5.30)a shows the first section and Fig. (5.30)c illustrates the tenth section which, for this example, is produced by adding some detail (Fig. (5.30)b) to the first section, such that

$$\mathbf{X}_{10} = \mathbf{X}_1 + 0.5\mathbf{X}_d$$

where $\mathbf{X}_{10}$ is the tenth section, $\mathbf{X}_1$ is the first section and $\mathbf{X}_d$ represents the detail.

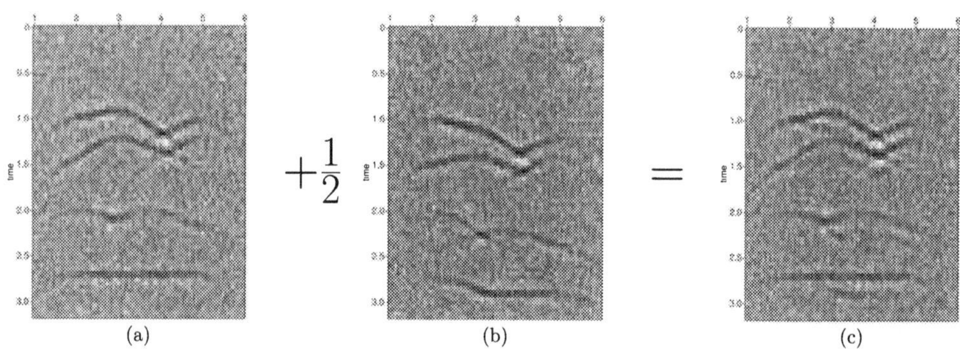

Figure 5.30: The poststack data example. (a) The first CMP, (b) The detail. (c) The tenth CMP which is the sum of the first CMP and the detail.

Figs. (5.31)a-c show the first two and the tenth sections respectively. Figs. (5.31)d-e demonstrate the first two eigensections ($\mathbf{u}_1$ and $\mathbf{u}_2$ respectively) and Figs. (5.31)f-h depict the third, sixth and ninth eigensections ($\mathbf{u}_i$ for $i = 3, 6$ and 9 respectively). Essentially all the detail which exists in the tenth section is extracted to the last eigensection, which is shown in Fig. (5.31)i. A very small contribution of energy is also present in the second eigensection [4]. For comparison, Fig. (5.31)j illustrates the detail, $\mathbf{X}_d$, that was used to create the tenth section. It is apparent that Figs. (5.31)i and j contain the same relevant information. A closeup of the tenth section and the actual and recovered detail is shown in Figs. (5.32)a-c, and accentuates the recovery of the detail information in the eigensection decompsoition.

5.3.8.4 Remarks

The study of eigenfaces has, as the ultimate goal, the task of recognizing and detecting human faces. The application of the eigenface approach to seismology (Kaplan and Ulrych, 2002) has, as an ultimate goal, the processing and analysis of seismic reflection data. Both eigenfaces and eigensections expand 3D data using the SVD, but the manner in which eigenfaces are used is quite different to that of eigensections. In order to recognize a face, the eigenface approach is localized in the coordinates of the projection of a face onto a basis of eigenfaces. Analyzing these coordinates is what gives the eigenface technique the ability to recognize and detect faces. On the other hand, the eigensection approach is concerned mainly in the actual projection of a section onto some subset of the eigensection basis. Depending on which subset one chooses for a basis, this projection can accomplish one of two things. A projection onto the first few eigensections filters incoherent noise from the CMP, providing an enhanced SNR in the prestack domain. A projection onto the last few eigensections extracts detail unique to a particular poststack section which could be used, for example, in the processing of 4D data or in the analysis of 3D poststack data volumes.

[4] The location of the detail information in the eigensection decomposition is partly dependent on the signal to noise ratio in the data.

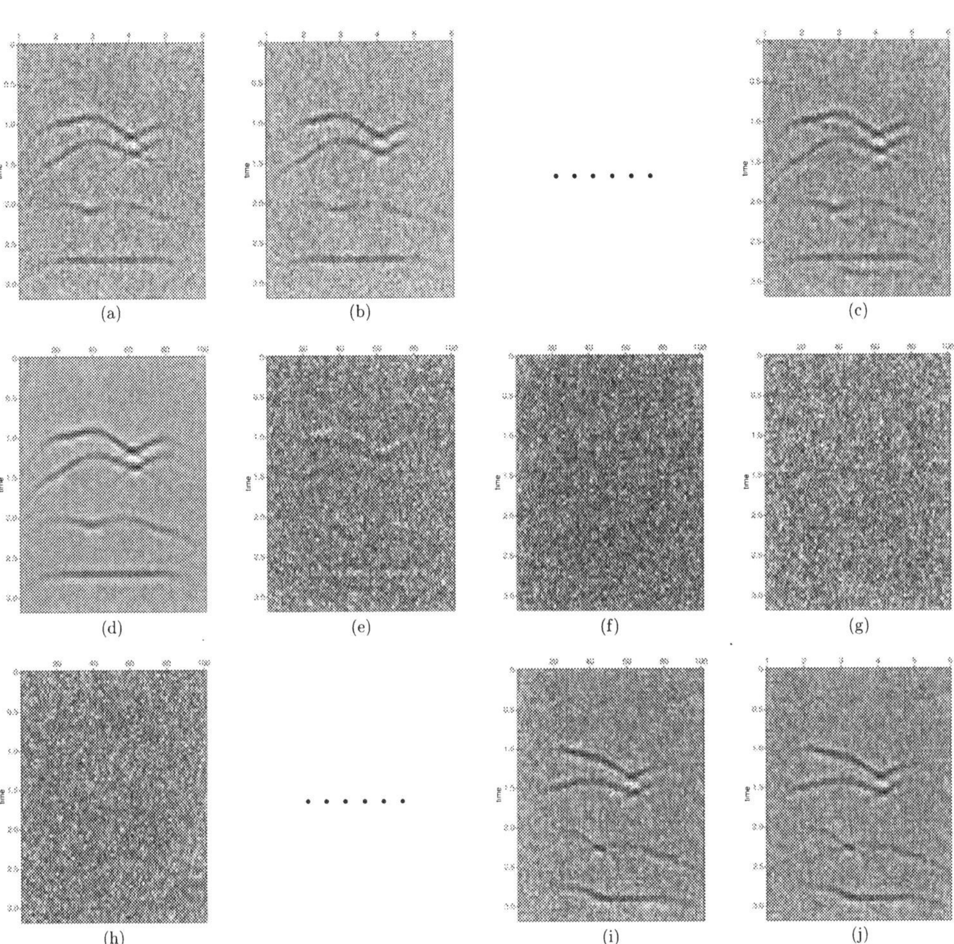

Figure 5.31: The poststack data example. (a-b) The first two seismic sections. (c) The tenth seismic section. (d-e) The first two eigensections. (f-h) The third, sixth and ninth eigensections, respectively. (i) The tenth eigensection. (j) The detail as in Fig. 5.30b.

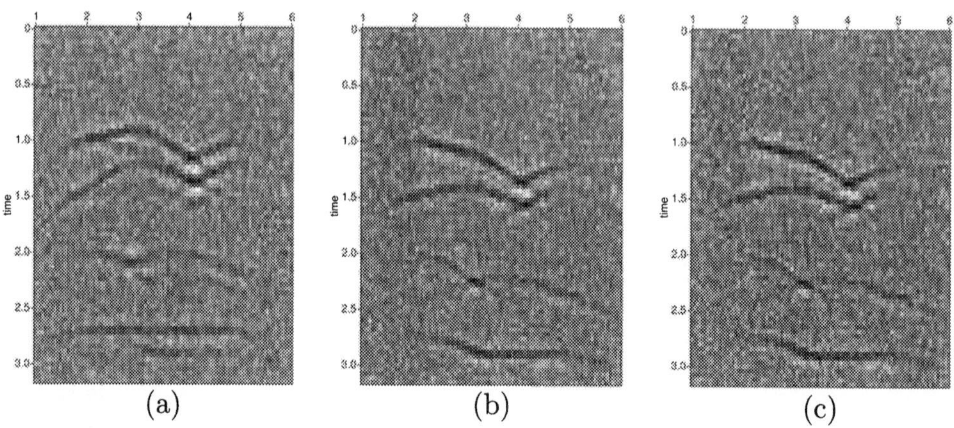

Figure 5.32: The poststack data example. (a) The tenth seismic section. (b) The actual detail. (c) The recovered detail.

5.3.9 Discussion

2D eigenimage and 3D eigensection reconstruction are non-linear and data dependent filtering methods that may be applied either in the real or complex domain to achieve a number of important processing objectives. Since seismic sections are in general highly correlated trace-to-trace, eigen reconstruction is a parsimonious representation in the sense that the data may be reconstructed from only a few images. A natural consequence of such a reconstruction is, as we have shown, an improvement of the SNR. Eigenimage reconstruction has a similar capacity of $f-k$ filtering of removing events which show an apparently different phase velocity on the section. The actual process by which this is accomplished is quite different to that which is entailed in $f-k$ filtering. In the former approach, events are removed which do not correlate well from trace to trace. In the latter, events are rejected which possess different $f-k$ signatures. One of the consequences of this difference is that eigenimage filtering is not subject to spatial aliasing in the sense of $f-k$ filtering. However, eigenimage reconstruction encounters similar difficulties as does the $f-k$ approach in separating events with similar dips. As we have mentioned, the linear moveout model is not essential in eigenimage processing. In fact, as discussed above, a hyperbolic moveout model may be used to estimate stacking velocities in shot gathers. A very fine example of SNR enhancement using eigenimages on stacked seismic sections which contain highly curved events has been presented by Al-Yahya (1991) who applied the filter for selected dips and then combined the results for all dips to produce the composite enhanced section. Eigensection reconstruction is applicable to 3D volumes of data. As such, an important application, as mentioned above, is in the 4D monitoring of reservoir properties.

Many people concerned with SNR enhancement employ a spectral matrix rather than an eigenimage approach. It is of interest, therefore, to consider the correspondence between these two techniques. The spectral matrix is formed in the frequency domain as the complex covariance matrix at each frequency. For linear events in time, each frequency contains a sum of L harmonics associated with the L events on the section. Separation of the events is then achieved by means of averaging in frequency and/or space using window functions which depend on the input data. This is possible because of the redundancy of information which exists in the frequency-space domain. Unlike the eigenimage technique, prior information concerning the dips of the events is not required. However, as shown by the work of Rutty and Jackson (1992), this entails certain penalties which manifest themselves as edge effects and the necessity that events to be separated have different energies. On the other hand, unlike the eigenimage approach, amplitudes of the events are recovered. An interesting possibility might be to combine the spectral matrix and eigenvector methods in order to develop an approach which reflects the best of both techniques.

5.4 Radon Transforms

An approach that is very often successful in attenuating noise, is to map the data into a new domain where random and/or coherent noise can be isolated from the signal. The Radon transform, RT, is an example of such a transformation and has found considerable application in the processing of seismic data. Seismic sections may be considered, depending on the scale of the data window and preprocessing steps, to be composed of events that obey linear, parabolic or hyperbolic signatures. Each type of event demands a different RT and we outline the theoretical details and present illustrative examples.

5.4.1 The Linear Radon Transform (LRT)

We begin with the definition of the linear Radon transform (LRT), also called the slant stack transform in geophysics. Let $d(h,t)$ represent a seismic signal. Throughout this section, the variable t designates the time and h the offset or range. Using a continuous formulation for convenience for the moment, we define the slant stack by means of the following transformation

$$m(p,\tau) = \mathcal{L}\, d(h,t) = \int d(h, t = \tau + hp)dh$$

where p and τ denote the slope, or ray parameter, and the intercept time, respectively and $m(p,\tau)$ is used to designate the signal in the transform, $\tau-p$, domain. The adjoint transform, $\mathcal{L}^*$, is

$$\tilde{d}(h,t) = \mathcal{L}^*\, m(p,\tau) = \int m(p, \tau = t - hp)dp$$

In the frequency domain, the pair of transformations is given by

$$M(p,f) = \int D(h,f)e^{i2\pi fph}dh \tag{5.82}$$

and

$$\tilde{D}(h, f) = \int M(p, f) e^{-i2\pi f p h} dp \tag{5.83}$$

In as much as we can now map our data from (h, t) to (τ, p), we require an inverse transformation that will allow us to return to the (h, t) domain. Substituting, Eq. (5.82) into Eq. (5.83) yields

$$\tilde{D}(h, f) = \int D(h', f) \int e^{-i2\pi f p (h-h')} dp\, dh'$$

It is clear that this equation is a convolution integral that can be written as

$$\tilde{D}(h, f) = D(h, f) * \rho(h, f) \tag{5.84}$$

where

$$\rho(h, f) = \int e^{-i2\pi f p h} dp \tag{5.85}$$

Making the substitution $z = -2\pi f p$, Eq. (5.85) becomes

$$\rho(h, f) = \int \frac{1}{2\pi |f|} e^{ihz} dz = \frac{1}{|f|} \delta(h)$$

The convolution operator is a delta function with respect to h. Since $\int f(x)\delta(x)dx = f(0)$, we can write

$$\tilde{D}(h, f) = \frac{1}{|f|} D(h, f) * \delta(h) \tag{5.86}$$

$$\tag{5.87}$$

$$= \frac{1}{|f|} D(h, f)$$

and the inverse transformation becomes

$$D(h, f) = |f| \tilde{D}(h, f) \tag{5.88}$$

The inverse transform is computed in two stages. First, the adjoint is used to evaluate $\tilde{D}(h, f)$ in Eq. (5.88). Secondly, $\tilde{D}(h, f)$ is multiplied by the frequency response of the ρ filter. The slant stack pair in the frequency domain results takes the form

$$M(p, f) = \int D(h, f) e^{i2\pi f p h} dh$$

$$D(h,f) = |f| \int M(p,f) e^{-i2\pi f p h} dp$$

Let the ray parameter, p, be defined in the interval $p \in [-P, P]$. The resulting ρ filter is

$$\begin{aligned} \rho(h,f) &= \int_{-P}^{P} e^{-i2\pi f p h} dp \\ &= 2P \frac{\sin(2\pi f P h)}{2\pi f P h} \end{aligned} \tag{5.89}$$

Substituting Eq. (5.89) into Eq. (5.84)

$$\tilde{D}(h,f) = 2P \int D(h',f) \frac{\sin(2\pi f P(h-h'))}{2\pi f P(h-h')} dh'$$

Evidently, the data may be recovered after solving a deconvolution problem. Spatial deconvolution is required since the infinite range of the variable p is truncated to a finite range (Zhou and Greenhalgh, 1994).

(5.90)

5.4.2 The Inverse Slant Stack Operator

The definition of the forward and adjoint slant stack operators may be changed to construct another slant stack pair. In this case we begin by defining the forward operator as follows

$$d(h,t) = \mathcal{L}^* m(h,t) = \int m(p, t = \tau - hp) dp$$

Similarly, we define the adjoint operator

$$\tilde{m}(p,\tau) = \mathcal{L} d\,(h,t) = \int d(h, t = \tau + hp) dh, \tag{5.91}$$

The pair of transformations can be written in the frequency-offset domain

$$D(h,f) = \int M(p,f) e^{-i2\pi f p h} dp \tag{5.92}$$

and

$$\tilde{M}(p,f) = \int D(h,f) e^{i2\pi f p h} dh$$

Substituting Eq. (5.91) into Eq. (5.92) yields

$$\tilde{M}(p,f) = \int M(p',f) \int e^{-i2\pi f h(p'-p)} dh\, dp' \tag{5.93}$$

where now, the convolution is with respect to p. The convolutional operator is given by

$$\gamma(p,f) = \int \frac{1}{2\pi|f|} e^{ihp} dh = \frac{1}{|f|}\delta(p)$$

The γ filter is a delta function with respect to p and, hence, Eq. (5.93) becomes

$$\tilde{M}(p,f) = \frac{1}{|f|} M(p,f)$$

or equivalently

$$M(p,f) = |f|\tilde{M}(p,f)$$

From the above derivation, it is clear that the ρ and the γ filters have the same frequency response. Finally, the slant stack pair becomes

$$\begin{aligned} M(p,f) &= |f| \int D(h,f) e^{i2\pi f p h} dh, \\ D(h,f) &= \int M(p,f) e^{-i2\pi f p} dp \end{aligned} \tag{5.94}$$

Assuming a finite aperture defined by $h \in [-H, H]$, the γ filter has the following structure

$$\gamma(p,f) = \int_{-H}^{H} e^{i2\pi f p h} dh = 2H \frac{\sin(2\pi f H p)}{2\pi f H p} \tag{5.95}$$

Hence, $M(p,f)$ may be calculated by solving

$$\tilde{M}(p,f) = 2H \int V(p',f) \frac{\sin(2\pi f H(p-p'))}{2\pi f H(p-p')} dp' \tag{5.96}$$

After a comparison of the slant stack pairs, given by Eqs. (5.94) and (5.95), it is clear that a deconvolution procedure is required in both cases. In the slant stack transform, the deconvolution is necessary to recover the data from the $\tau - p$ transform. In the inverse slant stack transformation, the deconvolution process is required to compute the $\tau - p$ transform. Truncation effects in p may be alleviated by choosing the proper region of support for the transform. The truncation in h is associated with the resolution of the transform and cannot be alleviated by simple means. Generally, both h and p are truncated and deconvolution should be applied in both the forward and inverse transforms. However, the range of p may be chosen in such a way that most of the energy in the signal lies within this range.

5.4.3 The Sampling Theorem for Slant Stacks

Assuming that the wavefield is evenly sampled, $D(n\Delta h, f)$ $n = 0, \pm 1, \pm 2, \ldots$, the relationship between the $\tau - p$ and the $h - t$ spaces is given by

$$D(n\Delta h, f) = |f| \int M(p, f) e^{-i2\pi f p n \Delta h} dp$$

where $M(p, f)$ denotes the slant stack corresponding to the continuous wavefield $D(h, f)$. The integration domain can be decomposed into small subdomains as follows

$$\begin{aligned} D(n\Delta h, f) &= |f| \sum_k \int_{(2k-1)\frac{\pi}{f\Delta h}}^{(2k+1)\frac{\pi}{f\Delta h}} M(p, f) e^{-i2\pi f p n \Delta h} dp \\ &= |f| \sum_k \int_{-\frac{\pi}{f\Delta h}}^{\frac{\pi}{f\Delta h}} M(p + 2k\frac{\pi}{f\Delta h}, f) e^{-i2\pi f (p + 2k\frac{\pi}{f\Delta h}) n \Delta h} dp \end{aligned} \quad (5.97)$$

Since $e^{-i2\pi nk} = 1 \ \forall \ n$ and k, Eq. (5.97) may be written as

$$U(n\Delta h, f) = |f| \int_{-\frac{\pi}{f\Delta h}}^{\frac{\pi}{f\Delta h}} M_d(p, f) e^{-i2\pi f p n \Delta h} dp$$

where the relationship between the slant stack of the continuous signal and the one corresponding to the sampled wavefield, $M_d(p, f)$, is given by

$$M_d(p, f) = \sum_k M(p + 2k\frac{\pi}{f\Delta h}, f)$$

Thus, the discrete signal has an $f - p$ representation with support in the range $p \in [-\frac{\pi}{f\Delta h}, \frac{\pi}{f\Delta h}]$. The components with slope $p - 2\frac{\pi}{f\Delta h}, p + 2\frac{\pi}{f\Delta h}, p - 4\frac{\pi}{f\Delta h}, p + 4\frac{\pi}{f\Delta h}, \ldots$ will appear to have slope p and every slope outside the range $(-\frac{\pi}{f\Delta h}, \frac{\pi}{f\Delta h})$ will have an alias inside this range. If the continuous signal has all the components inside that range, the aliased components do not exist and, therefore, we can write $M_d(p, f) = M(p, f)$. It is clear from the above discussion that spatial sampling must be chosen so as to avoid the aliasing effect. If $P = P_{max} = -P_{min}$, the following relationship guarantees the absence of alias

$$\Delta h \leq \frac{1}{2P f_{max}}$$

5.4.4 Discrete Slant Stacks

Discrete versions of the continuous Radon pair are obtained by discretizing the offset and ray parameter axes. Consider a seismic image that contains N traces at offset positions $h_0, h_1, \ldots, h_{N-1}$. Sampling the p axis at M locations, $p_0, p_1, \ldots, p_{M-1}$, the LRT becomes

$$m(p_k, \tau) = \sum_{l=0}^{N-1} d(h_l, \tau + p_k h_l)$$

Similarly, we define the adjoint of the LRT operator as follows

$$\tilde{d}(h_l, t) = \sum_{k=0}^{M-1} m(h, t - h_l p_k)$$

Taking the Fourier transform with respect to the temporal variables, yields

$$\begin{aligned} M(p_k, f) &= \sum_{l=0}^{N-1} D(h_l, f) e^{2\pi i f h_l p_k} \\ \tilde{D}(h_l, f) &= \sum_{k=0}^{M-1} M(p_k, f) e^{-2\pi i f h_l p_k} \end{aligned}$$

Using matrix notation it is possible to rewrite the LRT and its adjoint as follows (f is omitted for convenience)

$$\mathbf{m} = \mathbf{L}^H \mathbf{d} \tag{5.98}$$

$$\tilde{\mathbf{d}} = \mathbf{L}\mathbf{m} \tag{5.99}$$

where

$$\mathbf{L} = \begin{pmatrix} e^{i2\pi f p_0 h_0} & e^{i2\pi f p_1 h_0} & \ldots & e^{i2\pi f p_{M-1} h_0} \\ e^{i2\pi f p_0 h_1} & e^{i2\pi f p_1 h_1} & \ldots & e^{i2\pi f p_{M-1} h_1} \\ e^{i2\pi f p_0 h_2} & e^{i\,2\pi\, f\, p_1\, h_2} & \ldots & e^{i2\pi f p_{M-1} h_2} \\ \vdots & \vdots & \vdots & \vdots \\ e^{i2\pi f p_0 h_{N-1}} & e^{i2\pi f p_1 h_{N-1}} & \ldots & e^{i2\pi f p_{M-1} h_{N-1}} \end{pmatrix}$$

An inversion formula for the LRT may be derived by substituting Eq. (5.98) into Eq. (5.99)

$$\tilde{\mathbf{d}} = \mathbf{L}\mathbf{L}^{\mathbf{H}}\mathbf{d} \tag{5.100}$$

Eq. (5.100) is invertible provided that $\det(\mathbf{L}\,\mathbf{L}^{\mathbf{H}}) \neq \mathbf{0}$ and we may write

$$\begin{aligned}\mathbf{d} &= (\mathbf{L}\mathbf{L}^H)^{-1}\tilde{\mathbf{d}} \\ &= \mathbf{G}^{-1}\tilde{\mathbf{d}}\end{aligned}$$

The $N \times N$ matrix $\mathbf{G} = \mathbf{L}\mathbf{L}^H$ represents a discrete version of the ρ filter. The pair of transformations which map a signal from $f-h$ to $f-p$, and vice-versa, are given by

$$\mathbf{m} = \mathbf{L}\mathbf{d}$$

and

$$\mathbf{d} = \mathbf{G}^{-1}\mathbf{L}\mathbf{m}$$

The vector $\mathbf{m}$ is computed by means of a simple matrix times vector multiplication. Both expressions constitute an inverse pair when the inverse of $\mathbf{G}$ exits. It is important to stress that this formulation of the RT does not permit adequate modelling of the signal when additive noise is present which maps into the Radon domain. We deal with this shortcoming below.

5.4.5 Least Squares Inverse Slant Stacks

Assume that the data, $\mathbf{d}$, result from a Radon (slant stack) operator applied to $\mathbf{m}$, and that $\mathbf{d}$ is contaminated by noise, $\mathbf{n}$

$$\mathbf{d} = \mathbf{L}\mathbf{m} + \mathbf{n}$$

We form an objective function

$$J = ||\mathbf{d} - \mathbf{L}\mathbf{m}||^2$$

which is minimized to obtain the LS solution

$$\mathbf{m} = (\mathbf{L}^H\mathbf{L})^{-1}\mathbf{L}^H\mathbf{d}$$

a stabilized form of which is

$$\mathbf{m} = (\mathbf{L}^H\mathbf{L} + \mu\mathbf{I})^{-1}\mathbf{L}^H\mathbf{d} \tag{5.101}$$

Eq. (5.101) describes the conventional manner of computing the slant stack pair. Other techniques to improve the resolution of these operators were proposed by Sacchi and Ulrych (1995a). It is important to stress that the system represented by Eq. (5.101) can be efficiently solved using Levinsons recursion (Kostov, 1990). We return to this issue when discussing the numerical implementation of the RT.

5.4.6 Parabolic Radon Transform (PRT)

A clever modification of the LRT to implement the attenuation of multiple reflections is seismic records, was proposed by Hampson (1986). In the LRT, the integration path is given by

$$t = \tau + ph$$

In the PRT, the path is modified to

$$t = \tau + qh^2$$

This is a good approximation to reflections in CMP gathers after normal moveout, NMO, correction, where the moveout of the primaries is zero (we describe this correction below).

To derive the parabolic approximation we first assume a CMP composed of two events, a primary and a multiple with equivalent intercept time, T_0. The travel-time curve for the primary is given by

$$T_p = \sqrt{(T_0^2 + h^2/V_p^2)}$$

and the expression for the multiple is

$$T_m = \sqrt{(T_0^2 + h^2/v_m^2)}$$

where V_p and V_m are the primary and multiple velocities, respectively. If the multiple is generated by a shallow layer or by the water column (in the marine case) we can consider $V_p > V_m$. Suppose that we now apply the NMO correction to the CMP gather with a NMO law that uses the velocity V_{NMO}. The NMO correction entails applying the following time shift to the data

$$\Delta T_{NMO} = T_0 - \sqrt{(T_0^2 + h^2/V_{NMO}^2)}$$

and, therefore, the time of the primary after NMO is

$$T_{p[NMO]} = T_p + \Delta T_{NMO}$$

It is clear that the NMO velocity is equal to the velocity of the primary and that the travel-time of the primary reflection becomes

$$T_{p[NMO]} = T_0$$

The travel-time of the multiple after NMO is given by

$$T_{m[NMO]} = T_m + T_0 - \sqrt{(T_0^2 + h^2/v_{NMO}^2)}$$

or, after replacing T_m

$$T_{m[NMO]} = T_0 + \sqrt{(T_0^2 + h^2/v_m^2)} - \sqrt{(T_0^2 + h^2/v_{NMO}^2)}$$

The two square roots in the above equation can be expanded in a Taylor series and, keeping only second order terms, we have

$$T_{m[NMO]} \approx T_0 + \frac{1}{2T_0 v_m^2}h^2 - \frac{1}{2T_0 v_{NMO}^2}h^2$$

which can be re-written as

$$T_{m[NMO]} \approx T_0 + qh^2$$

where

$$q = \frac{1}{2T_0}(\frac{1}{V^2} - \frac{1}{V_{NMO}^2})$$

Hence, a transform with a parabolic integration path can be constructed by a simple interchange of h with h^2 in the LRT. Sometimes, it is preferable to parameterize the parabola in terms of q, the residual moveout time at far offset, in which case

$$t = \tau + q\frac{h^2}{h_{max}^2}$$

where the q has units of time.

5.4.7 High Resolution Radon Transforms

We have met problems associated with the issue of resolution before, an excellent example being that of computing the power spectrum of a time limited signal. Aperture always limits resolution and, not surprisingly therefore, the RT suffers from the same malaise. But things are not all black, however. It turns out, as we now explore, that one can devise a high resolution version of the RT that is much superior to the LS transform that we have described thus far. The least squares inverse, LSIRT, for the RT, is obtained by minimizing a cost function of the form

$$J_{GG} = \frac{1}{\sigma_n^2}||\mathbf{Lm} - \mathbf{d}||^2 + \frac{1}{\sigma_m^2}||\mathbf{m}||^2$$

The first term in this equation is the data misfit, the second term, often called the regularizer, is used to stabilize the solution by keeping the elements of the vector of unknown parameters small (please see Section 4.2.2). Minimizing J_{GG} in standard fashion, leads to the damped LS solution

$$\mathbf{m}_{LS} = (\mathbf{L}^H \mathbf{L} + \mu \mathbf{I})^{-1}\mathbf{L}^H\mathbf{d}$$

where the trade-off parameter is given by

$$\mu = \frac{\sigma_n^2}{\sigma_m^2}$$

As described in Section 4.5.12.1, $\mathbf{m}_{LS}$ can also be obtained as the MAP (maximum a posteriori) solution in a Bayesian sense, assuming that both the noise and model parameters are normally distributed (Sacchi and Ulrych, 1995a).

It is important to understand the effect of μ in the solution. First, we notice that large values of μ lead to the following form

$$\mathbf{m}_{LS} \approx \frac{1}{\mu}\mathbf{L}^H\mathbf{d}$$

which is a scaled version of the low resolution RT obtained by operating on the data with the adjoint $\mathbf{L}^H$. This is the best we can do in extremely low SNR situations. We also notice that μ is model independent, in other words, all the model parameters are assumed to have the same variance. If the data consist of a limited number of events (linear or parabolic events) the vector of unknowns, $\mathbf{m}$, should exhibit some degree of sparseness (most elements equal to zero, with a few elements different to zero at the parameter positions, p or q, where waveforms exist). It is clear that we need a regularization method capable of retrieving sparse models. This idea has been extensively studied in Chapter 4 and will be exploited in the Chapter 5.

Following Sacchi and Ulrych (1995a), we define the Cauchy norm with scale parameter β of our (complex) vector of unknown parameters as

$$J_C = \sum_{i=1}^{M} \log(1 + \frac{|m_i|^2}{\beta^2})$$

and form the objective function for our problem

$$J_{GC} = \frac{1}{\sigma_n^2}||\mathbf{Lm} - \mathbf{d}||^2 + J_C$$

where the subscript GC stands for a Gauss-Cauchy model and emphasizes that the errors are modelled using a Gaussian distribution whereas the model parameters are modelled according to a Cauchy distribution (Section 4.5.12.2). Taking derivatives of J_{GC} and equating to zero leads to

$$\mathbf{m} = (\mathbf{L}^H\mathbf{L} + \mathbf{Q}(\mathbf{m}))^{-1}\,\mathbf{L}^H\mathbf{d}$$

where

$$\mathbf{Q}_{ii} = \frac{\sigma_n^2}{\beta^2(1 + \frac{|m_i|^2}{\beta^2})} \tag{5.102}$$

As previously discussed, this equation must be solved using some type of a nonlinear optimization scheme. We have found that, in general, good results are obtained using IRLS (iterative re-weighted LS) (Scales et al., 1988). In IRLS, the unknown vector $\mathbf{m}$ is retrieved iteratively

$$\mathbf{m}_k = (\mathbf{L}^H\mathbf{L} + \mathbf{Q}_{k-1})^{-1}\mathbf{L}^H\mathbf{d} \tag{5.103}$$

where $\mathbf{Q}_{k-1}$ indicates the matrix of weights computed from the solution $\mathbf{m}_{k-1}$ at iteration $k-1$. We now compare the performance of the LS LRT and PRT with the high resolution version of these transforms. Results for the linear case are shown in Figs. 5.33 and 5.34.

It is clear that the LS solution produces Radon panels dominated by smearing and truncation artifacts. The high resolution Radon Transform, on the other hand, has managed to collapse the waveforms to isolated wavelets in the transform domain. The same features are observed in Figs. 5.35 and 5.36 that illustrate results for parabolic events. The importance of achieving high resolution at the time of filtering waveforms of different curvature in the Radon panel is quite apparent from these simple examples.

5.4.8 Computational Aspects

Computations involved in the various RTs that we have discussed are not trivial, particularly if large volumes of data need to be processed and time and memory requirements are of the essence. We concentrate on the computational aspects of the LS inverse RT and of the high resolution RT, the HRRT. We also discuss various computational issues concerning the PRT, issues that apply equally to the LRT.

5.4.8.1 Least Squares Radon Transform

The LSIRT entails the solution of the following system

$$\begin{aligned}\mathbf{m} &= (\mathbf{L}^H\mathbf{L} + \mu\mathbf{I})^{-1}\mathbf{m}_{adj} \\ &= (\mathbf{R} + \mu\mathbf{I})^{-1}\mathbf{m}_{adj}\end{aligned}$$

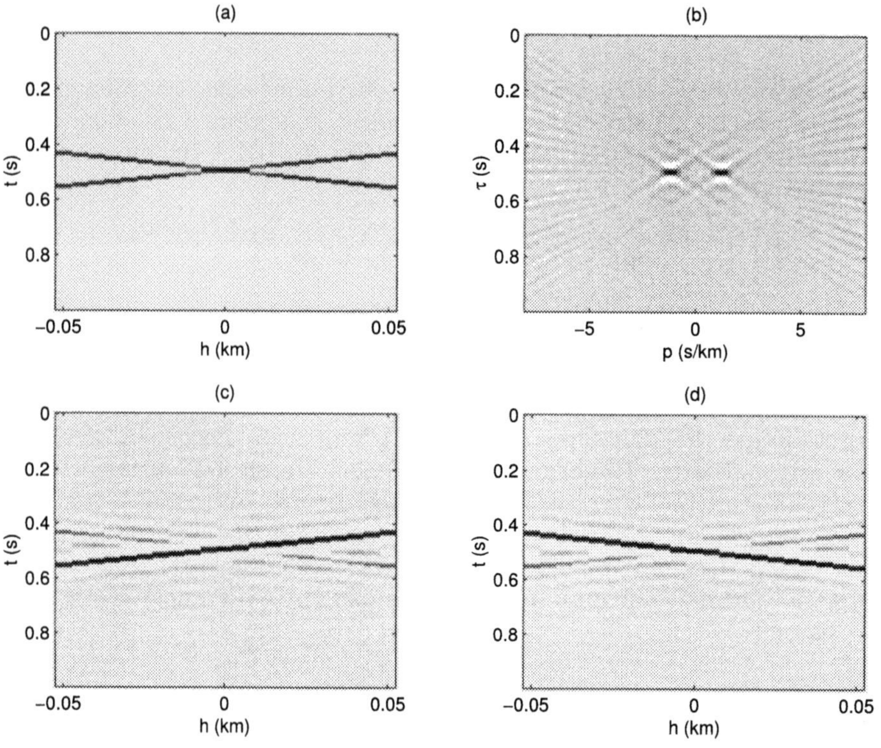

Figure 5.33: Linear RT synthetic example. (a) Events in $t-x$. (b) $\tau-p$ panel computed using the LS linear RT. (c) and (d) Separated waveforms using the $\tau-p$ panel.

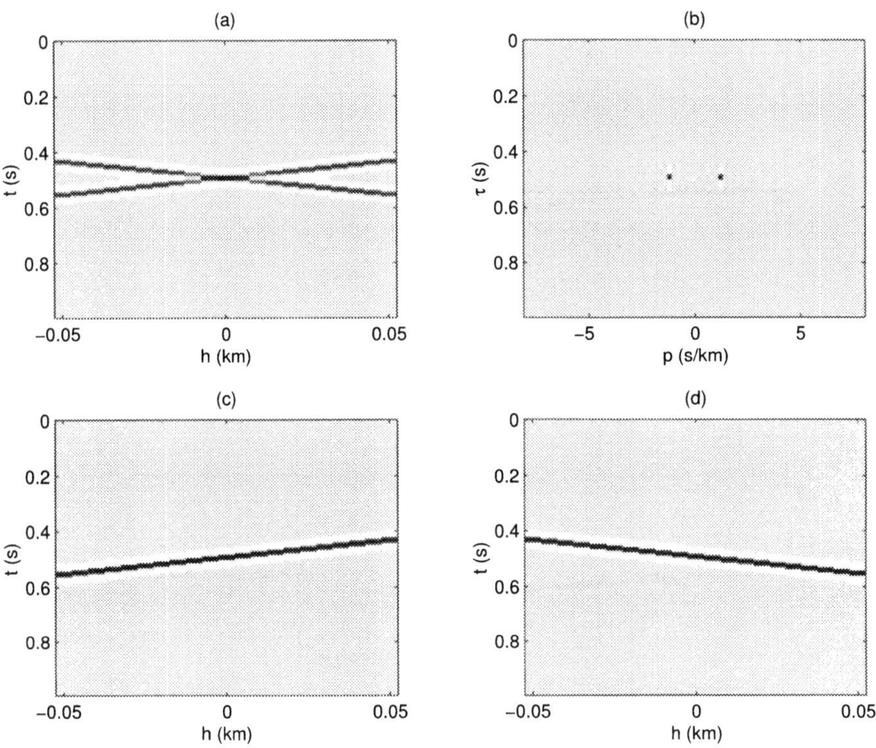

Figure 5.34: High resolution linear RT synthetic example. (a) Events in $t - x$. (b) Radon panel computed using the high resolution linear RT. (c) and (d) Separated waveforms using the $\tau - p$ panel (to be compared with Fig. 5.33).

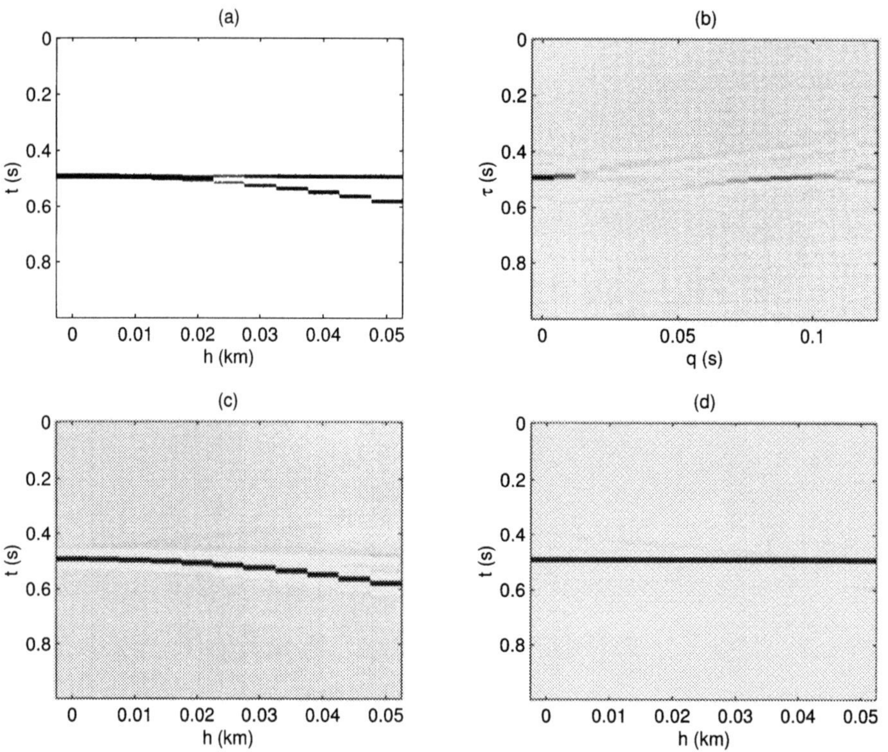

Figure 5.35: PRT synthetic example. (a) Events in $t - x$. (b) $\tau - q$ panel computed using the LS parabolic RT. (c) and (d) Separated waveforms using the $\tau - q$ panel.

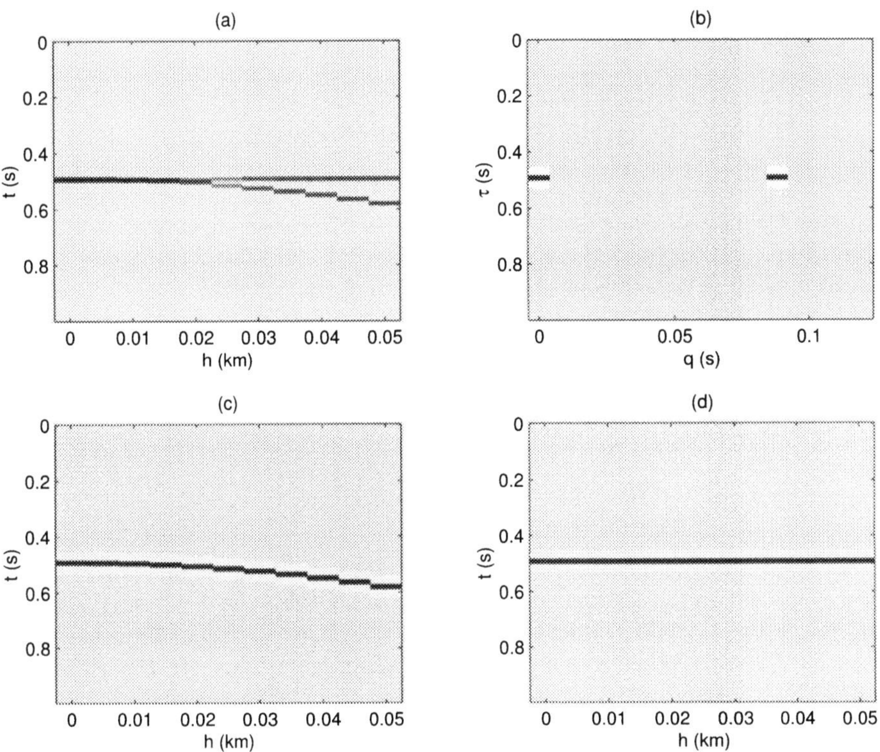

Figure 5.36: High resolution parabolic RT synthetic example. (a) Events in $t-x$. (b) Radon panel computed using the high resolution parabolic RT. (c) and (d) Separated waveforms using the $\tau - q$ panel (to be compared with Fig. 5.35).

where

$$\mathbf{m}_{adj} = \mathbf{L^H d}$$

and defines the adjoint model as described by Eq. (5.99). We note a number of important points.

$$\mathbf{R} = \mathbf{L}^H \mathbf{L} + \mu \mathbf{I}$$

is Toeplitz (Kostov, 1990), with elements given by

$$\{\mathbf{R} + \mu \mathbf{I}\}_{lm} = \sum_{k=1}^{N} e^{-i2\pi f \Delta q(l-m)h_k^2} + \mu \delta_{lm}$$

Solving this system using the Levinson recursion requires approximately $4M^2 + 7M$ operations, and storage of only the first row of the matrix (Marple, 1987). This feature yields a very efficient algorithm to compute the PRT.

5.4.8.2 High Resolution Parabolic Radon Transform

In the high resolution PRT, **m** is computed by solving

$$(\mathbf{R} + \mathbf{Q})\mathbf{m} = \mathbf{m}_{adj} \tag{5.104}$$

R is a diagonal matrix with elements that depend on **m**. This leads to an iterative algorithm where **Q** is bootstrapped from the result of a previous iteration. In general, the iterative procedure is not required if we are able to design **Q** from a priori information. We will return to this point when discussing a non-iterative implementation of the high resolution RT. The weight matrix, **Q**, is diagonal with elements

$$\{\mathbf{Q}\}_{lm} = Q_l \, \delta_{lm}, \quad l, m = 1, 2, \ldots, M$$

The diagonal elements of $\mathbf{R} + \mathbf{Q}$ are given by

$$\{\mathbf{R} + \mathbf{Q}\}_{lm} = \sum_{k=1}^{N} e^{-i2\pi f \Delta q(l-m)h_k^2} + w_l^2 \delta_{lm}$$

It is clear that the addition of a diagonal matrix with non constant elements destroys the Toeplitz structure of the operator. The above matrix can be inverted using the Cholesky method in a number of operations proportional to M^3. From the computational point of view, it is more convenient to compute the RT using a constant diagonal regularization (Sacchi and Porsani, 1999). However, if we want to estimate a high resolution Radon operator, the regularization term must be diagonal with non constant elements. To solve Eq. (5.104), we adopt the method of conjugate gradients as follows: Begin with an initial solution, $\mathbf{m_0}$

$$
\begin{aligned}
\text{set } \mathbf{p}_0 &= \mathbf{r}_0 = \mathbf{m}_{\text{adj}} - (\mathbf{R}+\mathbf{Q})\mathbf{m}_0 \\
\text{for } i = &\ 1:K \\
\alpha_{i+1} &= (\mathbf{r}_i, \mathbf{r}_i)/(\mathbf{p}_i, (\mathbf{R}+\mathbf{Q})\mathbf{p}_i) \\
\mathbf{m}_{i+1} &= \mathbf{m}_i + \alpha_{i+1}\mathbf{p}_i \\
\mathbf{r}_{i+1} &= \mathbf{r}_i - \alpha_{i+1}(\mathbf{R}+\mathbf{Q})\mathbf{p}_i \\
\beta_{i+1} &= (\mathbf{r}_{i+1}, \mathbf{r}_{i+1})/(\mathbf{r}_i, \mathbf{r}_i) \\
\mathbf{p}_{i+1} &= \mathbf{r}_{i+1} + \beta_{i+1}\mathbf{p}_i \\
\text{end} &
\end{aligned}
$$

The cost of the conjugate gradient, CG, algorithm is dominated by the cost of multiplying a matrix by a vector. In general, a matrix times vector multiplication requires $O(M^2)$ operations. In our problem we will use the Toeplitz structure of $\mathbf{R}$ to find a fast method to compute this product. $(\mathbf{R}+\mathbf{Q})\mathbf{x}$ can be decomposed into two products, $\mathbf{Rx}+\mathbf{Qx}$. The first product can be efficiently computed using the Fast Fourier Transform (FFT), while the second involves only $2M$ operations (M products plus M additions) and does not substantially increase the computational cost of the inversion. The first product, $\mathbf{y}=\mathbf{Rx}$, is evaluated by augmenting the system as follows

$$
\begin{bmatrix} \mathbf{y} \\ \mathbf{y}' \end{bmatrix} = \mathbf{R}_{aug} \begin{bmatrix} \mathbf{x} \\ \mathbf{0} \end{bmatrix}
$$

where $\mathbf{R}_{\text{aug}}$ is the original Toeplitz matrix but properly folded to become a circulant matrix (Strang, 1988, and Section 1.9). The vector $[\mathbf{y}, \mathbf{y}']^T$ is computed by multiplying the FT of the first row of $\mathbf{R}_{aug}$ by the FT of the vector $[\mathbf{x}, \mathbf{0}]^T$, and taking the inverse FT of this product. Now our matrix times vector takes $O(M' \log M')$ operations, where M' is the size of augmented matrix ($M' = 2M$). We have found that the CG algorithm converges after a few iterations. The inversion, therefore, becomes a $O(KM' \log(M'))$ process. This is more efficient than the direct inversion of Eq. 5.104 using Cholesky.

5.4.8.3 Non-iterative High Resolution Radon Transform

An interesting modification to the high resolution RT has been proposed by Herrmann et al. (2000), who computed the weighting operator, $\mathbf{Q}$, from a previous frequency estimate of $\mathbf{m}$. One of the advantages of this procedure is that an iterative scheme to compute the weights is avoided. Moreover, since low frequency estimates of the RT are used to constraint high frequency estimates, the procedure operates as an anti-alias operator.

The original algorithm by Sacchi and Ulrych (1995a) can be summarized as follows:

- For all frequencies, f, solve

$$
[\mathbf{R}(f) + \mathbf{Q}(f)_{k-1}]\mathbf{m}(f)_k = \mathbf{L}^H \mathbf{d}(f) \quad k = 1, 2, \ldots.
$$

The non iterative algorithm of Herrmann et al. (2000) leads to:

- For all frequencies, f, solve

$$[\mathbf{R}(f) + \mathbf{Q}(f - \Delta f)]\mathbf{m}(f) = \mathbf{L}^H\mathbf{d}(f)$$

where $\mathbf{Q}(f - \Delta f)$ indicates the weights computed from the solution $\mathbf{m}$ at temporal frequency $f - \Delta f$.

An alternative algorithm is to use a two pass process:

- Compute the LS solution $M(q, f)$
- Determine the q-spectrum $S(q) = \sum_{f\in B} |M(q, f)|^2$
- Determine the frequency independent weights $Q(q) = \frac{\alpha}{\beta^2+S(q)}$
- For all frequencies, f, solve

$$[\mathbf{R}(f) + \mathbf{Q}]\,\mathbf{m}(f) = \mathbf{L}^H\mathbf{d}(f)$$

In the last algorithm, the low resolution LS solution is used to estimate the q-spectrum. It is clear that a combination of all these strategies is a plausible path to developing stable, high resolution Radon operators.

Figs. 5.37 and 5.38 illustrate a real data example. The Radon operator is computed using the LS procedure. A mask in the Radon domain is used to isolate the multiple energy. The estimated multiples are subtracted from the data to obtain an estimate of the primaries. It is clear that in the LS PRT there is not sufficient separation of events in the transform domain to properly design a mask function (Fig. 5.38). When the HR PRT is implemented (in this case we used the two passes algorithm described above) primaries and multiples are properly modelled and are isolated in the Radon panel. The resulting demultipled data and the Radon panel are shown in Figs. 5.39 and 5.40.

5.5 Time variant Radon Transforms

We will discuss in this section the computation of time variant operators that can be used as an alternative to the parabolic Radon transform. The parabolic Radon transform is a time invariant operator, therefore it can be implemented in the frequency domain. This trick permits one to solve several small problems, one at each frequency, instead of a large problem involving all the time-offset-velocity samples at the same time. It is clear that in the case of the parabolic Radon transform time invariance is achieved by means of an approximation. In situations where the parabolic approximation is not properly satisfied (travel-times especially at far offsets are not properly modelled) we can construct a Radon transform with integration along hyperbolic paths.

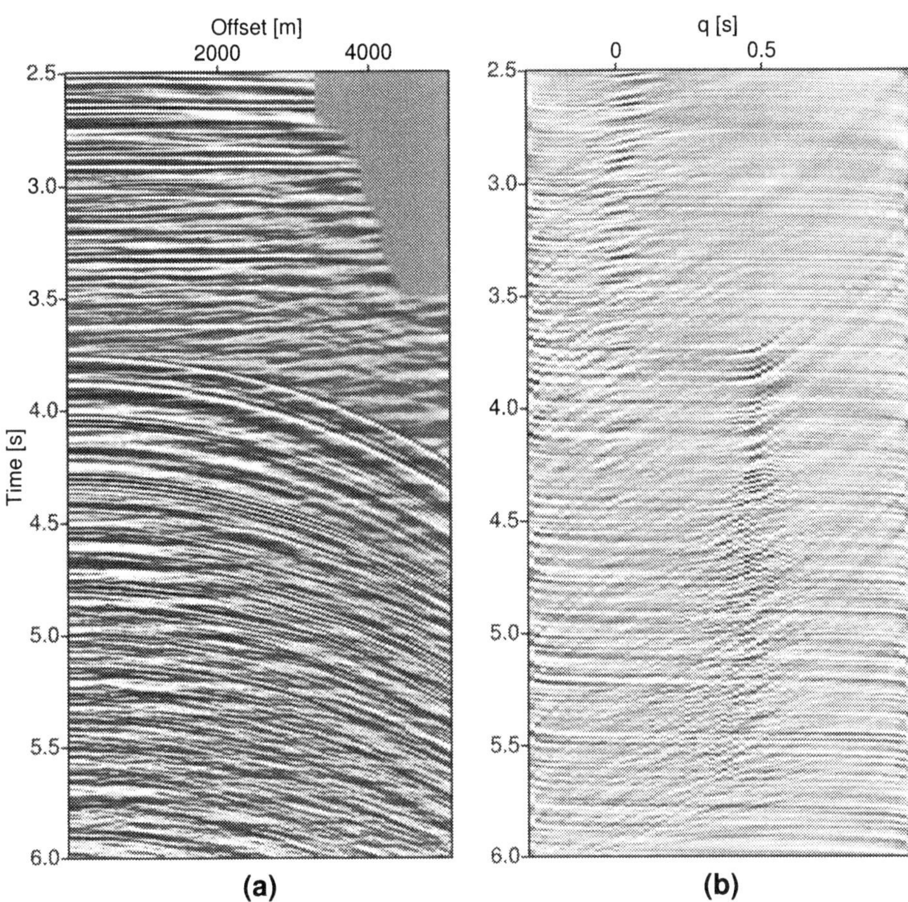

Figure 5.37: (a) Marine gather after NMO correction. (b) Radon panel computed via the least squares parabolic Radon transform (LSPRT).

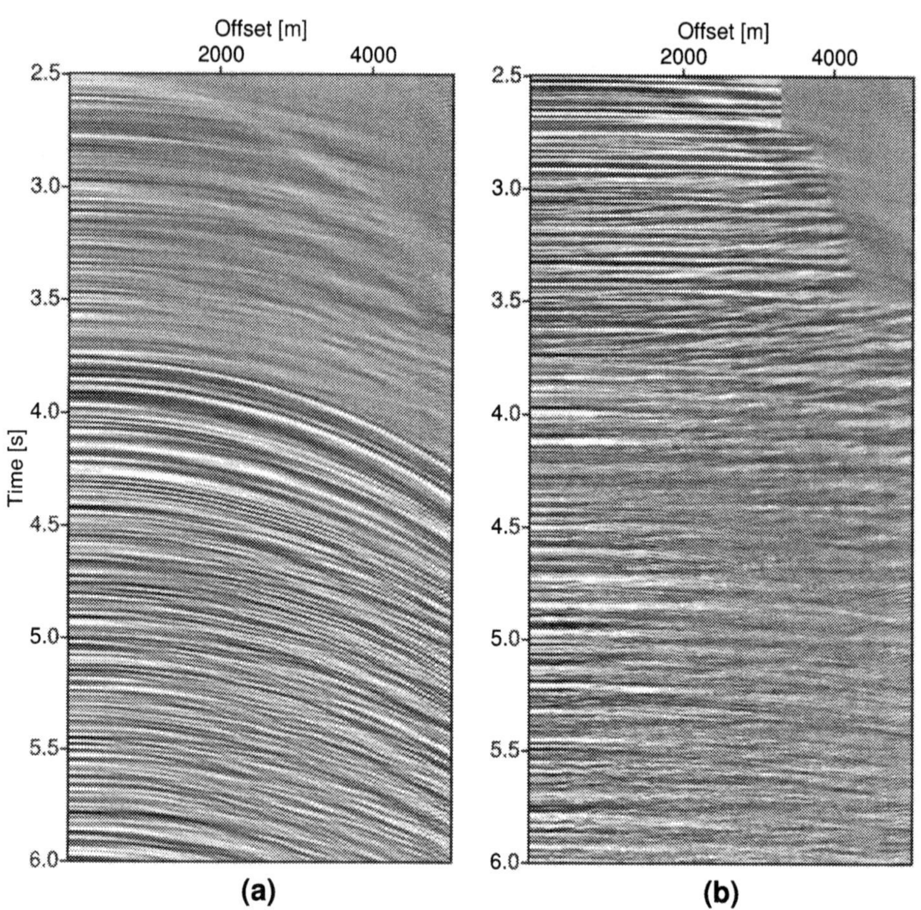

Figure 5.38: (a) Multiples estimated by muting and mapping back to data space the Radon panel obtained via the LSPRT. (b) Primaries obtained by subtracting the estimated multiples from the original data.

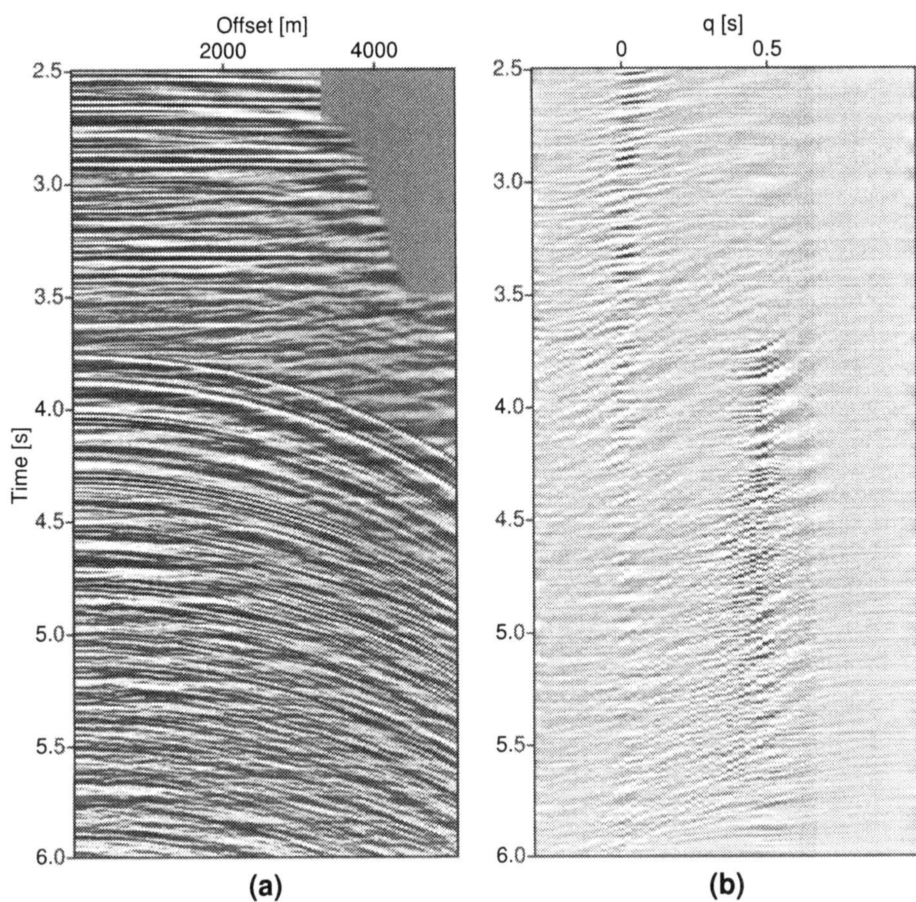

Figure 5.39: (a) Marine gather after NMO correction. (b) Radon panel computed via the high resolution parabolic Radon transform (two pass algorithm described in the text).

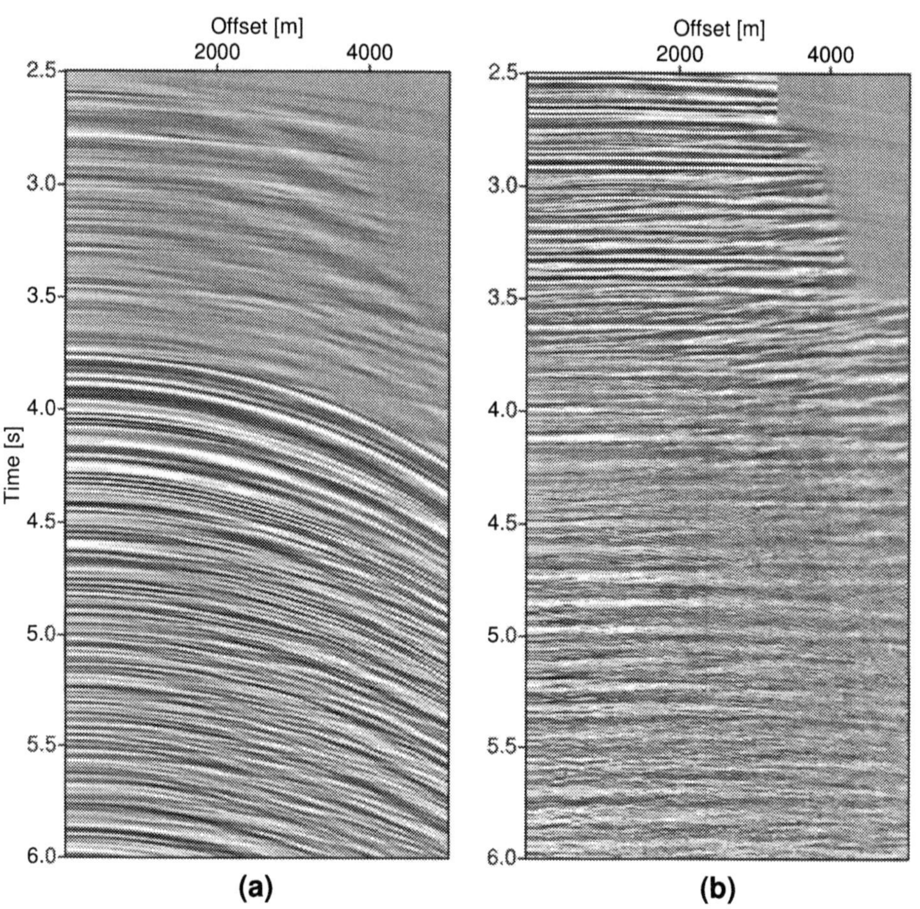

Figure 5.40: (a) Multiples estimated by muting and mapping back to data space the Radon panel obtained via the high resolution Radon Transform. (b) Primaries obtained by subtracting the estimated multiples from the original data.

The time variant hyperbolic Radon transform operator is defined in terms of summation along Dix hyperbolas. If $m(\tau, v)$ is used to designate the Radon panel and $d(t, h)$ the CMP gather then,

$$d(t,h) = \int m(\tau = \sqrt{t^2 - h^2/v^2}, v)\, dv \tag{5.105}$$

where h is source-receiver offset, t is two-way travel-time, v is the RMS velocity, and τ is two-way vertical travel-time. After discretization and lexicographic arrangement, Eq. (5.105) can be written as

$$\mathbf{d} = \mathbf{L}\mathbf{m}$$

The vectors $\mathbf{d}$ and $\mathbf{m}$ have $nt \times nh$ and $n\tau \times nv$ elements, respectively. The dimension of the operator $\mathbf{L}$ is $(nt \times nh) \times (nt \times n\tau)$. The forward or modelling operator, $\mathbf{L}$, picks a wavelet in velocity space and produces a hyperbola in data space. The transpose operator $\mathbf{L}^T$ is capable of undoing the aforementioned process. It is important to stress, however, that the transpose operator is a bad approximation to the inverse of $\mathbf{L}$, and therefore a procedure is required to find an inverse operator. We consider the problem

$$\text{Minimize } \phi = ||\mathbf{L}\,\mathbf{m} - \mathbf{d}||_2^2$$

Differentiating ϕ with respect to $\mathbf{m}$ yield the least-squares solution

$$\hat{\mathbf{m}} = (\mathbf{L}^T\,\mathbf{L})^{-1}\mathbf{L}^T\,\mathbf{d} = (\mathbf{L}^T\,\mathbf{L})^{-1}\mathbf{m}_{adj} \tag{5.106}$$

$\mathbf{m}_{adj}$ in Eq. (5.106) is the low resolution velocity-stack computed by means of the adjoint or transpose operator. The $\tau - v$ panel computed after inversion, $\hat{\mathbf{m}}$, possesses more resolution that $\mathbf{m}_0$. Unfortunately, the computation of $\hat{m}$ involves the inversion of $\mathbf{L}^T\mathbf{L}$. If we assume a typical CMP gather of 48 channels and 1000 samples per trace. In addition, suppose that 48 traces and 1000 samples were used to discretize the velocity axis, v , and the intercept time, τ, respectively. In this case $\mathbf{L}^T\mathbf{L}$ has dimension 24000 × 24000. It is evident that direct methods cannot be applied to this type of problems. The trick here is to use a semi-iterative technique, like the conjugate gradient, CG, method, to find an approximate solution to our problem. The advantage of the CG algorithm is that the matrix $\mathbf{L}$ is replaced by an operator acting on a vector. Details pertaining to the implementation of the hyperbolic Radon transform can be found in Trad et al. (2002). In addition, this reference describes a strategy to efficiently manipulate large inverse transformations by incorporating sparse matrix operations. In Fig. (5.41) we provide an example that summarizes our discussion. In this example we have used the adjoint operator to compute the low resolution $\tau - v$ panel (Fig. 5.41 b) and the least squares solution using CG (Fig. 5.41 c). It is important to note that, like in the processing of seismic data with the parabolic Radon transform, filtering and mapping back to the original data space is required to estimate a noise model (multiples). It is also important to mention that sparseness constraints can also be included in the formulation of the hyperbolic Radon transform, a number of authors have investigated this problem (Thorson and Claerbout (1985), Cary (1998), and Trad et al. (2002)). A survey that introduces several algorithms is provided in Trad et al. (2003).

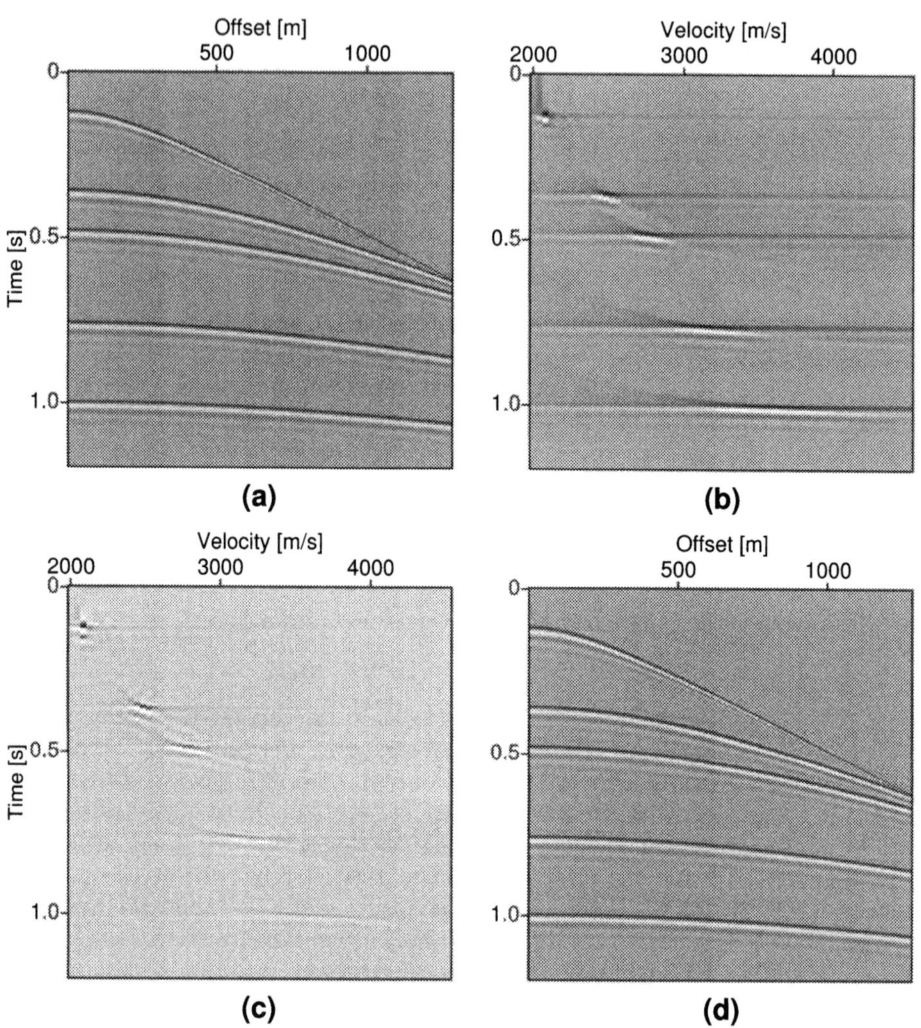

Figure 5.41: (a) Synthetic CMP gather. (b) $\tau - v$ panel obtained with the adjoint operator ($\mathbf{L}^T$). (c) $\tau - v$ panel obtained using the method of conjugate gradients to invert the hyperbolic Radon operator. (d) Data predicted by applying the modelling operator ($\mathbf{L}$) to panel (c).

5.5.1 Discussion

We emphasized in the introduction to this chapter that one of the difficult tasks facing the seismic processor, is the elimination of coherent noise, the term we designated as $\mathbf{n}_c$. The most common form of $\mathbf{n}_c$ in seismic sections occurs as ground-roll and multiples. The latter are of particular significance since, unless identified and attenuated, they may lead to erroneous and extremely costly drilling programs. We have presented a popular approach to multiple elimination by means of the RT. Of special significance in the design of suitable Radon operators is the resolution properties of the Radon panel where both identification and attenuation is carried out.

Chapter 6

Deconvolution with Applications to Seismology

6.1 Introduction

We have explored the concept of convolution in depth in Chapter 1. Specifically, we showed (Section 1.7.6) that convolution is ubiquitous because it is the process by which a linear and time invariant system transforms input into output. Section 1.7.10 introduced the inverse process of deconvolution, particularly stressing the importance of minimum phase in stable deconvolution filter design. In this chapter we consider the problem in much more depth. In particular we are concerned with the problem of deconvolving and inverting zero offset seismograms. This problem as we will see requires, essentially, the solution of one equation with two unknowns, a daunting task indeed. As a first step, we derive the convolution model by assuming that our seismic data have been acquired at normal incidence.

6.2 Layered Earth Model

The layered earth model has a very rich history in exploration seismology. We have explored some details in Chapter 2, specifically in Appendix A.1 in relationship to the AR model. The reasons for the importance of this geologically ideal model lies in the fact that, due to its simplicity, it lends itself to a rigorous analysis that, in turn, teaches much about the earth. At normal incidence, this 1D earth model allows, as we have seen, a complete inversion of the surface wavefield. i.e., the primary wavefield can be reconstructed from the recorded seismogram that contains *all* the multiple reflections that are entailed. Although the material that is the topic of the next two sections is covered in various texts (see, for example, Claerbout, 1976; Robinson and Treitel, 2002), it is of such a basic nature to the problem at hand, that we feel it is important to treat it here as well.

6.2.1 Normal Incidence Formulation

Consider a plane wave impinging at angle $i = 0$ with respect to the normal, Fig. 6.1. In this case we have three waves:

- Incident wave: $\downarrow$ in medium 1
- Reflected wave: $\uparrow$ in medium 1
- Transmitted wave: $\downarrow$ in medium 2

Assuming an incident wave of unit amplitude, the amplitude of the reflected wave is r, and the amplitude of the transmitted wave is t. At the boundary, the condition that must be satisfied due to the continuity of displacements is

$$1 + r = t \tag{6.1}$$

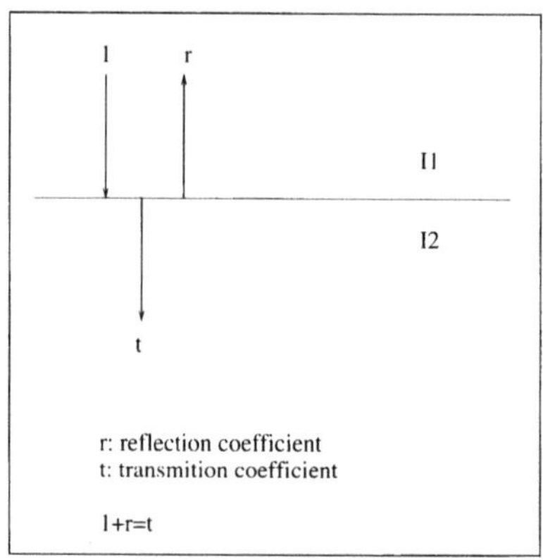

Figure 6.1: P-wave at normal incidence. The incident wave is propagating downwards.

This equation has two unknowns, and we require an additional equation in order to compute r and t. This equation is obtained by considering the conservation of energy which, in the acoustic normal incidence case, leads to

$$I_1 \times 1 + I_1 \times r^2 = I_2 \times t^2 \tag{6.2}$$

The quantities I_1 and I_2 are called *acoustic impedances*

$$I_1 = \rho_1 v_1$$

and

$$I_2 = \rho_2 v_2$$

where ρ_1 and ρ_2 are the densities of the material above and below the interface and v_1 and v_2 are the P-wave velocities, respectively[1]. After combining Eqs. (6.1) and (6.2), we obtain the following expressions

[1] P, as always, indicates the compressional wave

$$
\begin{aligned}
r &= \frac{I_2 - I_1}{I_2 + I_1}
\end{aligned}
$$

and

$$
\begin{aligned}
t &= \frac{2I_1}{I_2 + I_1}
\end{aligned}
$$

The above analysis is valid for an incident plane wave propagating downwards. Considering an incident wave propagating upwards, Fig. 6.2

- Incident wave: $\uparrow$ in medium 2
- Reflected wave: $\downarrow$ in medium 2
- Transmitted wave: $\uparrow$ in medium 1

the reflection and transmission coefficients are given by

$$
\begin{aligned}
r' &= \frac{I_1 - I_2}{I_2 + I_1}
\end{aligned}
$$

and

$$
\begin{aligned}
t' &= \frac{2I_2}{I_2 + I_1}
\end{aligned}
$$

which implies that

$$
r' = -r
$$

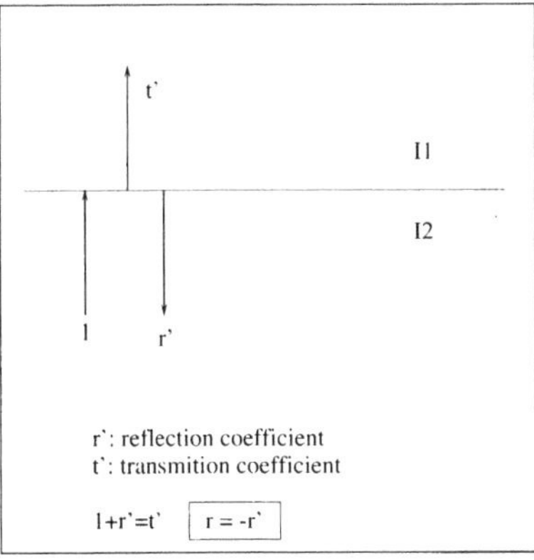

Figure 6.2: P-wave normal incidence. The incident wave is propagating upwards.

6.2.2 Impulse Response of a Layered Earth

Having established the essentials of reflection and transmission coefficients, we proceed to develop the expression for the Earth's impulse response. This issue has been treated in some detail in Appendix A.1, Chapter 2, and we deal here only with certain aspects, that are required for our treatment of deconvolution.

Let us assume that we have conducted a zero offset experiment in a stratified earth composed of 4 layers plus a half-space of impedances given by $I_1, I_2, I_3, I_4$and I_5. (Figure 6.3). At $t = 0$, a delta-like source propagates energy into the Earth that is transmitted and reflected from the various layers. If multiple reflection are not considered, our seismogram will consist of 4 arrivals (4 reflections). To simplify the problem, we compute the amplitude of the wave recorded at the surface that has been reflected at interface 4. First, we compute the amplitude transmitted to each one of the layers before layer 4. This is given by the product of the transmission coefficients of each layer. In Fig. 6.3, the transmission coefficients, t, are replaced by their equivalent expressions, $(1 + r)$. The amplitude of the wave reaching layer 4 is

$$1 \times t_1 \times t_2 \times t_3 \quad = \quad (1 + r_1)(1 + r_2)(1 + r3)$$

Following reflection from layer 4, the amplitude must be multiplied by the reflection coefficient of interface 4

$$1 \times t_1 \times t_2 \times t_3 \times r_4 \quad = \quad (1 + r_1)(1 + r_2)(1 + r3)r_4$$

The transmission coefficients for the upward propagating wave are given by terms of the form

$$1 + r' = 1 - r$$

The final amplitude after propagating to the receiver at the surface is given by

$$\underbrace{(1 + r_1)(1 + r_2)(1 + r_3)}_{\text{Transmission} \downarrow} \times \underbrace{r_4}_{\text{Reflection}} \times \underbrace{(1 - r_1)(1 - r_2)(1 - r_3)}_{\text{Transmission} \uparrow}$$

and the expression for the amplitude of the wave reflected at interface 4 becomes

$$(1 - r_1^2)(1 - r_2^2)(1 - r_3^2)r_4$$

Reflections occur at all the layers, of course, and the respective amplitudes are

Amplitude of the reflection generated at interface 1

$$A_1 \quad = \quad r_1$$

Amplitude of the reflection generated at interface 2

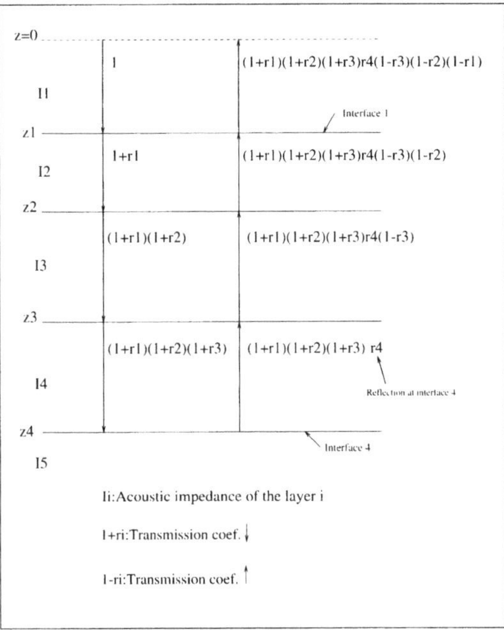

Figure 6.3: Amplitude of a plane wave propagating in a layered medium. Analysis of the wave reflected at interface 4.

$$A_2 = (1 - r_1^2)r_2$$

Amplitude of the reflection generated at interface 3

$$A_3 = (1 - r_1^2)(1 - r_2^2)r_3$$

Amplitude of the reflection generated at interface 4

$$A_4 = (1 - r_1^2)(1 - r_2^2)(1 - r_3^2)r_4$$

A general expression for the amplitude of a reflection at the k-th interface can, consequently, be inferred

$$\begin{aligned} A_1 &= r_1 \\ A_k &= \prod_{i=1}^{k-1}(1 - r_i^2)r_k \quad k = 2, 3, 4, \ldots \end{aligned}$$

Let us now interpret these results. Assuming that the Earth is excited with a delta function, and neglecting the presence of multiples, the zero-offset seismogram will be a collection of delta functions (we will usually call these, spikes) at arrival times given by the two-way travel time formula. The strength of each arrival will be proportional to the amplitude A_k. In the real world, however, it is impossible to have a source that resembles a delta. The source signature, called a wavelet, is a finite length time function that we denote by $w(t)$. In this case, the seismogram is composed of a superposition of wavelets arriving at different times, with amplitude proportional to A_k. The seismogram that describes our 4 interface model (Fig. 6.3) will have 4 arrivals of amplitudes A_1, A_2, A_3 and A_4, and can be expressed as

$$x(t) = A_1 w(t - t_1) + A_2 w(t - t_2) + A_3 w(t - t_3) + A_4 w(t - t_4)$$

where t_1, t_2, t_3 and t_4 are the arrival times of each reflection. Neglecting transmission effects, the amplitude A_i can be replaced by the reflection coefficient r_i and allows us to express the seismogram as a convolution between two time series, a wavelet and the reflectivity sequence

$$x_k = w_k * q_k \tag{6.3}$$

6.3 Deconvolution of the Reflectivity Series

Deconvolution is not a new subject at this stage. We have met it before in various applications, in particular in Sections 1.7.10 and 1.8.5 where, we met inverse dipole filters and optimum spiking filters. In these applications, we considered known seismic signatures. Our particular problem in this instance is much more complicated, and has been outlined before in Section 2.6. Reiterating here for clarity, our task consists of the deconvolution of the desired reflectivity, q_k, from the recorded data, x_k, as expressed by Eq. (6.3).

The simple model of Eq. (6.3), must be modified to include the presence of random noise, n_k

$$x_k = w_k * q_k + n_k \tag{6.4}$$

where we emphasize again (as we did in Section 2.6) that q_k represents the primary reflectivity, i.e., the reflectivity of the layered earth assuming that multiple events have been attenuated (an easy assumption to make on paper and extremely difficult to realize in practice). We are faced with (we reiterate), a problem of one equation (one observable, the seismogram) and two unknowns (the wavelet and the reflectivity) which we will attempt to solve as two sub-problems,

I. Wavelet estimation

II. Operator design

Wavelet estimation refers to methods of estimating the source signature from the seismic trace. In general, these methods are statistical techniques that explore some properties of the remaining unknown, the reflectivity, which are used as a priori information.

6.3.1 The Autocovariance Sequence and the White Reflectivity Assumption

We have seen (Section 1.8.5) that the design of a Wiener filter involves the inversion of a Toeplitz autocovariance matrix. This matrix arises from the fact that we have represented our convolution model as a matrix- vector multiplication. To clarify, we assume a 3 point wavelet and compute the autocovariance matrix. The convolution matrix is

$$\mathbf{C} = \begin{pmatrix} w_0 & 0 \\ w_1 & w_0 \\ w_2 & w_1 \\ 0 & w_2 \end{pmatrix}$$

and the autocovariance matrix becomes

$$\mathbf{R} = \mathbf{C}^{\mathbf{T}}\mathbf{C} = \begin{pmatrix} r_0 & r_1 \\ r_1 & r_0 \end{pmatrix}$$

Writing the autocovariance coefficients in terms of the wavelet w_k

$$r_0^w = w_0^2 + w_1^2 + w_2^2$$

and

$$r_1^w = w_0 w_1 + w_1 w_2$$

The first coefficient is the zero-lag correlation coefficient[2], also a measure of the energy of the wavelet. The second coefficient, r_1^w, is the first lag of the correlation sequence. The coefficients can be written using the following expression

$$r_j^w = \sum_k w_k w_{k+j}\,, \qquad j = 0, \pm 1, \pm 2 \ldots$$

In the Wiener filter, $\mathbf{R}$ is an $N \times N$ matrix where N is the length of the filter, and the N autocovariance coefficients are computed by

$$r_j^w, \qquad j = 0, 1, \ldots, N-1$$

The design of a Wiener or spiking filter requires knowledge of the wavelet, which is generally unknown. To solve this problem we use the white reflectivity assumption (Section 2.6). Under this assumption the seismic reflectivity (the *geology*) is considered to be a zero mean, white process (Robinson and Treitel, 2002). Such a process is uncorrelated, i.e., if r_j^q is the autocovariance function of the reflectivity, then

[2]The superscript, w, is used to indicate that the autocovariance is that of the wavelet

$$r_j^q = \begin{cases} P_q & j = 0 \\ 0 & j = \pm 1, \pm 2, \pm 3, \ldots \end{cases}$$

As we know, the autocovariance is a measure of similarity of a time series with itself. The zero lag coefficient is an estimate of the power of the signal, P_q, whereas the $j = 1$ coefficient is a measure of the similarity of the signal with a one-sample shifted version of itself. If the reflectivity is a zero mean, white noise process, we have seen, when dealing with MA time series (Section 2.3), that the following remarkable property is true

$$r_j^x = P_q r_j^w$$

In other words: *the autocovariance function of the trace is an estimate (within a scale factor) of the autocovariance of the wavelet.* We can estimate the autocovariance of the unknown wavelet from the autocovariance of the observed seismic trace. Hence, we are able to compute an estimate of the amplitude spectrum of the wavelet by Fourier transformation. Associating a suitable phase function has been fully described in Section 2.3.1 and we will not repeat the steps here. Suffice it to say that we have obtained an estimate of the wavelet, a very special, minimum phase, estimate to be sure.

6.3.2 Deconvolution of Noisy Seismograms

We begin with our 1D model, Eq. (6.4), written here again for convenience

$$x_k = w_k * q_k + n_k \tag{6.5}$$

In order to recover q_k from x_k, a filter f_k must be computed such that $f_k * w_k = \delta_k$. Generally, of course, only an estimate, $\hat{f}_k$, can be computed, and $\hat{f}_k * w_k = a_k$, where a_k is called the averaging function that approximates a delta function only in the ideal case. Applying $\hat{f}_k$ to both sides of Eq. (6.5) yields the estimated output of the deconvolution process

$$\begin{aligned} \hat{q}_k &= a_k * q_k + \hat{f}_k * n_k \\ &= q_k + (a_k - \delta_k) * q_k + \hat{f}_k * n_k \end{aligned} \tag{6.6}$$

Since our main requirement is to estimate a model $\hat{q}_t$ that is close to the true reflectivity, it is important to design a filter such that the error terms in Eq. (6.6) are as small as possible. Consequently, one seeks a solution with the following properties

$$a_k = w_k * f_k \approx \delta_k$$

and

$$f_k * n_k \approx 0$$

The last two expressions can also be written in matrix form

$$\mathbf{C}_w \mathbf{f} \approx \mathbf{d}$$

and

$$\mathbf{C}_n \mathbf{f} \approx \mathbf{0}$$

where $\mathbf{C}_w$ and $\mathbf{C}_n$ denote the convolution matrices for the wavelet and the noise, respectively. Both equations are honored when the following objective function is minimized

$$J = ||\mathbf{C}_w \mathbf{f} - \mathbf{d}||^2 + \beta ||\mathbf{C}_n \mathbf{f}||^2$$

where β is a tradeoff parameter. The second term in the last equation can be written as

$$||\mathbf{C}_n \mathbf{f}||^2 = \mathbf{f}^T \mathbf{C}_n^T \mathbf{C}_n \mathbf{f}$$

where $\mathbf{C}_n^T \mathbf{C}_n$ is the noise autocovariance matrix. If the noise is uncorrelated, $\mathbf{C}_n^T \mathbf{C}_n$ can be replaced by its estimator

$$E[\mathbf{C}_n^T \mathbf{C}_n] = \sigma_n^2 \mathbf{I}$$

where σ_n^2 is an estimate of the variance of the noise. J is now given by

$$J = ||\mathbf{C}_w \mathbf{f} - \mathbf{d}||^2 + \mu ||\mathbf{f}||^2$$

where $\mu = \sigma_n^2 \times \beta$. This is the objective function used to design an inverse filter, with solution given by

$$\mathbf{f} = (\mathbf{R}_w + \mu \mathbf{I})^{-1} \mathbf{C}_w^T \mathbf{d}$$

Figs. (6.4), (6.5) and (6.6) illustrate the performance of the LS inversion when dealing with noise free and noisy data. It is clear that the pre-whitening parameter plays a key role in the deconvolution of noisy data.

6.3.3 Deconvolution in the Frequency Domain

A procedure similar to the one outlined in the previous section can be used to deconvolve data in the frequency domain. Taking the DFT of Eq. (6.6) yields

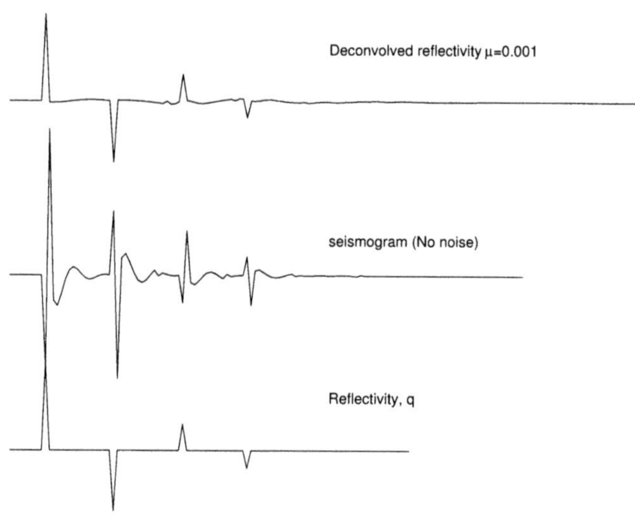

Figure 6.4: Deconvolution of a clean seismogram.

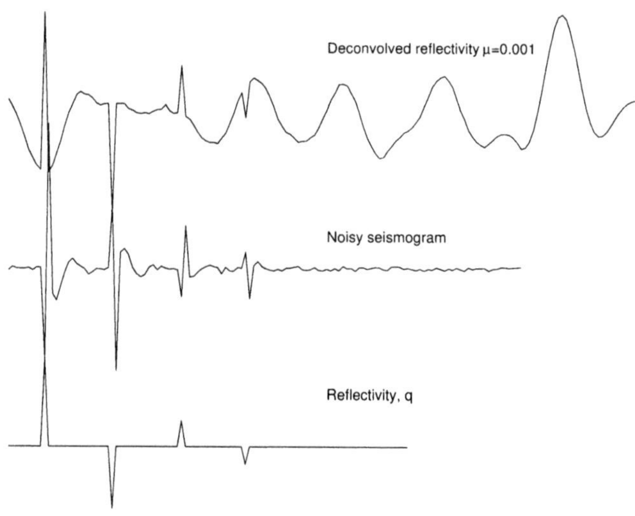

Figure 6.5: Deconvolution of a noisy seismogram. The tradeoff parameter is too small and the result is unstable.

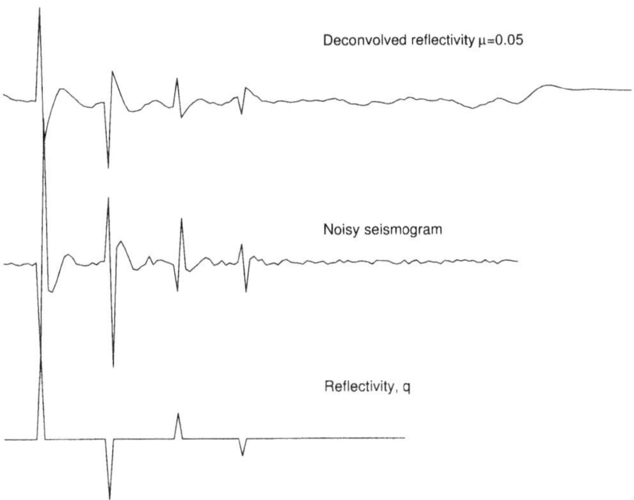

Figure 6.6: Deconvolution of a noisy seismogram. The tradeoff parameter has been increased to stabilize the solution

$$\hat{Q}_k = Q_k + (A_k - 1)Q_k + \hat{F}_k N_k \tag{6.7}$$

Since a_k is to be a good approximation to a delta function, it turns out that the filter design should satisfy the following requirement

$$W_k F_k = A_k \approx 1 \qquad \forall k$$

Furthermore, in order to maintain the noise at a small level

$$F_k N_k \approx 0 \qquad \forall k$$

These two requirements may be combined into one. In order to achieve this, we construct the following objective function

$$J = \sum_k |A_k - 1|^2 + \alpha \sum_k |F_k N_k|^2$$

that, when minimized with respect to the filter coefficients leads to

$$\hat{F}_k = \frac{W_k^*}{|W_k|^2 + \alpha |N_k|^2}$$

Finally, the reflectivity estimate is given by

$$\begin{aligned}\hat{Q}_k &= D_k \frac{W_k^*}{|W_k|^2 + \alpha|N_k|^2} \\ &= D_k \frac{W_k^*}{|W_k|^2 + \mu}\end{aligned}$$

Since the noise has a flat spectrum ($|N_k|^2 = \sigma_n^2$) we can replace $\alpha|N_k|^2$ by another constant, μ. An estimate of the variance of the reflectivity estimate in the frequency domain is given by

$$\mathrm{Var}(\hat{Q}_k) = |\hat{F}_k|^2 {\sigma_n}^2$$

and, following a little algebra, we end up with

$$\mathrm{Var}(\hat{Q}_k) = \frac{|W_k|^2 {\sigma_n}^2}{(|W_k|^2 + \mu)^2}$$

When $\mu = 0$, the variance can be extremely high at the frequencies at which the wavelet power is small. We can also find an expression for the norm of the reflectivity estimate in the frequency domain

$$\begin{aligned}N_r &= \textstyle\sum_k |\hat{Q}_k|^2 \\ &= \tfrac{1}{{\sigma_n}^2} \textstyle\sum_k |S_k|^2 \mathrm{Var}(\hat{Q}_k)\end{aligned}$$

and for the misfit function

$$\begin{aligned}\Phi &= \sum_k |S_k - W_k \hat{Q}_k|^2 \\ &= \frac{1}{{\sigma_n}^2} \sum_k |S_k|^2 \left(\frac{\mu}{|W_k|^2 + \mu}\right)^2\end{aligned}$$

The norm, N_r, and the misfit, Φ, clearly indicate that there is a trade-off between honoring the data and the stability of the reflectivity estimate. We also note that the trade-off parameter, μ, basically avoids division by zero.

We continue the analysis in order to quantify the errors associated with the final estimate of the reflectivity. If E_k, the deviation of the filter from the true inverse filter, is defined by

$$E_k = 1 - \hat{F}_k W_k$$

we can write Eq. (6.7) as follows

$$\begin{aligned}\hat{Q}_k &= \hat{F}_k W_k Q_k + \hat{F}_k N_k \\ &= (1 - E_k) Q_k + \frac{1 - E_k}{W_k} N_k\end{aligned}$$

Then, the difference between the true reflectivity, Q_k, and the estimate, $\hat{Q}_k$, is given by

$$\hat{Q}_k - Q_k = \underbrace{-E_k Q_k}_{\text{RE}} + \underbrace{\frac{1 - E_k}{W_k} N_k}_{\text{NAE}}$$

where RE and NAE signify regularization and noise amplification errors, respectively. NAE is independent of the data and can be expressed as a function of the wavelet

$$\text{NAE} = \frac{W_k^*}{|W_k|^2 + \mu} N_k$$

It is clear that the more the filter resembles the inverse wavelet, W_k^{-1}, the larger this error will be. RE, on the other hand, introduces data-dependent degradation (i.e., ringing).

6.4 Sparse Deconvolution and Bayesian Analysis

The deconvolution operator is usually stabilized by adding a small perturbation to the diagonal of the autocovariance matrix, that is equivalent to zero order quadratic regularization (Section 4.5.12.1). Regularization is used to estimate a unique and stable solution by introducing some type of prior information. Our task, here, is to examine different regularization strategies that may be used to improve the deconvolution of seismic records. Specifically, we use the Huber and the Cauchy criteria to retrieve a sparse reflectivity sequence. These criteria are related to long tail probability distributions (Huber, 1981) which, in a Bayesian context (please see Section 4.5), are used as a prior distribution for the reflectivity sequence. The seismic wavelet is assumed known or to be have been accurately estimated.

Re-weighting strategies have been used in conjunction with LS type estimators to diminish the influence of outliers in inverse problems (Scales et al., 1988). In robust statistics, the influence function (Huber, 1981) that measures the influence of the residuals on the estimators is constructed so as to attenuate outliers. A similar construction can be applied to the regularization function. In this case the goal is to attenuate the side-lobe artifacts which are introduced in the deconvolution process. In this context, the regularization strategy is used to drive the output of the deconvolution to a prescribed part of model space, in contrast to classical applications in robust statistics where the chief goal is to attenuate the influence of gross errors. When the problem is properly regularized, the resolution of close seismic arrivals can be significantly enhanced. The described procedure is used to overcome the poor resolution associated with quadratic regularization strategies. We have to realize, however, that only when the seismogram is composed of a finite superposition of seismic wavelets (a sparse reflectivity assumption) may these techniques provide a substantial improvement with respect to conventional deconvolution.

6.4.1 Norms for Sparse Deconvolution

Restating the deconvolution problem, we have

$$x_k = \sum_j w_j q_{k-j}$$

where q_k, $k = 1, 2, \dots, M_q$ and x_k, $k = 1, 2, \dots, M_x$ are the input and output to the convolution process, respectively, and w_k, $k = 1, 2, \dots, M_w$ is the *blurring function* or source wavelet. In time domain deconvolution, the goal is to find $\hat{q}_k$ such that the residuals, ϵ_k, given by

$$\epsilon_k = x_k - \sum_j w_j \hat{q}_{k-j}$$

are minimized. In the LS approach, the objective function is

$$J = \sum_k \rho(\frac{\epsilon_k}{\sigma_k})$$

where

$$\rho(u) = \frac{1}{2}u^2$$

The residuals are weighted according to the data precision that is given by the inverse of the standard error of each observation, σ_k. For simplicity, we shall assume that $\sigma_k = \sigma_n$, $k = 1, 2, \dots, M_x$. The minimization of J is accomplished by solving the system

$$\frac{\partial J}{\partial q_l} = \sum_k \psi(\frac{\epsilon}{\sigma_n}) w_{k-l} = 0 \,, \qquad l = 1, 2 \dots, M_q$$

where

$$\psi(u) = \frac{d\rho(u)}{du} \tag{6.8}$$

and is called the influence function. This function measures the influence of the residuals on the estimates of the parameters The minimization of J leads to the system of normal equations

$$\sum_k \sum_j w_{k-j} w_{k-l} q_j = \sum_k w_{k-l} s_k$$

or, in matrix form

$$\mathbf{Rq} = \mathbf{g} \tag{6.9}$$

In usual fashion, Eq. (6.9) is stabilized by adding a small perturbation to the diagonal of $\mathbf{R}$, which is equivalent to minimizing the modified objective, or cost, function

$$J = \sum_k \rho(\frac{\epsilon_k}{\sigma_n}) + \sum_i \rho(\frac{q_i}{\sigma_q}) \tag{6.10}$$

where

$$J_q = \sum_k \rho(\frac{q_k}{\sigma_q})$$

The second term in the cost function is the regularizer of the problem. The minimum of Eq. (6.10) is reached at the point

$$\hat{\mathbf{q}} = (\mathbf{R} + \mu \mathbf{I})^{-1} \mathbf{g} \tag{6.11}$$

where $\mu = \sigma_n^2 / \sigma_q^2$ is the damping parameter (also the pre-whitening parameter, Robinson and Treitel, 2002).

6.4.1.1 Modifying J

A standard procedure in robust statistics is based on redesigning the influence function, Eq. (6.8), in order to attenuate the effect of outliers. A similar modification can be used to design the regularization term in J. The data misfit is modelled using the functional ρ, while the following modification is introduced for the regularization term

$$\rho_H(u) = \begin{cases} u^2/2 & \text{if } |u| \leq a \\ a|u| - a^2/2 & \text{if } |u| > a \end{cases}$$

The deconvolution problem is now solved by minimizing

$$J_H = \sum_k \rho(\frac{\epsilon_k}{\sigma_n}) + \sum_i \rho_H(\frac{q_i}{\sigma_q})$$

The influence function for ρ_H becomes

$$\psi_H(u) = \begin{cases} u & \text{if } |u| \leq a \\ a \,\text{sign}(u) & \text{if } |u| > a \end{cases}$$

The function ρ_H behaves identically to ρ for small values of u. When $|u| > a$, ρ_H defines a line and its associated influence function becomes a constant.

We can define another function with similar behavior

$$\rho_C(u) = \log(\frac{u^2}{2} + 1)$$

When u is small $\rho_C \to \rho$. The influence function corresponding to ρ_C is given by

$$\psi_C(u) = \frac{u}{\frac{u^2}{2} + 1}$$

If ρ_C is adopted, the deconvolution problem is solved by minimizing function designated by J_C

$$J_C = \sum_k \rho(\frac{\epsilon_k}{\sigma_n}) + \sum_i \rho_C(\frac{q_i}{\sigma_q})$$

J_C can be derived using Bayes rule (Section 4.4.1.1) by assuming Gaussian errors and a Cauchy prior probability to model the unknown parameters (Sacchi and Ulrych, 1995a). We have studied this type of solution in the context of High Resolution Fourier Transforms in section (4.5.12.2). Fig. 6.7 shows the functions ρ, ρ_H and ρ_C, while Fig. 6.8 displays the corresponding influence functions ψ, ψ_H and ψ_C. The functions ρ_H and ψ_H were calculated for $a = 1$ and $a = 2$. When the parameter a is small compared to the normalized variable u, the width of the transition zone $-a < u < a$ becomes very narrow. In this case the cost function $\sum_i \rho_H(u_i)$ behaves like a l_1 norm, $l_1 = \sum_i |u_i|$.

The solution to the LS problem with zero order quadratic regularization expressed in equation (6.11) can be modified by introducing the functionals ρ_H (the Huber criterion) and/or ρ_C (the Cauchy criterion). In this case the system of normal equations is obtained by equating the gradient of the objective function J_H or J_C, to zero,

$$(\mathbf{R} + \mu \mathbf{Q})\mathbf{q} = \mathbf{g} \tag{6.12}$$

If the problem is regularized with ρ_H, $\mathbf{Q}$, which we designate as $\mathbf{Q}_H$, has the following diagonal elements

$$Q_{H_{ii}} = \begin{cases} 1 & \text{if } |\frac{q_i}{\sigma_q}| \le a \\ a/|\frac{q_i}{\sigma_q}| & \text{if } |\frac{q_i}{\sigma_q}| > a \end{cases}$$

When ρ_C is adopted, $\mathbf{Q}$ in Eq. (6.12), which is denoted by $\mathbf{Q}_C$, has the following diagonal elements

$$Q_{C_{ii}} = \frac{1}{1 + \frac{q_i^2}{2\sigma_q^2}}$$

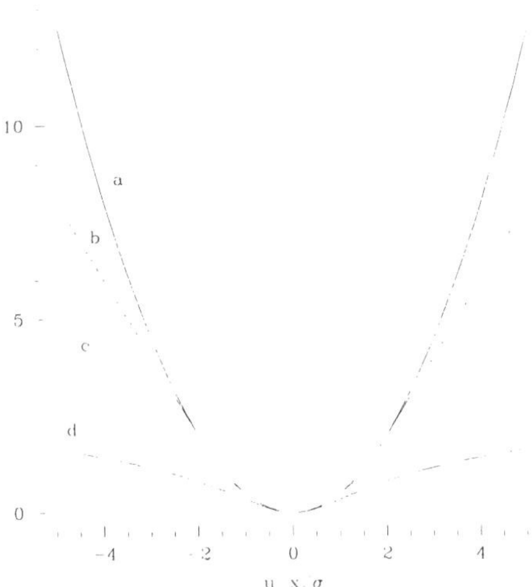

Figure 6.7: Cost functions. (a) $\rho(u)$. (b) $\rho_H(u)\, a = 1$. (c) $\rho_H(u)\, a = 2$. (d) $\rho_C(u)$.

We can make an analogy with the zero order quadratic regularization to understand the effect of $\mathbf{Q}$ on solving the system expressed by Eq. (6.12). In the zero order quadratic regularization, the damping term μ in Eq. (6.11) corresponds to the ratio of two variances, $\mu = \sigma_n^2/\sigma_q^2$. When ρ_H or ρ_C are used, the damping term becomes a ratio of the variance of the noise to a model dependent variance which we designate by $\sigma^2(q_i)$. This variable takes the following form when the problem is regularized with ρ_H

$$\sigma^2(q_i) = \begin{cases} \sigma_q & \text{if } |\frac{q_i}{\sigma_q}| \leq a \\ \sigma_q|q_i|/a & \text{if } |\frac{q_i}{\sigma_q}| > a \end{cases}$$

The last equation shows that, above a threshold, the variance of x_i is proportional to $|x_i|$. A similar argument leads to the variance for the regularization with ρ_3

$$\sigma^2(q_i) = \sigma_q^2 + \frac{q_i^2}{2}$$

in which case the variance has a parabolic growth with the amplitude of q.

Eq. (6.12) is solved using the following iterative scheme

1. Start with an initial reflectivity sequence $\mathbf{q}^0$
2. Select the hyper-parameters of the problem, σ_n, σ_q, and a (the Huber criterion) or σ_n and σ_q (the Cauchy criterion).
3. Compute $\mu = \sigma_n^2/\sigma_q^2$, $\mathbf{Q}^{(0)}$, and the source autocovariance matrix $\mathbf{R}$.

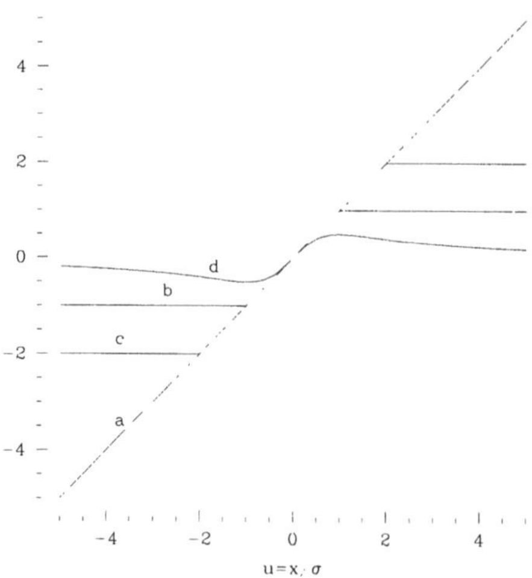

Figure 6.8: Influence functions. (a) $\psi(u)$. (b) $\psi_H(u)\, a = 1$. $\psi_H(u)\, a = 2$. (d) $\psi_C(u)$.

4. Iteratively solve Eq. (6.12) using the following algorithm

$$\mathbf{q}^{(k)} = (\mu \mathbf{Q}^{(k-1)} + \mathbf{R})^{-1} \mathbf{g}$$

   where k is the iteration number.

5. The procedure is stopped when the following tolerance criterion is satisfied

$$2 \frac{|J^{(k)} - J^{(k-1)}|}{(|J^{(k)}| + |J^{(k-1)}|)} < tolerance$$

   where $J = J_H$ or J_C depending on the regularization criterion.

6. Compute the data misfit. Select new hyperparameters if the misfit is not satisfactory (the strategy for hyperparameter selection is discussed below).

Each iteration demands one matrix inversion. The procedure can be accelerated by using an iterative solver like the conjugate gradient, CG, algorithm. The advantage of using CG is that an approximate solution can be computed by truncating the number of iterations.

The effect of $\mathbf{Q}$ can be summarized as follows. In each iteration, the nonlinearity produces a solution that has the minimum amount of structure or maximum sparseness. The validity of this type of solution is subject to the validity of the sparse reflectivity assumption.

The determination of the parameter μ in Eq. (6.12) is crucial but, unfortunately, cannot be performed a priori. A wrong selection of μ may yield a solution that is unreasonable. Assuming that the variance of the noise, σ_n^2, is known, we can make the following observations. If the Cauchy criterion is adopted, only one independent parameter σ_x must be determined ($\mu = \sigma_n^2/\sigma_q^2$). When the Huber criterion is used, two independent parameters are needed, σ_q and a. The parameter a is assigned as $a = c \times \sigma_q$, where c is a scalar $(0.1 \leq c \leq 1)$. If c is large $(c > 2)$ the Huber criterion behaves like the standard quadratic form ρ.

We adopt the discrepancy principle which determines the parameter σ_q from the requirement that the data misfit matches the power of the noise. Since we have assumed that the noise is normally distributed, the data misfit obeys a χ^2 statistic.

$$\chi^2 = \frac{1}{\sigma_n^2} \sum_{k=1}^{M_x} \epsilon_k^2$$

The expected value of the χ^2 statistic is used as a target misfit, $E[\chi^2] = M_x$ (M_x is the number of observations), where the largest acceptable value at 99% confidence limit is $\approx M_x + 3.3\sqrt{M_x}$.

Fig. 6.9a illustrates a simulated reflectivity impulse response for a simple earth model, the seismogram and the seismic source. Gaussian noise was added to the synthetic seismogram with standard deviation $\sigma_n = 5 \times 10^{-2}$. This represents a relative noise amplitude of 17.5%. The relative noise magnitude is specified by a percentage of the maximum noise-free seismogram amplitude whose standard deviation represents $\sigma_n/\max(x_k) \times 100$. The deconvolution was carried out using zero order quadratic regularization , the Huber criterion ρ_H (minimizing J_H) and the Cauchy criterion ρ_C (minimizing J_C). The estimated impulse responses are displayed in Figures 6.9b, c, and d together with the reconstructed seismograms and residuals (original minus reconstructed data). The parameter σ_s was selected according to the χ^2 criterion. The solution with ρ (zero order quadratic regularization) does not allow us to properly identify each arrival. The re-weighted strategy, on the other hand, yields highly resolved estimates of the position and amplitude of each seismic reflection. In general, about $5-10$ iterations are sufficient to find a good approximation to the minimum of the cost function. The performance of the re-weighted deconvolution procedure when dealing with field data, is tested on the portion of a stacked seismic section shown in Fig. 6.10a. The stacked section is obtained by applying a NMO correction, and summing traces from CMP gathers. The data consist of 24 traces which delineate several seismic horizons. In particular, we are interested in the coherent events at $\approx$ 0.95sec which may represent a thin layer. It is important to stress that, unlike many deconvolution scenarios, seismic deconvolution implies that the kernel function (i.e., the source wavelet) is unknown. In this particular example, the source wavelet was retrieved using a combined cepstrum-cumulant approach, a technique that is discussed in Section 6.6.

The deconvolved data are shown in Fig. 6.10. In this example, the problem is regularized by means of the Cauchy criterion (ρ_C). The χ^2 criterion was used to estimate the parameter σ_x. A relative noise amplitude of 2% was assumed. The latter was used to estimate the standard error of the noise σ_n. Similar results were obtained using Huber's weights.

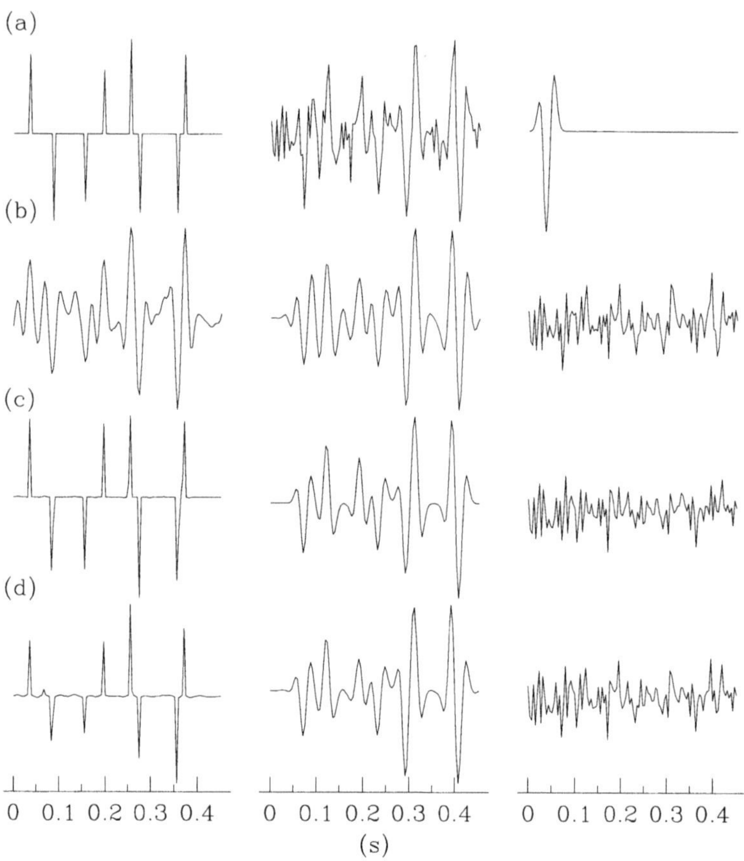

Figure 6.9: A synthetic deconvolution example. (a) Forward model. (Left) Synthetic impulse response. (Center) Seismogram. (Right) Source wavelet. (b) Deconvolution using zero order quadratic regularization: (Left) Estimated impulse response. (Center) Reconstructed seismogram. (Right) Residuals (original minus reconstructed data). (c) Deconvolution by means of the Huber criterion ($a = 1$). (d) Deconvolution by means of the Cauchy criterion.

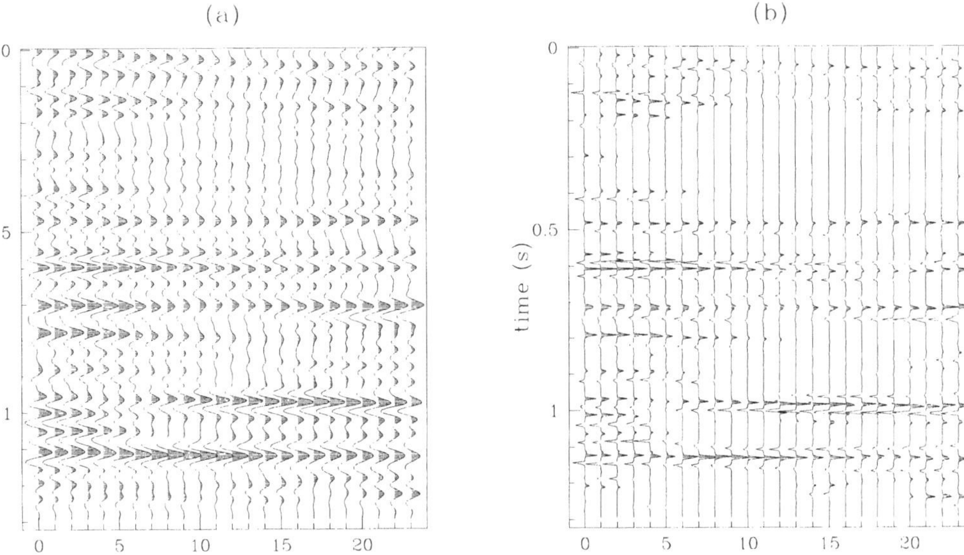

Figure 6.10: Deconvolution of field data. (a) Original seismic section. (b) Deconvolved seismic section using the Cauchy criterion to regularize the inversion. The source wavelet was retrieved using a combined cepstrum-cumulant approach.

6.5 1D Impedance Inversion

The object of 1D impedance inversion is the recovery of the acoustic impedance (defined in Section 6.2.1) from band-limited reflection seismograms. This is a difficult task, indeed, and we will develop various approaches that explore many issues of interest. We define the problem in some detail.

Since, as always, we adopt the convolutional model for the seismogram, it is required that initial processing has removed multiples and has recovered true amplitudes as well as possible. We have, therefore

$$x_t = w_t * q_t \tag{6.13}$$

where the q_t are the *primary* reflection coefficients[3]. Even considering this simple model, the recovery of the acoustic impedance is a goal not easily attained. This may

[3]We stress, once again, that this model is a simplification. It assumes, in particular, that w_t is a

be most easily seen in the frequency domain. Taking FTs in Eq. (6.13)

$$X(f) = W(f)Q(f)$$

In general, $Q(f)$ is full band but $W(f)$ is limited to a band of approximately 5-70 Hz. Consequently, $Q(f)$, the acoustic impedance, cannot be recovered directly, from the seismogram. The loss of the low frequencies, as we will show below, is particularly crucial for the correct inversion of the acoustic impedances. The loss of the high frequencies is important as far as the resolution of thin layers is concerned.

We refer to the reflectivity following deconvolution in the allowable bandwidth, as the band-limited reflectivity, given by

$$\hat{q}_t = q_t * a_t \tag{6.14}$$

where a_t, the averaging function defined in Eq. (6.6), is zero phase with constant amplitude in the frequency range $[\omega_L, \omega_H]$ and zero amplitude outside this interval. Estimating q_t from $\hat{q}_t$ is a nonunique linear inverse problem, a fact that, neglecting the noise term, may be easily confirmed by taking the FT of Eq. (6.14)

$$\hat{Q}(\omega) = A(\omega)Q(\omega) \tag{6.15}$$

We see that $Q(\omega)$ can take any value at those frequencies at which $A(\omega)$ vanishes. The knowledge of $\hat{Q}(\omega)$ is not enough to estimate the portion of $Q(\omega)$ outside the non-zero band of $A(\omega)$. Hence, there exist an infinite number of models, q_t, that satisfy equation (6.15). In other words, $\hat{Q}(\omega)$ gives no information about the part of $Q(\omega)$ which belongs to the null space.

There are two paths that one can follow. The first is to live with a_t, but use the nonuniqueness of the deconvolution problem in such manner, that the solution is a useful estimate of the full band reflectivity. The second is to attempt to obtain our full band estimate directly from the solution of Eq. (6.3). We deal with the first path below, and present various approaches to the problem of band-limited extrapolation. The second path is dealt with in Sections 6.7.1 and 6.7.3.

6.5.0.2 Acoustic Impedance

Following our discussion in Section 6.2.1, if q_k is the reflection coefficient of the kth layer in our layered earth model, the acoustic impedance in the kth layer is

$$\xi_k = \rho_k v_k$$

where ρ_k and v_k are the density and velocity in the kth layer, respectively. The relationship between q_k and ξ_k is given by

$$q_k = \frac{\xi_{k+1} - \xi_k}{\xi_{k+1} + \xi_k}$$

---

non dispersive pulse and that the earth is made up of plane parallel layers having constant material properties.

which, after rearranging, yields

$$\xi_{k+1} = \xi_k \left(\frac{1+q_k}{1-q_k} \right) = \xi_1 \prod_{j=1}^{k} \left(\frac{1+q_j}{1-q_j} \right) \tag{6.16}$$

and depicts the inverse relationship between the impedance and the reflection coefficients.

We may now establish a procedure for estimating the acoustic impedance from the seismogram. First, we find an inverse filter f_t for the wavelet w_t which is assumed "known" (we deal with this "known" issue in the following sections)

$$f_t * w_t = a_t \tag{6.17}$$

where a_t is the averaging function, explored in Section 6.3.2, and, owing to the band limited nature of $w(t)$, only approximates the desired delta. In fact, a_t is zero phase and band-limited. The reflection coefficients, therefore, are derived as averages of the true values

$$\begin{aligned} f_t * x_t &= f_t * w_t * q_t \\ &= a_t * q_t \\ &= < q_t > \end{aligned}$$

Substituting $< q_t >$ into Eq. (6.16), we obtain the estimated acoustic impedances which are denoted by $\widehat{\xi}_t$.

$$\widehat{\xi}_{k+1} = \widehat{\xi}_k \left(\frac{1+ < q_k >}{1- < q_k >} \right)$$

The relationship between the estimated and the true acoustic impedances may be best seen by considering the continuous case. As shown by Oldenburg et al. (1983), so long as the reflectivity coefficients are less than about 0.3, the relationship between $q(t)$ and $\xi(t)$ may be written as

$$q(t) = \frac{1}{2} \frac{d[\log \xi(t)]}{dt}$$

Rewriting Eq. (6.18) in terms of the surface impedance, $\xi(0)$, we obtain

$$\xi(t) = \xi(0) e^{2 \int_0^t q(u) du} \tag{6.18}$$

Defining

$$\eta(t) = \log \left(\frac{\xi(t)}{\xi(0)} \right) \tag{6.19}$$

and substituting $< q(t) >$ in place of $q(t)$ in Eq. (6.19), we have

$$\widehat{\eta}(t) = \ln\left(\frac{\widehat{\xi}(t)}{\widehat{\xi}(0)}\right) = 2\int_0^t < q(u) > du \tag{6.20}$$

It is clear from Eq. (6.20) that the logarithm of the estimated normalized impedance, $\widehat{\eta}(t)$, which, in fact, is in general the average of the true logarithm of the normalized acoustic impedance, cannot have information at those frequencies which are absent in $< q(t) >$. Hence the low- and the high- frequency information about the acoustic impedance cannot be obtained from the seismogram by direct methods.

We now examine the crucial role that the low frequencies play in the determination of $\widehat{\eta}(t)$. Consider, for the sake of clarity, the problem of estimating the impedance for the case of one layer, with reflectivity $q(t) = q_1\delta(t-\tau)$. We know, from Eq. (6.20), that the ideal impedance is $q_1\,H(t-\tau)$, where $H(\cdot)$ is the Heaviside function. The estimated impedance, on the other hand, is given by

$$\widehat{\eta}(t) = 2\int_0^t q_1\delta(u-\tau) * a(u)du = 2\int_0^t q_1 a(u-\tau)du$$

If $A(0)$ is the value of the FT of $a(t)$ at zero frequency

$$A(0) = \int a(t)dt = 0$$

since low frequencies are absent. Consequently

$$\widehat{\eta}(t) \rightarrow 0 \text{ as } t \rightarrow \infty$$

Fig. 6.11 illustrates the above discussion in toy fashion. The desired impedance, Fig. 6.11d, is achieved by the integration of the full band reflectivity shown in Fig. 6.11a. Panel (e) is the recovered impedance obtained if no deconvolution is performed (not a practice adopted in processing) and Panel (f) illustrates the recovered impedance using the best conventional deconvolution shown in Panel . For this example, Fig. 6.11c is, in fact, the averaging function defined by Eq. (6.6). Given that, in general, velocity increases with depth, the behavior of the impedance depicted in Fig. 6.11f, is highly undesirable. The way to overcome this problem is, clearly, to obtain a full band deconvolved reflectivity or, at least, a reflectivity containing low frequencies. We now present some algorithms which attempt to do, precisely, this.

6.5.1 Bayesian Inversion of Impedance

The first approach to the estimation of a *blocky* impedance model from normal incidence data that we discuss is by means a Bayesian methodology. We assume that impedance constraints are provided at different time levels. Using Eq. (6.20), and dropping the $\widehat{}$ notation for simplicity, we have

$$\eta_k = \frac{1}{2}\log(\frac{\xi_k}{\xi_0}) = \sum_{i=1}^{N_k} q_i \tag{6.21}$$

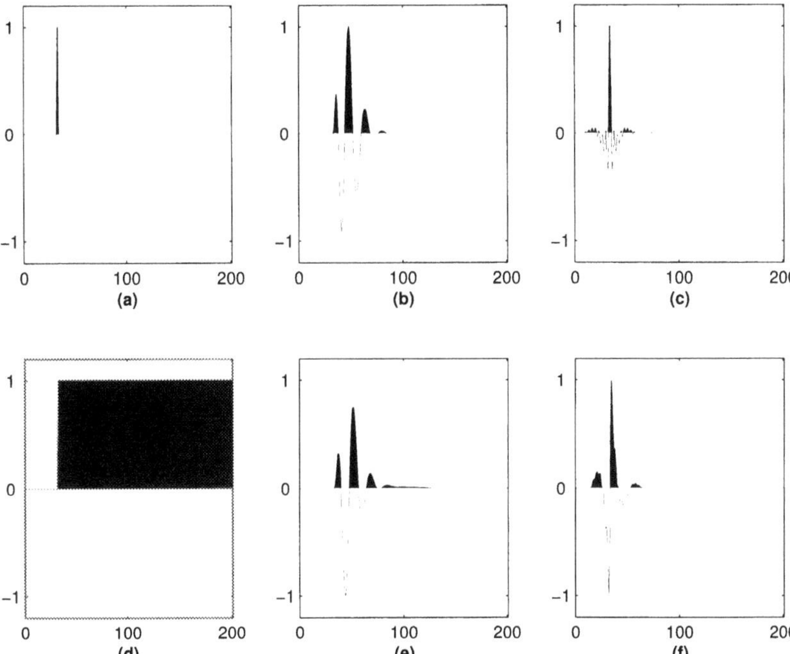

Figure 6.11: A toy example: impedance recovery for a single layer. Panels (a), (b) and (c) show the reflectivity, the recorded seismogram and the best conventional deconvolution. Panels (d), (e) and (f) show the respective recovered impedances.

Using matrix notation, we write Eqs. (6.3) and (6.21) as

$$\mathbf{W}\mathbf{q} = \mathbf{x} + \mathbf{e} \tag{6.22}$$

and

$$\mathbf{C}\mathbf{q} = \boldsymbol{\eta} + \boldsymbol{\epsilon} \tag{6.23}$$

where $\mathbf{W}$ and $\mathbf{C}$ correspond to the wavelet and impedance constraint matrices, respectively, and $\boldsymbol{\epsilon}$ represents the uncertainties in the constraints. The matrix $\mathbf{W}$ contains the wavelet properly padded with zeros in order to express discrete convolution in matrix form (Section 1.9). The matrix $\mathbf{C}$ is a simple integrator operator. Bayes rule (Section 4.4.1.1) is used to incorporate the prior probability of the unknown reflectivity $\mathbf{q}$ into the problem. In applications where a blocky impedance profile is to be estimated, a long tailed distribution may be used to estimate a sparse reflectivity sequence. Noise in the trace, $\mathbf{e}$, is modelled by means of the usual Gaussian assumption. The uncertainties of the constraints are also assumed Gaussian (note that we are using the variable η which can take positive and negative values and not ξ which is strictly positive).

Bayes rule is used to define the posteriori probability of the reflectivity sequence.

The MAP solution is computed by maximizing the posterior probability, and is equivalent to minimize the following cost function

$$J = \underbrace{\alpha J_q}_{1} + \underbrace{\frac{1}{2}||\frac{1}{\sigma}(\mathbf{Wq}-\mathbf{x})||^2}_{2} + \underbrace{\frac{1}{2}||\mathbf{S}^{-1}(\mathbf{Cq}-\boldsymbol{\xi})||^2}_{3} \tag{6.24}$$

where σ^2 is the variance of the noise in the seismic trace and $\mathbf{S}$ is a diagonal matrix with elements

$$S_{ii} = \sigma_{c_i}$$

Eq. (6.24) specifies three different features that the solution must satisfy

- 1 - The solution must be sparse.
- 2 - The solution must honor the data.
- 3 - The solution must honor a set of impedance constraints.

α is the weighting or hyperparameter that determines the relative amount of sparseness that can be introduced. J_q is derived using four different priors which induce the associated regularization criteria for sparse spike inversion. In section (6.4.1.1) we have discussed the application of the Huber and Cauchy criteria for high resolution (sparse) deconvolution. Now, we will incorporate other criteria that can also be used to reach the desire resolution. The four sparseness criteria that we have studied are the L_p, *Cauchy*, *Sech* and *Huber* criteria, all of which have been used in robust statistics to diminish the influence of outliers in estimation problems (Huber, 1981). In our application, these criteria are used to impose sparseness into the reflectivity estimate. J_q in Eq. (6.24) is given by one of the following regularization terms

$$\begin{aligned} J_p &= \frac{1}{p}\sum_i |q_i|^p \\ J_C &= \frac{1}{2}\sum_i \log(1+\frac{q_i^2}{\sigma_q^2}) \\ J_S &= \sum_i \log(\cosh\frac{q_i}{\sigma_q}) \end{aligned}$$

and

$$J_H = \sum_i \begin{cases} q_i^2/2 & \text{if } |q_i| \le q_c \\ a\,|q_i| - q_c^2/2 & \text{if } |q_i| > q_c \end{cases}$$

A nonlinear CG algorithm is used to minimize the cost function. α is continuously adapted to satisfy a misfit criterion. The misfit target that was used is

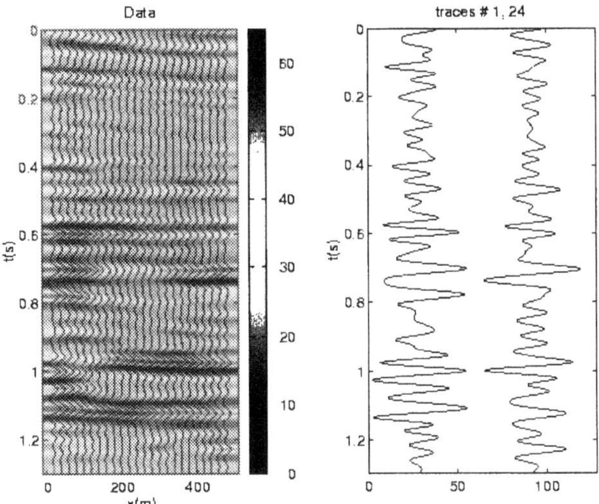

Figure 6.12: A window (and two trace detail) of a seismic section pre-processed for impedance inversion.

χ^2 = number of traces + number of constraints. Fig. 6.12 shows a window of a seismic section (and a detail for two traces) pre-processed for impedance inversion. The inversion was carried out using the *Huber* and L_p criteria ($p = 1.1$). The reflectivity estimates are shown in Figures (6.13) and (6.15). The constrained inversion of the impedance profile is depicted in Figures (6.14) and (6.16).

6.5.2 Linear Programming Impedance Inversion

In this section we discuss the classical approach to impedance inversion proposed by Oldenburg et al. (1983). The assumptions in this approach are identical to those in Section 6.5.1. The procedure proposed to recover the reflectivity is based on the minimization of the l_1 cost function of the problem. Instead of using a CG technique to minimize the cost function (as we did in the Bayesian approach of Section 6.5.1) we adopt the linear programming approach. The cost function of the problem is defined as

$$J = \alpha|\mathbf{q}|_1 + |\mathbf{e}|_1$$

where the variables are defined in Eqs. (6.22) and (6.23), and the constrained minimization problem is set up as

$$\begin{aligned} \text{Minimize} \quad & J = \alpha|\mathbf{q}|_1 + |\mathbf{e}|_1 \\ \text{subject to} \quad & \mathbf{Wq} = \mathbf{x} + \mathbf{e} \\ \text{and} \quad & \eta_l < \mathbf{Cq} < \eta_u \end{aligned}$$

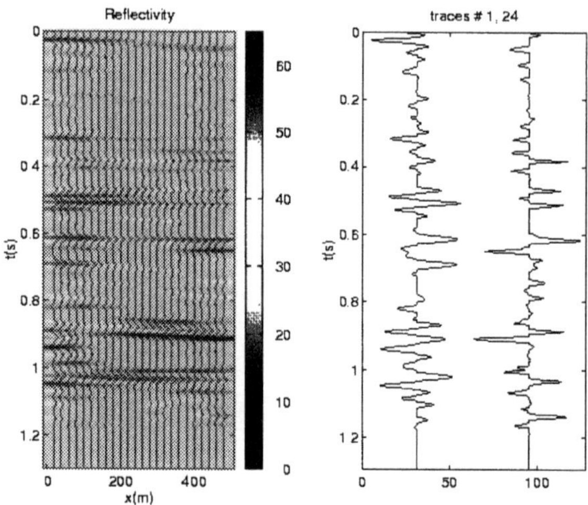

Figure 6.13: Reflectivity inversion using the L_p norm, $p = 1.1$

and is solved using the linear programming approach. α is the tradeoff parameter and is assigned a value as a percentage of the l_1 norm of the wavelet. This is analogous to the pre-whitening parameter in spiking deconvolution which is given as a percentage of the zero lag autocovariance coefficient.

6.5.3 Autoregressive Recovery of the Acoustic Impedance

The approach described here is quite different to the two previous techniques that we have dealt with. It is of much interest because it combines many principles that we have presented in the previous chapters, in particular, autoregressive, AR, modelling (Sections 2.4 and 5.2). Our model for the reflectivity is

$$q(t) = \sum_k q_k \delta(t - \tau_k)$$

where τ_k is the two-way travel time in the kth layer of the plane layered earth model. Transforming into the frequency domain

$$Q(f) = \sum_{k=1}^{N-1} \exp(-i2\pi k f) \tag{6.25}$$

and shows that the frequency domain representation of the reflectivity function consists of a sum of complex sinusoids. The AR method and the extensions detailed later are based on the representation of Eq. (6.25) as a complex AR process. In actuality, since noise must be taken into account, the AR representation is only an approximation as detailed briefly below and explored in Sections 2.5.1 and 5.2.

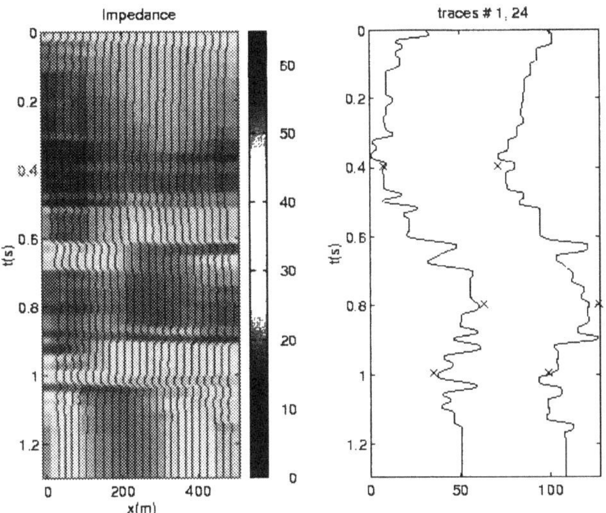

Figure 6.14: Constrained impedance inversion using the L_p norm, $p = 1.1$.

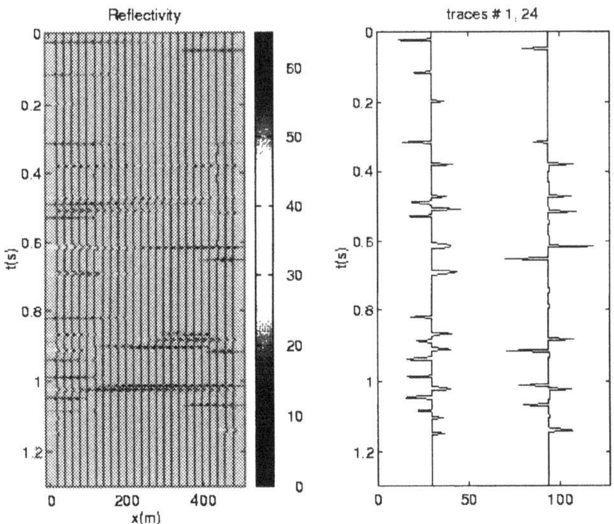

Figure 6.15: Reflectivity inversion using the *Huber* norm.

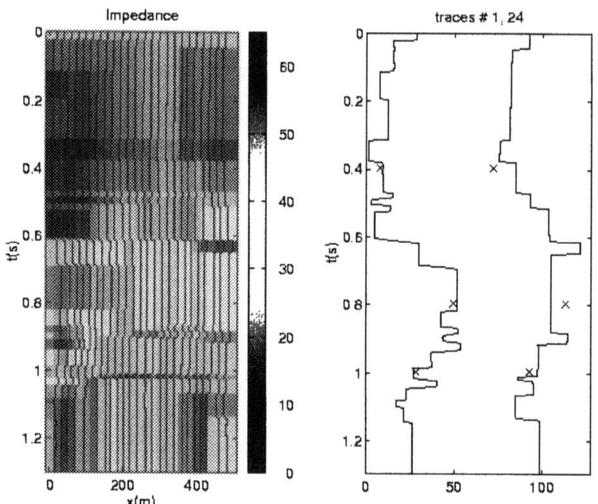

Figure 6.16: Constrained impedance inversion using the *Huber* norm.

As we have shown in Section 2.5.1, p harmonics with additive white noise n_t may be modelled as

$$y_t = x_t + n_t$$

where

$$x_t = \sum_{m=1}^{2p} a_m x_{t-m}$$

is the $2p$th order difference equation with real coefficients whose roots are located in conjugate pairs on the unit circle in the z plane. As we show in Section 2.5.1, the appropriate model in terms of the PEO, $\mathbf{g}$, is

$$\sum_{k=0}^{2p} g_k y_{t-k} = \sum_{k=0}^{2p} g_k n_{t-k} \tag{6.26}$$

and shows that the correct representation for a process consisting of a sum of p harmonics with additive white noise is a special ARMA process of order $(2p, 2p)$, the coefficients of which are given by the eigenproblem

$$\mathbf{R_{yy}g} = \sigma^2 \mathbf{g} \tag{6.27}$$

where $\mathbf{R_{yy}}$ is the $2p$ autocorrelation matrix of the data and σ^2 is the noise variance. This model can be applied in the complex form to the band-limited FT of the deconvolved seismogram. In the frequency domain

$$Y_k = W_k Q_k + N_k \tag{6.28}$$

Since we have assumed that deconvolution with a known wavelet has been performed, $|W_k|$ in Eq. (6.28) is a box car of unit amplitude in the band of frequencies of the seismic wavelet. $W_k Q_k$ is just a sum of complex sinusoids limited to the band of frequencies of W_k. Consequently, the real and imaginary parts of Y_k may be predicted by a complex version of Eq.(6.27).

Although the eigenvector representation models the noisy data correctly, prediction with the eigenvector corresponding to the minimum eigenvalue is very sensitive to the SNR in the data . A far more robust and economical process is to approximate the ARMA process of Eq. (6.26) by a finite order AR process (please see Section 5.2). The AR approach to spectral extension adopted here is to model Y_k in Eq. (6.28) as

$$Y_k = -\sum_{n=1}^{p} g_k Y_{n-k} + E_k \tag{6.29}$$

where the g_ks are complex. E_k is the complex innovation series and p is the order of the process. The well known solution to Eq. (28), which we have discussed at length in Section 2.4, takes the form

$$\mathbf{R_{yy}g} = \sigma_E \mathbf{u} \tag{6.30}$$

where $\mathbf{R_{yy}}$ is the Toeplitz, Hermitian autocovariance matrix of the Y_k, σ_E is the innovation variance and $\mathbf{u} = [1, 0, \ldots, 0]$. It is important to stress that different methods of estimating $\mathbf{R_{yy}}$ exist (see Section 3.4) and should be allowed for when implementing the algorithm.

6.5.3.1 AR Gap Prediction

The data which form the input to the prediction algorithm have an important symmetry which must be taken into account. Specifically, the symmetry is contained in the Hermitian property of the frequency domain data. In theory, one should be able to predict the low-frequency spectrum from the positive frequency band, through zero frequency and match the data in the negative frequency band. The approach that is adopted here, makes use of the gap filling prediction technique of Fahlman and Ulrych (1982) that we have met before in Section 5.2.6 when dealing with the recovery of near offset traces. A brief outline is presented below, details may be found in Walker and Ulrych (1983). We consider those data that are missing from the spectrum between the low cut-off frequencies f_L in the positive frequency band and $-f_L$ in the negative frequency band as a gap to be filled. For this purpose, we assume that any point $\widehat{Y}_k$ in the gap may be represented as

$$\widehat{Y}_k = \widehat{Y}_k^f W_k^f + \widehat{Y}_k^b W_k^b \tag{6.31}$$

where $\widehat{Y}_k^f$ is the forward prediction based on the negative frequency band and $\widehat{Y}_k^b$ is the backward prediction based on the positive frequency band. W_k^f and W_k^b are weights which are optimally chosen by minimizing the prediction error for each point in the gap.

The first step is to determine the prediction error operator, **g**, from the known data by means of Eq. (6.30). The second step is to minimize the sum of the squared forward and backward prediction errors given by

$$S = \sum_k \{|E_k^f|^2 + E_k^b|^2\} \tag{6.32}$$

The resulting system of equations may be written as

$$\mathbf{G}^H \widehat{\mathbf{y}} = \mathbf{b} \tag{6.33}$$

where **G** is the complex Toeplitz matrix consisting of the transient autocorrelation values of the prediction error operator **g**, with elements $r_m = \sum_{j=0}^{p-m} g_j g_{j-m}^* = r_{-m}^*$, $\widehat{\mathbf{y}}$ is the vector of missing data points and **b** is a vector of known terms. The AR filter may be recomputed from the known and the predicted data iteratively.

6.5.3.2 Gap Prediction with Impedance Constraints

Since the sum of the predicted low-band values, $\widehat{Y}_k$, and the known mid-band values, Y_k, are the estimates of Q_k, we may rewrite Eq. (6.20) as

$$\begin{aligned} \widehat{\eta}(t) &= 2\int_0^{t_k} (\hat{y}_u + y_u)du \\ &= 2\int_0^{t_k} \sum_j (\widehat{Y}_j + Y_j)e^{i2\pi ju}du \\ &= 2\theta_k \end{aligned} \tag{6.34}$$

where θ_k is a constraint and is related to the logarithm of the relative impedance at time t_k. Expanding Eq. (6.34)

$$\sum_j \gamma_j \widehat{X}_j + \int_0^{t_k} x_u du = \theta_k$$

where $\gamma_j = \int_0^{t_k} \exp(i2\pi judu)$. If $\int_0^{t_k} = \mu_k$, we obtain the required constraint at time t_k

$$\sum_j \gamma_j \widehat{X}_j = \theta_k - \mu_k = \beta_k \tag{6.35}$$

The problem now is to minimize S in Eq. (6.32) subject to the constraint of Eq. (6.35). The general form of the solution for M constraints is given by the system of equations

$$\begin{bmatrix} \mathbf{G}^H & \mathbf{H} \\ \mathbf{H}^T & 0 \end{bmatrix} \begin{bmatrix} \hat{\mathbf{y}} \\ \boldsymbol{\lambda} \end{bmatrix} = \begin{bmatrix} \mathbf{b} \\ \boldsymbol{\beta} \end{bmatrix}$$

where $\mathbf{H}^T\hat{\mathbf{y}} = \boldsymbol{\beta}$ are the constraint equations and $\boldsymbol{\lambda}$ is the vector of Lagrange multipliers. If we denote the unconstrained solution by $\hat{\mathbf{z}}$, where $\hat{\mathbf{z}} = \mathbf{G}^{-H}\mathbf{b}$, the solution of the constrained problem is given by

$$\hat{\mathbf{y}} = \hat{\mathbf{z}} - \mathbf{G}^{-H}\mathbf{H}[\mathbf{H}^T\mathbf{G}^{-H}\mathbf{H}]^{-1}(\mathbf{H}^T\hat{\mathbf{z}} - \boldsymbol{\beta})$$

The constraints may also be introduced in the algorithm as inequality constraints (Ulrych and Walker, 1984).

6.5.3.3 Minimum Entropy Extension of the High Frequencies

Because of the position of the band of known frequencies in the spectrum, generally in the neighborhood of 5-70 Hz, AR prediction of the high frequencies is much more difficult than the AR prediction of the low frequencies. High frequency extension may be accomplished by means of the minimum entropy principle as discussed in Section 6.7.2.2. Examples of the application of this AR approach, to synthetic and real data, may be found in Walker and Ulrych (1983). We will come back to this issue in section (6.7.2) where the Minimum Entropy Deconvolution (MED) algorithm is formulated in the time and frequency domain.

6.6 Nonminimum Phase Wavelet Estimation

The crucial assumption in the previous sections was that the wavelet was either known or that an accurate estimate was available. Easier said than done. Wavelet estimation is quite an extraordinarily difficult task, but a task that is vital to much of seismic processing and interpretation. We have already dealt, in Section 2.3.1, with the estimation of minimum phase signatures. Here, we analyze the problem of estimating a nonminimum phase wavelet. We begin by discussing the cumulant approach, the technique that resulted in the wavelet estimate used to process the data in the preceding section.

6.6.1 Nonminimum Phase System Identification

The classical system identification literature has been primarily dominated by the least squares approach, an approach attractive because of its equivalence to maximum likelihood parameter estimation for Gaussian distributions. These identification procedures use the autocovariance function of the data which, as we have seen, annihilates the phase of the system (e.g., the wavelet) and, consequently, these techniques are useful primarily to identify minimum phase systems or systems where the phase can be specified a priori i.e., a zero phase assumption.

Wold's celebrated decomposition theorem, Section 2.2, states that a stationary, stochastic process, x_k, can always be written in terms of a linear transformation of a white process, ϵ_k. In other words

$$x_t = \sum_k \epsilon_k h_{t-k} \tag{6.36}$$

where, for simplicity, we have omitted the deterministic component of the process (please see the discussion concerning Eq. (2.1)) which is not needed in the seismic deconvolution scenario. Clearly, the Wold decomposition theorem dictates that a decomposition exists where the impulse response, h_k, is causal and minimum phase. One of the consequences of this theorem is that the description of the process is non-unique. One can model a stationary, linear process, x_k, by means of the MA representation in Eq. (6.36), in an infinite number of ways. We will use the following model to describe the seismic trace

$$s_k = q_k * w_k$$

and

$$x_k = s_k + n_k$$

where s_k is the noise-free seismic trace, and x_k is the trace corrupted by additive noise. As always, the source wavelet is denoted by w_k and the reflectivity sequence by q_k. We will assume that q_k is white and iid and that the noise, n_k, is zero-mean, Gaussian and independent of x_k. The transfer function of the system is given by

$$W(z) = |W(z)|e^{i\Phi(z)}$$

The seismic deconvolution problem is to reconstruct the magnitude and the phase of the transfer function from the output data x_k. Thus far, we have only assumed that q_k is white. To continue with the analysis the following comments are in order

1. If q_k is Gaussian and $W(z)$ is minimum phase, autocorrelation based methods will correctly identify both the amplitude and the phase of the system.

2. If q_k is Gaussian and $W(z)$ is nonminimum phase, no technique will correctly identify the phase of the system.

3. If q_k is nonGaussian and $W(z)$ is nonminimum phase, true magnitude and phase of the system transfer function can be recovered by knowing the actual distribution of q_k. For MA processes of order one, it has been demonstrated that a l_1 optimization provides an estimate of the amplitude and phase of the system when the driving noise of the process is nonGaussian (Scargle, 1977)

Statement 3 is very important. In fact, it suggests that we may have faith in stochastic wavelet estimation procedures. It is also clear that nonGaussianity plays a key role in nonminimum phase wavelet estimation and we will have more to say about this in Section 6.7.3.5. In fact, MED-type (where MED stands for minimum entropy deconvolution) estimators are an example where departure from Gaussianity is postulated. The non-uniqueness expressed by the Wold theorem is eliminated by restricting q_k to be nonGaussian. Higher-order spectra are defined in terms of higher-order cumulants and contain information regarding deviation from Gaussianity. Quite contrary to power spectral density estimates, higher-order spectra retain phase information which allows nonminimum phase system identification/estimation.

6.6.1.1 The Bicepstrum

If $W(z)$ is stable and if q_k is non-Gaussian, white, iid., with skewness $\beta \neq 0$ then the bispectrum of s_k is given by

$$B_s(z_1, z_2) = \sum_m \sum_n r_s^{(3)}(m,n) z_1^n z_2^m$$

where

$$r_s^{(3)}(m,n) = E[s_k s_{k+m} s_{k+n}]$$

is the third order moment of the data. Since the third order moment of a Gaussian signal is zero, we can write

$$B_x(z_1, z_2) = B_s(z_1, z_2)$$

The bispectrum can be written in terms of the transfer function, $W(z)$, as follows (Nikias and Raghuveer, 1987)

$$B_x(z_1, z_2) = \beta W(z_1) W(z_2) W(z_1^{-1}, z_2^{-1}) \tag{6.37}$$

Since the wavelet is, in general, nonminimum phase, the transfer function may be expressed as

$$W(z) = A z^l W^{min}(z) W^{max}(z^{-1}) \tag{6.38}$$

where A is a constant, l an integer associated with a linear phase shift, and $W^{min}(z)$, $W^{max}(z^{-1})$ are the minimum and maximum phase components of the wavelet, respectively. Substituting Eq. (6.38) into (6.37) the bispectrum becomes

$$\begin{aligned} B_x(z_1, z_2) = \quad & \beta W^{min}(z_1) W^{min}(z_2) \cdot \\ & W^{max}(z_1^{-1}) W^{max}(z_2^{-1}) \cdot \\ & W^{min}(z_1^{-1} z_2^{-1}) W^{max}(z_1 z_2) \end{aligned}$$

We can now define the bicepstrum as

$$\hat{b}(n,m) = \mathcal{Z}_{\in}{}^{-1}[\log B_x(z_1, z_2)]$$

where $\mathcal{Z}_{\in}{}^{-1}$ stands for the inverse 2D z transform. We note that

$$\begin{aligned}\log[B_x(z_1,z_2)] = \quad & \log[\beta] \\ + \quad & \log[W^{min}(z_1)] + \log[W^{min}(z_2)] \\ + \quad & \log[W^{max}(z_1^{-1})] + \log[W^{max}(z_2^{-1})] \\ + \quad & \log[W^{min}(z_1^{-1}z_2^{-1})] + \log[W^{max}(z_1 z_2)] \end{aligned} \tag{6.39}$$

The inversion of Eq. (6.39) yields a complete separation into minimumphase and maximumphase components. Since the homomorphic transform maps minimumphase components into the positive semi-axis of the cepstrum and maximumphase components into the negative semi-axis, it is easy to verify that

$$\hat{b}(m,0) = \hat{w}^{min}(m), \quad m > 0$$

and

$$\hat{b}(m,0) = \hat{w}^{max}(m), \quad m < 0$$

where $\hat{w}(n)$ indicates the cepstrum of the wavelet. It is clear that we can completely define the cepstrum of the wavelet (the value $\hat{w}(0)$ cannot be recovered but represents only a scale factor) with only two semi-axes of the bicepstrum. A similar approach can be used to compute the tricepstrum (the cepstrum of the fourth order cumulant).

6.6.1.2 The Tricepstrum

Defining the trispectrum as

$$T_s(z_1,z_2,z_3) = \sum_m \sum_n \sum_l r_s^{(4)}(n,m,l) z_1^n z_2^m z_3^l$$

where $r_s^{(4)}(n,m,l)$ is now the fourth order cumulant of s_k. The trispectrum can be written in terms of the transfer function of the system as follows

$$\begin{aligned}T_x(z_1,z_2,z_3) = \quad & \gamma A^4 W^{min}(z_1) W^{min}(z_2) W^{min}(z_3) \cdot \\ & W^{max}(z_1^{-1}) W^{max}(z_2^{-1}) W^{max}(z_3^{-1}) \cdot \\ & W^{min}(z_1^{-1}z_2^{-1}z_3^{-1}) W^{max}(z_1 z_2 z_3)\end{aligned}$$

After taking the logarithm of the trispectrum, we end up with

$$\begin{aligned}\log[T_x(z_1,z_2,z_3)] = \log[\gamma A^4] \quad + \quad & \log[W^{min}(z_1)] + \log[W^{min}(z_2)] + \log[W^{min}(z_3)] \\ + \quad & \log[W^{max}(z_1^{-1})] + \log[W^{max}(z_2^{-1})] + \log[W^{max}(z_3^{-1})] \\ & \log[W^{min}(z_1^{-1}z_2^{-1}z_3^{-1})] + \log[W^{max}(z_1 z_2 z_3)]\end{aligned}$$

The inversion of the last expression will map minimum and phase components into the tricepstrum domain according to

$$\hat{t}(m,n,l) = \begin{cases} \log[\gamma A^4] & m=n=l=0 \\ \hat{w}_{min}(m) & m>0, \quad n=l=0 \\ \hat{w}_{min}(n) & n>0, \quad m=l=0 \\ \hat{w}_{min}(l) & l>0, \quad m=n=0 \\ \hat{w}_{max}(m) & m<0, \quad n=l=0 \\ \hat{w}_{max}(n) & n<0, \quad m=l=0 \\ \hat{w}_{max}(l) & l<0, \quad m=n=0 \\ \hat{w}_{min}(m) & m=n=l<0 \\ \hat{w}_{max}(m) & m=n=l>0 \end{cases}$$

The origin of the tricepstrum, $\hat{t}(0,0,0) = \log(A^4\gamma)$ represents a scale factor and can be ignored. In general, the wavelet is estimated from 2 semi-axes of the tricepstrum

$$\hat{t}(m,0,0) = \hat{w}^{min}(m), \quad m>0$$

and

$$\hat{t}(m,0,0) = \hat{w}^{max}(m), \quad m<0$$

6.6.1.3 Computing the Bicepstrum and Tricepstrum

We first note that, if $z_1 = e^{-i\omega_1}$ and $z_2 = e^{-i\omega_2}$ we have

$$\begin{aligned} \hat{B}_x(\omega_1,\omega_2) &= \log[B_x(\omega_1,\omega_2)] \\ &= \log|B_x(\omega_1,\omega_2)| + i[2k\pi + \Phi(\omega_1,\omega_2)] \end{aligned}$$

Since the phase of the complex variable, $\hat{B}_x(\omega_1,\omega_2)$, is undefined, it appears that an unwrapping procedure is mandatory . Fortunately, the unwrapping can be omitted by defining the differential cepstrum, as follows

$$\frac{d\hat{B}_x(\omega_1,\omega_2)}{d\omega_1} = \frac{1}{B_x(\omega_1,\omega_2)} \frac{dB_x(\omega_1,\omega_2)}{d\omega_1} \tag{6.40}$$

(6.41)

where the derivatives are obtained as

$$\frac{dB_x(\omega_1,\omega_2)}{d\omega_1} = \mathcal{F}_2[-imr_x^{(3)}(m,n)]$$

(6.42)

and we can write Eq. (6.40) as

$$\hat{b}(m,n) = \frac{1}{m}\mathcal{F}_2^{-1}[\frac{\mathcal{F}_2[mr_x^{(3)}(m,n)]}{\mathcal{F}_2[r_x^{(3)}(m,n)]}] \tag{6.43}$$

(6.44)

A similar algorithm is used to estimate the tricepstrum

$$\hat{t}(m,n,l) = \frac{1}{m}\mathcal{F}_3^{-1}[\frac{\mathcal{F}_3[mr_x^{(4)}(m,n,l)]}{\mathcal{F}_3[r_x^{(4)}(m,n,l)]}] \tag{6.45}$$

6.6.1.4 Some Examples

The first simulation entails a nonminimum phase wavelet and an exponentially distributed sequence with skewness $\beta = 1$. The additive noise is white and Gaussian. The standard deviation of the noise represents 1% of the maximum amplitude of the signal. We used 6 records of 600 samples to estimate the third order cumulant and the bicepstrum was computed using Eq. (6.43). Fig. 6.17 shows the true cepstrum of the wavelet and the cepstrum extracted from the bicepstrum. Fig. 6.18 shows the true and the estimated wavelets together with the associated minimum and maximum phase components. The maximum phase component has a zero close to the unit circle that is manifested in the negative semi-axis of the cepstrum as a long decaying oscillatory signal.

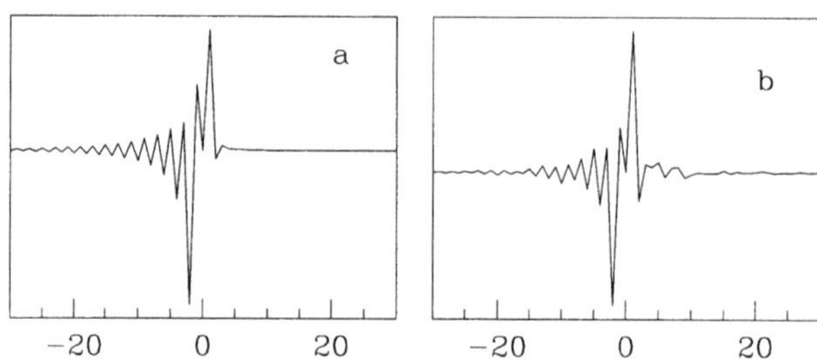

Figure 6.17: Synthetic example using an exponentially distributed sequence. (a) Cepstrum of the actual wavelet. (b) Cepstrum of wavelet derived from the bicepstrum.

The same wavelet was convolved with a white, nonGaussian reflectivity with non vanishing kurtosis. The fourth order cumulant was computed from 4 records of 1000 samples each. The cepstrum of the wavelet was estimated from the tricepstrum using Eq. (6.45). The results are shown in Fig. 6.19. The time signature and its maximum and minimum phase components are displayed in Fig. 6.20. The technique permits the recovery of the cepstral coefficients of the wavelet with an accuracy proportional to the accuracy of the estimation of the cumulant.

6.6.1.5 Algorithm Performance

We present a pictorial analysis of the performance of the algorithm, beginning with Fig. 6.21, which illustrates a schematic representation of the tricepstrum, Eq. (6.45), for a nonGaussian MA process. We simulate 20 realizations of a nonGaussian ($\gamma \neq 0$) process which is convolved with the source wavelet shown in Fig. 6.22. Fig. 6.23 illustrates the wavelet retrieved from the tricepstrum. The fourth order cumulant was

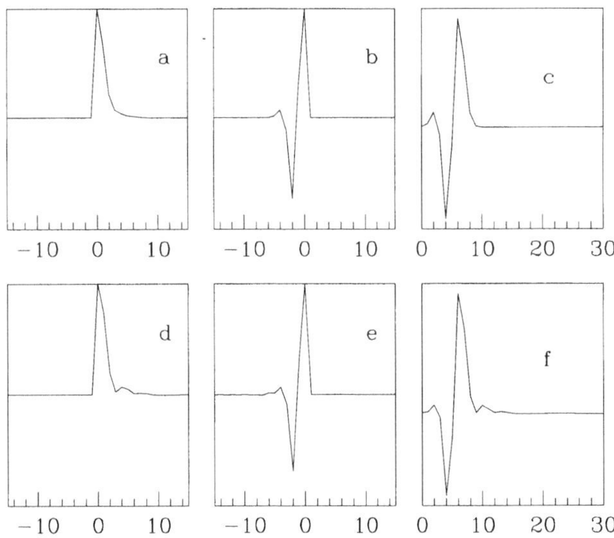

Figure 6.18: Synthetic example using an exponentially distributed sequence. (a) and (b) True minimum phase and maximum phase components of the wavelet (c) True wavelet. (d), (e), and (f) estimators of (a), (b), and (c) computed form the bicepstrum.

estimated using 4 segments of 1000 samples each. Figs. 6.24 and 6.25 depict results using 4 segments of 500 and 250 samples, respectively. These results indicate that a fairly good reconstruction of the amplitude and phase can be achieved for large data sets.

Fig. 6.26 portrays a segment of a seismic section pre-processed for impedance inversion. The segment is composed of 24 traces of 300 samples per trace. The fourth order cumulant is estimated from each trace and the average cumulant is used to identify the wavelet. Fig. 6.27 shows the cepstrum of the wavelet retrieved from one of the axes of the tricepstrum. The minimum and maximumphase components of the wavelet are shown in Fig. 6.28. The tricepstrum estimator of the wavelet is illustrated in Fig. 6.29. For comparison we also show the estimator of the wavelet computed using a cumulant matching approach (Velis and Ulrych, 1996). The latter uses a global optimization procedure (simulated annealing) to find the wavelet that best reproduces the cumulant of the data.

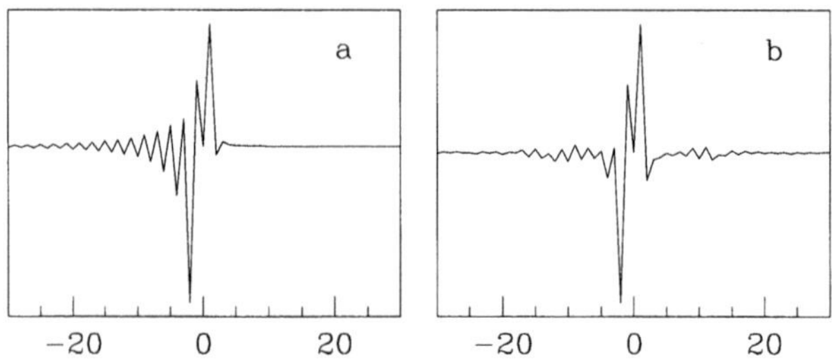

Figure 6.19: Synthetic example using a nonGaussian reflectivity with non zero kurtosis. (a) Cepstrum of the true wavelet. (b) Cepstrum of wavelet derived from the tricepstrum.

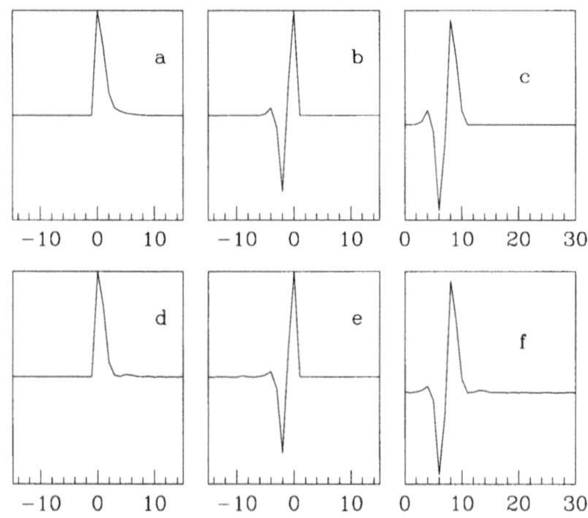

Figure 6.20: Synthetic example using a nonGaussian reflectivity with non zero kurtosis. (a) and (b) True minimumphase and maximum phase components of the wavelet. (c) True wavelet. (d), (e), and (f) estimators of (a), (b), and (c) computed from the tricepstrum.

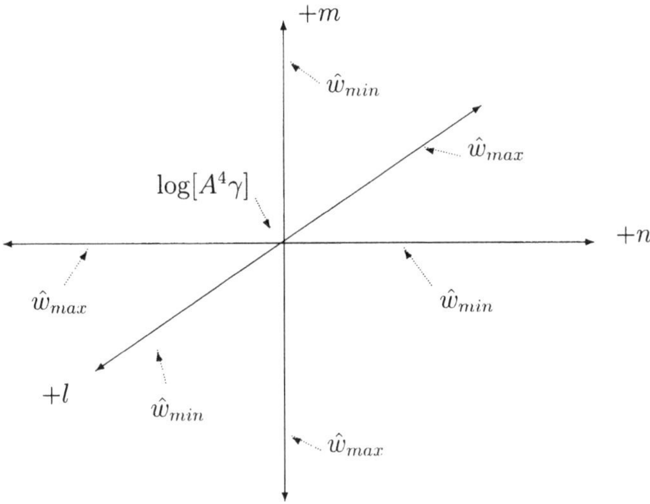

Figure 6.21: Schematic representation of the tricepstrum for a nonGaussian MA process.

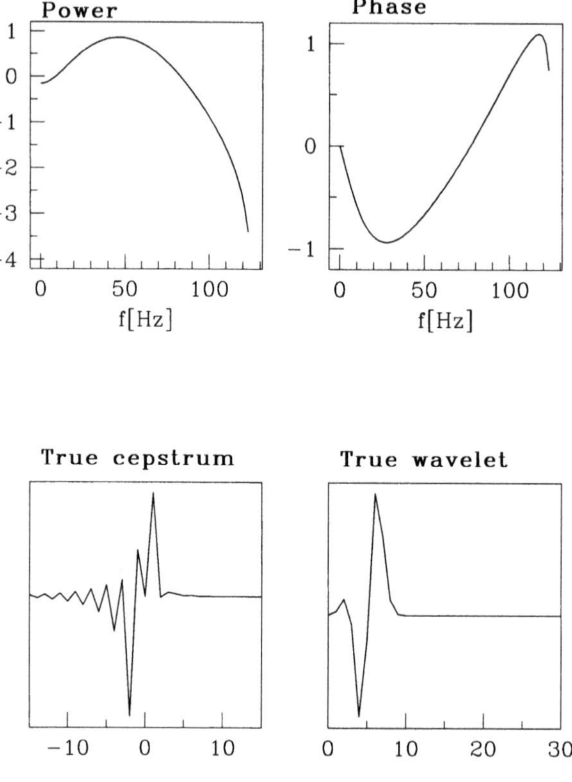

Figure 6.22: Synthetic wavelet.

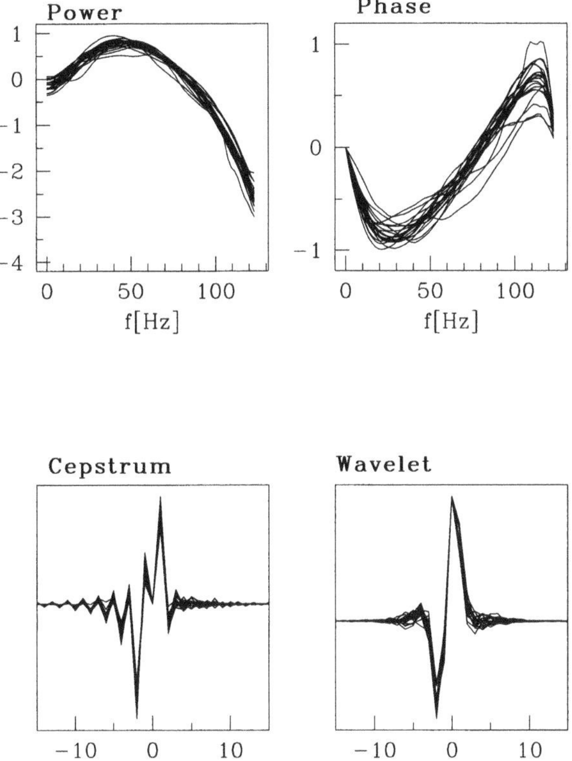

Figure 6.23: Wavelet estimation using the tricepstrum. The fourth order cumulant was estimated from 4 segments of 1000 samples per trace. The figures correspond to 20 realizations of the process.

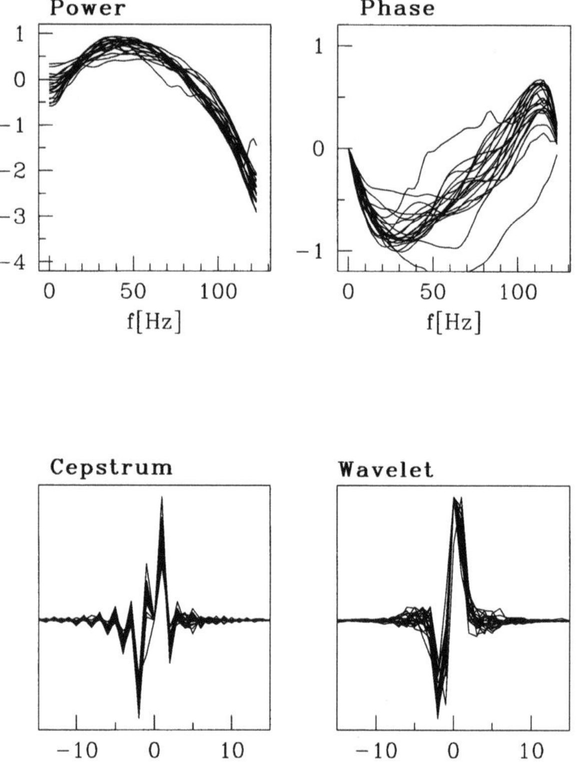

Figure 6.24: Wavelet estimation using the tricepstrum. The fourth order cumulant was estimated from 4 segments of 500 samples per trace. The figures correspond to 20 realizations of the process.

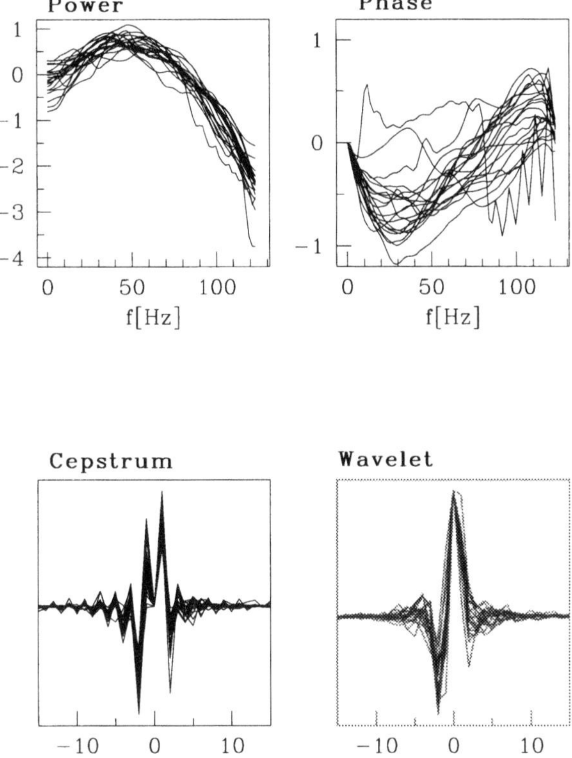

Figure 6.25: Wavelet estimation using the tricepstrum. The fourth order cumulant was estimated from 4 segments of 250 samples per trace. The figures correspond to 20 realizations of the process.

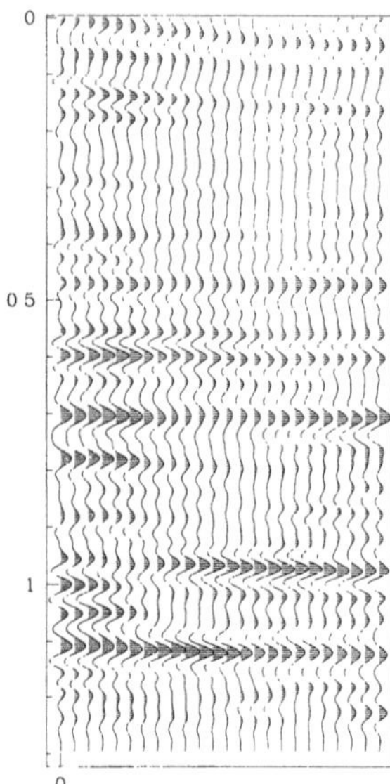

Figure 6.26: Segment of seismic section pre-processed for impedance inversion.

6.7 Blind, Full Band Deconvolution

In Section 6.5, we mentioned that impedance recovery can be approached in two ways. We will present two examples of the second approach, that of estimating a full band reflectivity by directly solving the nonunique problem posed in Eq. (6.3). Specifically, we describe the MED technique, minimum entropy deconvolution, proposed by Wiggins (1978), and an approach via independent component analysis, which we call ICBD, (independent component blind deconvolution). Both offer a very different approach to seismic deconvolution. While the classical methods such as spiking and predictive deconvolution seek to whiten the spectra, MED and ICBD seek a sparse spike solution consistent with the data.

6.7.1 Minimum Entropy Deconvolution, MED

Despite the differences between predictive deconvolution and MED outlined above, both methods constitute a linear approach to seismic deconvolution. While spiking and predictive filters are obtained by inverting a Toeplitz matrix, the MED filter is computed in an iterative procedure in which the Toeplitz matrix is inverted at each step.

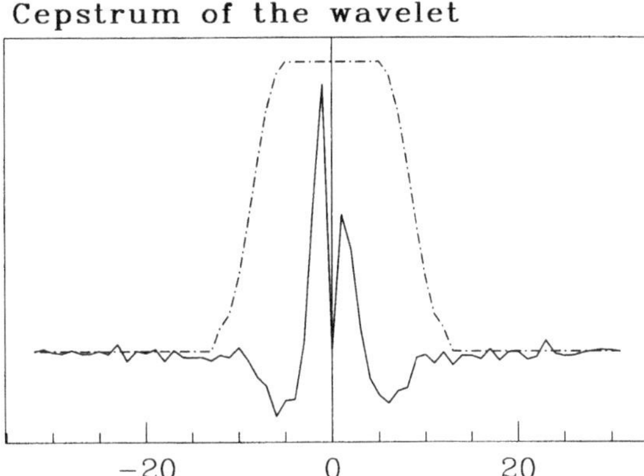

Figure 6.27: Cepstrum of wavelet estimated from the tricepstrum of the data. An average fourth-order cumulant derived from 24 traces was used to retrieve the tricepstrum.

6.7.1.1 Minimum Entropy Estimators

We begin, as always, with our canonical model, noiseless for the moment

$$s_t = w_t * q_t$$

Adopting a linear scheme to recover q_t from s_t, we look for an operator, f_t, such that

$$q_t = s_t * f_t \tag{6.46}$$

Importantly, note that if s_t is a band-limited signal, only a part of q_t can be recovered. Usually of course, the signal is contaminated with noise, and we measure

$$x_t = s_t + n_t = w_t * q_t + n_t$$

Following the discussion in Section 6.3.2, we compute the estimate of the filter, $\hat{f}_t$, and obtain the estimated reflectivity as

$$\begin{aligned} \hat{q}_t &= a_t * q_t + \hat{f}_t * n_t & (6.47) \\ &= q_t + (a_t - \delta_t) * q_t + \hat{f}_t * n_t. & (6.48) \end{aligned}$$

Eq. (6.48) shows that the task of $\hat{f}_t$ is not only to make the averaging function, a_t, close to a delta, but also to keep the noise level as small as possible. It follows that we are faced with the usual trade-off between resolution and statistical reliability.

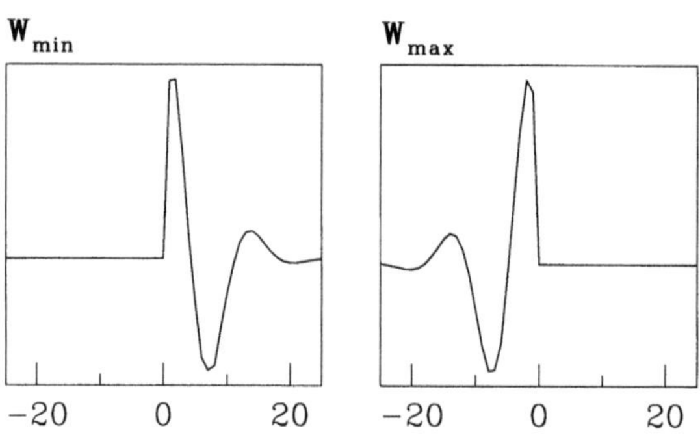

Figure 6.28: Minimum and maximum phase decomposition of the wavelet after cepstral liftering.

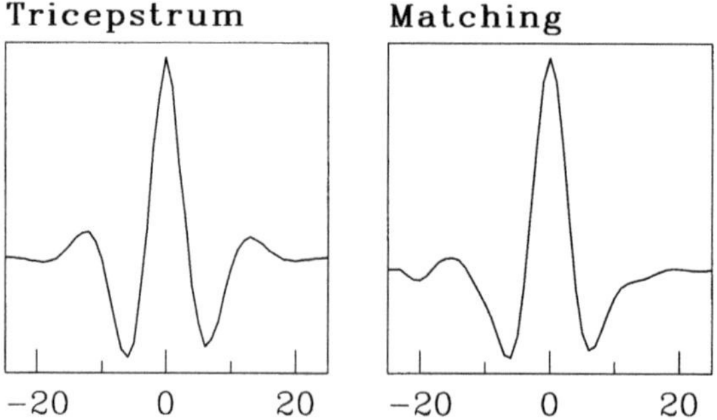

Figure 6.29: Wavelet estimates computed using the tricepstrum (Left) and cumulant matching plus nonlinear optimization (Right).

6.7.2 Entropy Norms and Simplicity

The problem that we have considered in the previous section is non-unique. An obvious approach, therefore, is to seek those particular solutions which contain reasonable features of the reflectivity. Certainly, parsimony or simplicity can be considered a required feature of an acceptable model (please see Section 3.6.3 for a discussion of this concept). Minimum structure or simple solution are terms often used for a model with a parsimonious characteristics. In Wiggins original approach, the term minimum entropy is used synonymously with maximum order, a term appropriate to describe the main difference between MED and spiking or predictive deconvolution. While the latter attempts to whiten the data (minimum order), the MED approach seeks a solution consisting mainly of isolated spikes. Wiggins norm was inspired by factor analysis techniques and can be regarded as a particular member of a broad family of norms of the form

$$V(\mathbf{q}) = \frac{1}{NF(N)} \sum_{i=1}^{N} \psi_i F(\psi_i) \tag{6.49}$$

where the vector $\mathbf{q}$ represents the reflection coefficient series of, length N, and ψ_i is an amplitude normalized measure given by

$$\psi_i = \frac{{q_i}^2}{\sum_k {q_k}^2/N}$$

$F(\psi_i)$ in Eq. (6.49) is a monotonically increasing function of ψ_i, often referred to as the entropy function (please see Section 3.2 for a discussion of entropy), and obeying the inequality

$$F(1)/F(N) \leq V(\mathbf{q}) \leq 1.$$

The normalization factor in Eq. (6.49) guarantees the same upper limit for any entropy function. For the simplest case, a series consisting of zeros and a single spike, the norm reaches the upper bound $V(\mathbf{q}) = 1$. When all the samples are equal $V(\mathbf{q})$ reaches the lower bound.

The original MED norm adopted by Wiggins (Wiggins, 1978), is obtained when $F(\psi_i) = \psi_i$, in which case

$$V(\mathbf{q}) = \frac{\sum_i \psi_i^4}{(\sum_i \psi_i^2)^2} \tag{6.50}$$

a norm that is a measure of the kurtosis, (Eq. (1.6)), of the reflectivity. In many synthetic examples we found that this norm is very sensitive to strong reflections. To avoid such a bias, we have tested other norms concluding that better results are achieved with the logarithmic norm in which $F(\psi_i) = \log(\psi_i)$

6.7.2.1 Wiggins Algorithm

A trivial solution to the problem stated in equation (6.46) is

$$\hat{f}_t = s_t^{-1}$$

implying

$$\hat{q}_t = \delta_t$$

where ${s_t}^{-1}$ stands for the inverse of s_t if it exists. To avoid such a solution, f_t must be assigned a particular length, L_f

$$q_n = \sum_{l=1}^{L_f} f_l s_{n-l} \tag{6.51}$$

The criterion for designing the operator f_k may be set as

$$\frac{\partial V}{\partial f_k} = 1/[NF(N)] \sum_i [F(\psi_i) + \psi_i \frac{\partial F(\psi_i)}{\partial \psi_i}] \frac{\partial \psi_i}{\partial f_k} = 0 \quad k = 1, 2, ..., L_f \tag{6.52}$$

From Eq. (6.51) it follows that

$$\frac{\partial q_n}{\partial f_k} = s_{n-k}$$

and, after some algebraic manipulations, Eq. (6.52) becomes

$$\sum_l f_l \sum_n s_{n-k} s_{n-l} = \sum_i b_i s_{i-k} \tag{6.53}$$

where

$$b_i = \frac{NG(\psi_i)\psi_i}{\sum_j G(\psi_j)\psi_j} \tag{6.54}$$

and

$$G(\psi_i) = F(\psi_i) + \psi_i \frac{\partial F(\psi_i)}{\partial \psi_i} \tag{6.55}$$

We can use two criteria, $F(\psi_i) = \psi_i$ (Wiggins entropy function) or $F(\psi_i) = \log(\psi_i)$ (logarithmic entropy function). Numerical experiments suggest that the second one appears to be a better choice for seismic deconvolution (Sacchi et al., 1994). Eq. (6.53) corresponds to the system used to design the well known Wiener or shaping filter (Robinson and Treitel, 2002), the task of which is to convert an input signal, $\mathbf{x}$, into a desired output, $\mathbf{d}$. In matrix notation, Eq. (6.53) becomes

$$\mathbf{Rf} = \mathbf{g}(\mathbf{f})$$

where $\mathbf{R}$ is the usual Toeplitz autocovariance matrix of the data and the vector $\mathbf{g}(\mathbf{f})$ is the crosscorrelation between $\mathbf{b}$ and $\mathbf{s}$. The system must be solved iteratively

$$\mathbf{f}^k = \mathbf{R}^{-1}\mathbf{g}(\mathbf{f})^{k-1}$$

where k denotes iteration number. The system is initialized with $\mathbf{f}^0 = [0, \cdots, 1, \cdots, 0]^T$ and the solution proceeds using Levinsons algorithm at each iteration. Note that, at each step, the system attempts to reproduce the features of the reflectivity series. If the proper length is chosen, the system leads to a useful maximum and the principal reflections can be estimated (Wiggins, 1978).

6.7.2.2 Frequency Domain Algorithm

It is possible to redevelop the MED approach in the frequency domain to extend missing high frequencies not recovered by deconvolution methods (Walker and Ulrych, 1983). In the frequency domain, the maximization of the entropy is subject to the following constraint

$$Q(\omega) = \hat{Q}(\omega) \quad \omega \in [\omega_L, \omega_H] \tag{6.56}$$

For practical purposes, we rewrite Eq. (6.15) in the discrete Fourier domain

$$\hat{Q}_k = A_k Q_k$$

where the index k denotes the frequency sample. The maximization of V with mid-band constraints can now be written as

$$\text{Maximize } V \text{ subject to } Q_k = \sum_{n=0}^{N-1} \hat{q}_n e^{-i2\pi kn/N} \qquad k = k_L, \cdots, k_H$$

where k_L and k_H are the samples that correspond to ω_L and ω_H, the low and high mid-band frequencies, respectively. The solution is achieved by considering the following system of equations

$$\frac{\partial V}{\partial Q_k} + \sum_{l=k_L}^{k_H} \lambda_l (Q_l - \sum_{n=0}^{N-1} \hat{q}_n e^{-i2\pi nl/N}) = 0$$

and

$$Q_k - \sum_{n=0}^{N-1} \hat{q}_n e^{-i2\pi nk/N} = 0 \qquad k = k_L, \cdots, k_H$$

where λ_l are the Lagrange multipliers of the problem. Taking derivatives, inserting Q_k into the constraint and substituting the derived multipliers, results in

$$Q_k = \begin{cases} \sum_{n=0}^{N-1} \left[\frac{G(\psi_n)\psi_n e^{-i2\pi kn/N}}{\sum_j G(\psi_j)\psi_j/N} \right] & k \notin [k_L, k_H] \\ \hat{Q}_k & k \in [k_L, k_H] \end{cases}$$

Eqs. (6.54) and (6.56) show that

$$Q_k = \begin{cases} B_k & k \notin [k_L, k_H] \\ \hat{Q}_k & k \in [k_L, k_H] \end{cases}$$

where B_k is the DFT of b_t (Eq. (6.54)). Since b_t is a nonlinear function of the reflectivity q_t, the problem must be solved as follows

1. The algorithm is initialized by letting $q_t = \hat{q}_t$
2. b_t and B_k are computed.
3. The missing low and high frequencies are replaced by B_k.
4. An estimate of the reflectivity is calculated from the inverse FT. The norm V is also evaluated to check convergence and a new iteration begins in step 2.

The algorithm is very simple and can be efficiently coded using the FFT. An application to real data can be found in Sacchi et al. (1994). A synthetic example portraying the method is provided in Fig. (6.30).

6.7.3 Blind Deconvolution via Independent Component Analysis

6.7.3.1 Introduction

As discussed in this chapter, a number of different schemes for deconvolution, each assuming some different a priori constraints[4], exist in the literature. Blind deconvolution, via Independent Component Analysis, ICA (which we call ICBD for short), assumes only independence. It is this topic that is covered here.

We rely much on the outstanding book Independent Component Analysis by Aapo Hyvärinen, Juha Karhunen and Erkki Oja, (Hyvärinen et al., 2001). We call this reference HKO since we refer to it constantly. We review, briefly, the central concepts and refer the reader to HKO for beautifully presented details.

[4]The most common in applied seismology being those of minimum phase and whiteness.

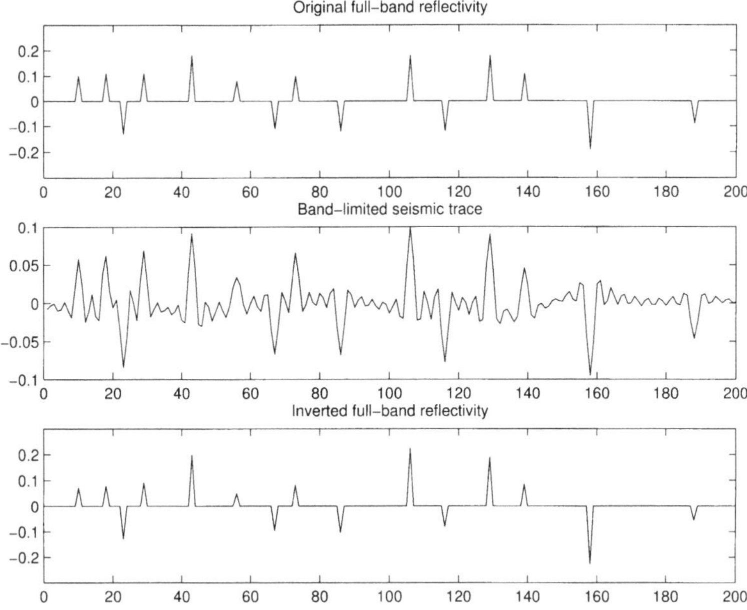

Figure 6.30: Frequency domain MED. (Top) True reflectivity. (Center) Band-limited reflectivity. (Bottom) Inverted reflectivity using FMED.

6.7.3.2 Blind Processing

Blind separation and blind deconvolution, the two main topics in blind processing, are integral parts of ICA. The former, that we are not concerned with here in any detail is, in the vernacular, called the Cocktail Party Problem, a name that defines the problem succinctly. The task of blind separation is to unmix mixtures. An example from HKO, that we have reprogrammed with somewhat different parameters, illustrates the problem. Fig. 6.31 illustrates quite clearly the potential power of ICA separation. The three mixtures, panels (d)-(f), that are almost the same by suitable choice of the mixing coefficients, show no sign whatsoever of the signals that have been added together in different proportions. The outputs, on the other hand, shown in panels (g)-(h) are, essentially, the inputs. We note that the amplitudes, both in size and sign, and the order of the signals, are different to the equivalent parameters in the input. The shapes, however, are essentially the same. How has this, seemingly miraculous, separation taken place? Of course, given that the problem essentially consists of 6 unknowns and 3 equations, some constraining criterion is required. This criterion is that of independence.

6.7.3.3 Independence

We reiterate here, in more detail, concepts that were introduced in Section 1.2.1. Since zero mean is assumed or imposed in all that follows, we will deal always with

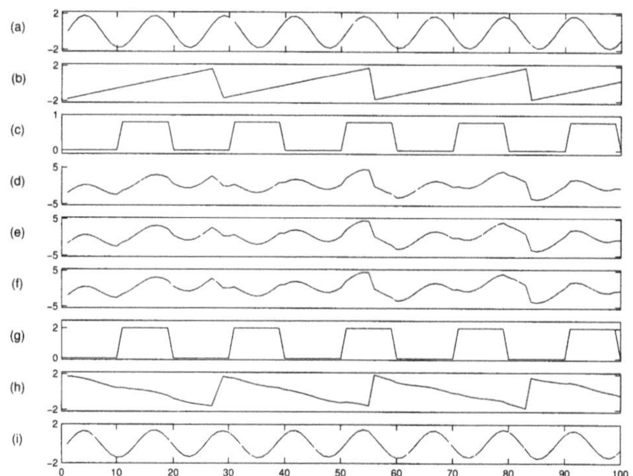

Figure 6.31: Blind separation. (a)-(c) input sources: (d)-(f) mixtures: (g)-(i) separated sources.

covariance rather than correlation. The covariance matrix for two zero mean vectors, $\mathbf{x}$ and $\mathbf{y}$, is defined as

$$\mathbf{C}_{\mathbf{xy}} = E[\mathbf{xy}^T]$$

and $\mathbf{C}_{\mathbf{xy}} = 0$ for uncorrelated vectors

When considering $\mathbf{x}$, we specify the case when the components x_i are mutually uncorrelated by

$$\mathbf{C}_{\mathbf{xx}} = \mathbf{D}$$

where $\mathbf{D}$ is a $N \times N$ diagonal matrix, with the variances of the individual components along the diagonal.

In keeping with the nomenclature of HKO, vectors that satisfy the condition

$$\mathbf{C}_{\mathbf{xx}} = \mathbf{I} \quad \text{(please remember that the mean vector, } \mathbf{m}_{\mathbf{x}} = 0)$$

are said to be *white*, a property that is preserved under orthogonal transformations.

Much of what we do in the processing of our data involves the property of uncorrelatedness, either by assumption (e.g., signal and noise are uncorrelated) or by design (e.g., Gram-Schmidt orthogonalization (Section 1.5.1). In fact, one may say that the property of uncorrelatedness is central in most of what we do. There is, however, a concept that not only embodies uncorrelatedness, but is in fact stronger. This is the concept of independence. Independence is stronger because, whereas all processes that are independent are also uncorrelated, processes which are uncorrelated are not necessarily independent. Defining $p(\mathbf{x})$ and $p(\mathbf{y})$ to be the marginal pdfs associated with $\mathbf{x}$ and $\mathbf{y}$ and $p(\mathbf{x}, \mathbf{y})$ to be their joint pdf, $\mathbf{x}$ and $\mathbf{y}$ are independent if, and only if

$$p(\mathbf{x}, \mathbf{y}) = p(\mathbf{x})p(\mathbf{y})$$

a property that generalizes to any number of vectors.

As will become apparent, lack of correlation is not sufficient in itself to accomplish blind deconvolution. This fact is not surprising. Our canonical model for the seismic trace, $x(t)$, given by Eq. (6.3), is one equation in two unknowns and, as we have pointed out, Enders Robinson required *two* constraints to solve it. The constraint of an uncorrelated reflectivity was not, in itself, sufficient.

A link between the concepts of uncorrelatedness and independence lives in nonlinearities. Consider the expectation of the nonlinear expression $\mathbf{f}(\mathbf{x})\mathbf{g}(\mathbf{y})$, where $\mathbf{f}(\mathbf{x})$ and $\mathbf{g}(\mathbf{y})$ are some general nonlinear transformations of the random vectors. In terms of the appropriate pdfs, we have

$$E[\mathbf{f}(\mathbf{x})\mathbf{g}(\mathbf{y})] = \int\int \mathbf{f}(\mathbf{x})\mathbf{g}(\mathbf{y})p(\mathbf{x}, \mathbf{y})d\mathbf{x}d\mathbf{y}$$

and, assuming independence

$$\begin{aligned} &= \int \mathbf{f}(\mathbf{x})p(\mathbf{x})d\mathbf{x} \int \mathbf{g}(\mathbf{y})p(\mathbf{y})d\mathbf{y} \\ &= E[\mathbf{f}(\mathbf{x})]E[\mathbf{g}(\mathbf{y})] \end{aligned} \tag{6.57}$$

We see, therefore, that independence implies the factorization of the nonlinear correlation as expressed by Eq. (6.57). This is a very important relationship since it explains, and emphasizes, the role of nonlinearities in ICA. It also says (by setting $f(x) = x$ and $g(y) = y$) that independence implies uncorrelatedness. That independence is, indeed, deeper than uncorrelatedness can be intuitively illustrated by means of a simple example. Consider the functional relationship between two scalar, random variables

$$y = x^2$$

then

$$\mathrm{cov}(x, y) = E[x^3] - E[x^2]E[x]$$

If $p(x)$ is symmetric about $x = 0$

$$E[x^3] = E[x] = 0$$

and

$$\text{cov}(x, y) = 0$$

x and y are uncorrelated but dependent. On the other hand, if x and y are independent, they are necessarily uncorrelated.

6.7.3.4 Definition of ICA

ICA is a linear model that describes the mixing of source components, $s_i(t)$, $i = 1, 2, \ldots, M$ by means of a mixing matrix, $\mathbf{A}$, to produce mixtures $x_i(t)$, $i = 1, 2, \ldots, L$. In matrix form, defining $\mathbf{s}$ and $\mathbf{x}$ to be the column vectors associated with the source and mixed signals, we have

$$\mathbf{x} = \mathbf{A}\mathbf{s} \tag{6.58}$$

Eq. (6.58) represents a very simple model but, in reality, its solution to obtain the desired $\mathbf{s}$ is anything but simple. The reason is, of course, that both $\mathbf{s}$ and $\mathbf{A}$ are unknown. As mentioned previously, simple whiteness is not sufficient to solve the ICA problem. It turns out, however that, the assumption that the components, s_i, $i = 1, 2, \ldots, M$, are independent is enough to allow the required decomposition. The discussion pertaining to nonlinear correlation allows us to see why, and we explore this concept in the sections to come.

It turns out that nonlinearities are central to the ICA decomposition. In turn, these nonlinear functions imply the use of higher-order statistics (Section 6.6.1). A simple, intuitive view, (HKO, page 37), is helpful. Algorithms often compute $g(\mathbf{u})$ where $\mathbf{u}$ contains the desired unknown parameters and $g(\cdot)$ is a nonlinearity. Consider $g(u) = \tanh(u)$. Since

$$\tanh(u) = u - \frac{1}{3}u^3 + \frac{2}{15}u^5 - \cdots$$

higher-order statistics of $\mathbf{x}$ are clearly involved. Further, since, inverting Eq. (6.58) obtains

$$\mathbf{y} = \mathbf{W}\mathbf{x} \tag{6.59}$$

where $\mathbf{y}$ is an estimate of $\mathbf{s}$ and $\mathbf{W}$ is a regularized estimate of the inverse of $\mathbf{A}$, the existence of higher-order cumulants implies that the sources must be nonGaussian.

6.7.3.5 Specifying Independence

As demonstrated by Eq. (6.57), independence occurs if the nonlinear correlation factors. To summarize briefly at this stage,

- Independence of the components of $\mathbf{s}$ is sufficient to solve the ICA decomposition. A discussion of this essential proposition is left to Section 6.7.3.6.
- Independence implies nonlinear uncorrelatedness.

- Nonlinearities are central in ICA decomposition. Nonlinearities imply higher order statistics which, in turn, imply nonGaussianity (a Gaussian pdf is fully parameterized by 2nd order statistics only).

Fortunately, nonGaussianity is the rule rather than the exception in physical situations. Fig. 6.32 illustrates a pdf computed from an actual sonic log and a Gaussian pdf of equal variance. The superGaussian character of the reflectivity is quite clear.

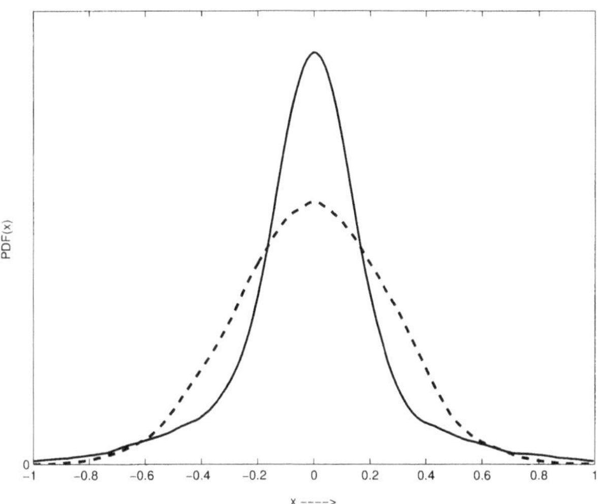

Figure 6.32: Gaussian and reflectivity pdfs.

We are now in a position to define suitable measures of independence by means of which we will formulate appropriate cost functions. All these measures will, in some fashion, specify the nonGaussian, and therefore independent, nature of $\mathbf{s}$. Here are our favorites.

Kurtosis.

In this regard, we remember the Wiggins norm defined by Eq. (6.7.2.1), and the discussion of MED in Section 6.7.1. The kurtosis, κ_4, is defined in Eq. (1.5). Fast gradient algorithms for maximizing κ_4 exist (HKO page 176). The problem, as discovered by the geophysical community, is that κ_4 is very sensitive to outliers.

Negentropy

Negentropy is defined as

$$J(\mathbf{x}) = H(\mathbf{x}_{gauss}) - H(\mathbf{x})$$

where $\mathbf{x}_{gauss}$ is a random vector with the same covariance matrix as $\mathbf{x}$. Clearly, maximizing negentropy will result in components that are as nonGaussian as possible.

Mutual Information

This very powerful concept has been elaborated in Section 3.6 and requires little elaboration here. The relationship between mutual information and entropy is expressed by Eq. (3.55), which is repeated here for convenience.

$$I(\mathbf{s},\mathbf{x}) = H(\mathbf{s}) - H(\mathbf{s}|\mathbf{x})$$

and

$$I(\mathbf{x},\mathbf{s}) = H(\mathbf{x}) - H(\mathbf{x}|\mathbf{s}) \qquad (6.60)$$

Eq. (6.60) represents the difference between the entropy of the output and the conditional entropy of the output and the signal, respectively. More specifically

$$H(\mathbf{x}|\mathbf{s}) = -\int\int p(x,s)\log p(x,s)dxds$$

and is a measure of the average loss of information in the transmission of the signal. The usefulness of this relationship lies in the fact that (avoiding, for the moment such issues as the divergence of entropy for continuous systems), maximizing I(**x**,**s**) with respect to a parameter w, say, clearly implies the maximization of the conditional entropy, a measure of the entropy of the output that is independent of that of the input. In other words, maximizing MI allows us to isolate the properties of the medium from those of the input signal[5].

6.7.3.6 Finally, the Reason to "Why Independence"?

We have alluded to the fact that independence of the source components is sufficient to solve the very nonunique ICA problem. That this is in fact so, can be seen by examining Eq. (6.59) and recalling the Central Limit Theorem, CLT. The essence of the CLT is that, if a random variable consists of the sum of independent random variables, then the distribution of this random variable tends to the Gaussian distribution. Intuitively, the reason for this is that, if a random variable z is the sum of independent random variables eg., $z = s_1 + s_2$, then $p(z) = p(s_1) * p(s_2)$. Since convolution is a smoothing operation, $p(z)$ is smoother than both $p(s_1)$ and $p(s_2)$. Of course, this simple view does not lead to the conclusion that $p(z)$ becomes Gaussian, but the CLT does.

We now turn to Eq. (6.59) which tells us that the estimated source vector $\mathbf{y}$ will be equal to a scaled version of the actual vector $\mathbf{s}$ (with arbitrary order of components) if $\mathbf{W}$ is full rank with row elements that are all zeros except for one element that is a constant (thus implying that no mixing of the components has occurred). Following the above discussion concerning the CLT, we see that $p(\mathbf{y})$, in the case described, must be more Gaussian than any other $p(\mathbf{y})$ resulting from a $\mathbf{W}$ that is less sparse. If, therefore, the components of $\mathbf{y}$ turn out to be independent, and the components of $\mathbf{s}$ are, as assumed, independent, then $\mathbf{y} \propto \mathbf{s}$ and truly represents the input sources.

[5]MI is also related to the Kullback-Leibler divergence, $I(1,2)$, defined in Section 3.3.

6.7.4 Blind Deconvolution

The ICA model is linear and, not surprisingly therefore, can be readily adapted to the problem of blind deconvolution. Here is how. Consider a source component, $s(t)$, discretized to yield s_t and which we will call q_t to signify that we will be dealing with the earths primary reflectivity. By primary, we mean that q_t consists only of primary events (the case of including multiples will be considered later). One thing that we know about q_t is that it can, fairly accurately, be assumed to be uncorrelated (we have discussed this issue in this chapter ad nauseam). The demultipled seismic trace is modelled as $x_t = b_t * q_t$ where b_t is the (stationary, for now) seismic wavelet (we call it b_t rather than w_t as previously, to differentiate it from the weighting coefficients). In matrix form

$$\mathbf{x} = \mathbf{B}\mathbf{q} \tag{6.61}$$

where $\mathbf{B}$ is the wavelet convolution matrix. Eq. (6.61) is now in ICA form, but to be useful for the purpose of BD, following HKO, we define a vector $\breve{\mathbf{q}}$ to be

$$\breve{\mathbf{q}} = [q_t, s_{t-1}, \ldots, s_{t-L+1}]^T$$

as one possibility. Here, L is the data-window length and $\breve{\mathbf{q}}$ consists of shifted versions of the source vector, $\mathbf{s}$. Applying the ICA equation gives rise to $\breve{\mathbf{x}}$

$$\breve{\mathbf{x}} = \breve{\mathbf{A}}\breve{\mathbf{q}} \tag{6.62}$$

or, in matrix form

$$\breve{\mathbf{X}} = \breve{\mathbf{A}}\breve{\mathbf{Q}} \tag{6.63}$$

Eq. (6.63), using different formulations of $\breve{\mathbf{X}}$, is used throughout the rest of this section.

6.7.4.1 The ICA Algorithm

We outline, here, some details concerning the nonlinear optimization algorithm that we use. We follow HKO. The first step in the algorithm is to preprocess the data by whitening $\mathbf{x}$ (our zero mean input vector). In the context used here, whitening implies that the covariance matrix of the whitened data, $\mathbf{C_{zz}} = \mathbf{I}$. It turns out that this step is a great simplification in the computation of the independent components. Whitening is accomplished by means of linear principal component analysis, LPCA, which we desribed in Section 5.3. The procedure is to perform a linear transformation on $\mathbf{x}$ to obtain $\mathbf{z}$

$$\mathbf{z} = \boldsymbol{\Gamma}\mathbf{x} \tag{6.64}$$

where $\boldsymbol{\Gamma}$ is a suitably computed orthogonal matrix. The LPCA approach is to compute $\boldsymbol{\Gamma}$ as

$$\boldsymbol{\Gamma} = \boldsymbol{\Lambda}^{-1/2}\mathbf{U}^{\mathrm{T}} \tag{6.65}$$

where $\mathbf{\Lambda}$ and $\mathbf{U}$ are obtained from the spectral decomposition of the data covariance matrix as

$$\mathbf{C_{xx}} = \mathbf{U \Lambda U^T} \tag{6.66}$$

That $\mathbf{z}$ is indeed white, may be seen as follows: Considering the covariance matrix of $\mathbf{z}$, we obtain, using Eqs. (6.64) and (6.65)

$$\begin{aligned} \mathbf{C_{zz}} &= E[\mathbf{zz}^T] \\ &= \mathbf{\Lambda} E[\mathbf{xx}^T]\mathbf{\Lambda}^T \\ &= \mathbf{\Lambda}^{-1/2}\mathbf{U^T} E[\mathbf{U\Lambda U^T}]\mathbf{U\Lambda^{-1/2}} \\ &= \mathbf{I} \end{aligned}$$

and the proof is complete.

The fact that $\mathbf{C_{zz}}$ is an orthogonal matrix simplifies the problem to a search for orthogonal matrices only. This is not the only advantage to LPCA preprocessing, however. Often, the input data, $\mathbf{x}$, are redundant, in the sense that $\mathbf{\Lambda}$ in Eq. (6.66) contains eigenvalues very close to zero. In such a case, $\mathbf{x}$ may be reduced in dimension with considerable algorithmic simplification. The latter advantage of LPCA preprocessing does not apply in the BD situation considered here.

Our BD model, using Eqs. (6.59) and (6.62), now becomes

$$\check{\mathbf{y}} = \mathbf{W}\check{\mathbf{z}} = \mathbf{W\Gamma}\check{\mathbf{x}} \tag{6.67}$$

The whiteness of $\check{\mathbf{z}}$ may be transferred to the estimated $\check{\mathbf{y}}$ by imposing constraints on $\mathbf{W}$. Using the above equation

$$\begin{aligned} E[\check{\mathbf{y}}\check{\mathbf{y}}^\mathbf{T}] &= E[(\mathbf{W}\check{\mathbf{z}})(\mathbf{W}\check{\mathbf{z}})^\mathbf{T}] \\ &= \mathbf{W}E[\check{\mathbf{z}}\check{\mathbf{z}}^\mathbf{T}]\mathbf{W^T} \\ &= \mathbf{WIW^T} \end{aligned}$$

Consequently, if $\mathbf{W}$ is constrained to be orthogonal, the estimated independent components will be white.

The cost function to be optimized is nonlinear and has multiple local minima. Of interest is to find these minima and, thus, recover the independent components. The constraints introduced in the previous section must also be taken into account. We first consider an algorithm for finding one independent component such that $y_i = \mathbf{w}^T\mathbf{z}$, where $\mathbf{w}$ is a row of $\mathbf{W}$. A simple extension of this algorithm allows for the simultaneous solution for all the independent components.

The idea for the optimization scheme used in finding one independent component, that is one local minimum, is illustrated, for $\mathbf{w} \in \mathcal{R}^2$, in Fig. 6.33. This is a modified gradient descent algorithm that, defining $\Phi(\mathbf{w})$ to be a suitable nonlinear function of the model weights, proceeds by

$$\mathbf{w}^{(i+1)} = \mathbf{w}^{(i)} + \alpha\nabla\Phi\left(\mathbf{w}^{(i)}\right)$$

$$\mathbf{w}^{(i+1)} \leftarrow \frac{\mathbf{w}^{(i+1)}}{||\mathbf{w}^{(i+1)}||_2}$$

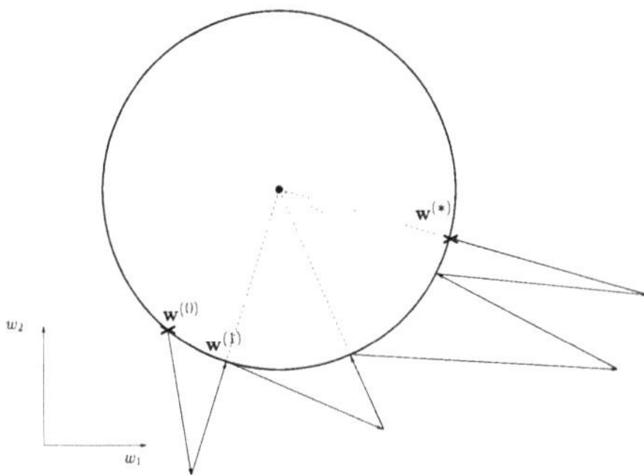

Figure 6.33: The basic optimization scheme for ICA. Given a vector, $\mathbf{w}^{(0)}$ on the unit circle, a step is taken following the gradient of the cost function producing a new point which is projected back onto the unit circle giving $\mathbf{w}^{(1)}$. This is repeated until $||\mathbf{w}^{(i+1)} - \mathbf{w}^{(i)}||_2^2 < \epsilon$ where ϵ is some prescribed tolerance.

where α is some user defined constant, $\mathbf{w}^{(i)}$ is the model at iteration i and $\mathbf{w}^{(i+1)}$ is the model at iteration $i+1$. That is, a step away from the unit circle follows the gradient of the cost function and the result is projected back onto the unit circle. Examination of the cost function shows that at a local minimum, $\mathbf{w}^{(j)} = \mathbf{w}^{(*)}$, where $\mathbf{w}^{(*)}$ is the minimum of the cost function, and the gradient is perpendicular to the unit circle. Consequently, executing an iteration of this algorithm gives $\mathbf{w}^{(j+1)} = \mathbf{w}^{(j)} = \mathbf{w}^{(*)}$, and hence, the algorithm converges.

The extension of this algorithm to find all independent components simultaneously is achieved by taking into account the constraint on $\mathbf{W}$. That is, find $\mathbf{w}_1$ as above, then choose $\mathbf{w}_2$ such that $\mathbf{w_2} \perp \mathbf{w}_1$; choose $\mathbf{w}_3$ such that it is orthogonal to both $\mathbf{w}_1$ and $\mathbf{w}_2$ and so on. This orthogonality is easily achieved using Gram-Schmidt orthogonalization. An example of $\Phi(\mathbf{w}^{(i)})$ based on the maximization of negentropy is (Hyvärinen, 1998)

$$\Phi\left(\mathbf{w}_i\right) = \frac{1}{2}\sum_i c_i^2$$

where $c_i = E\left[g\left(y_i\right)\right]$, y_i is the ith component of $\mathbf{y}$ and $g\left(y\right)$ is some nonlinear function. Further refinements of the optimization algorithm use a Lagrange multiplier, $\mathbf{w}^T\mathbf{w} = 1$, for stability and a fixed iteration scheme for efficiency yielding the update rule, (Hyvärinen et al., 2001)

$$\mathbf{w} \leftarrow E[\mathbf{z}g'(\mathbf{w^T z})] - E[g''(\mathbf{w^T z})]\mathbf{w} \tag{6.68}$$

where $g'(\cdot)$ and $g''(\cdot)$ are, respectively, the first and second derivatives of $g(\cdot)$. A typical non linearity is

$$g(y) = -exp(-y^2/2)$$

Eq. (6.68) represents one of many algorithms used in ICA. It maximizes the negentropy of y_i and, hence, maximizes its nonGaussian statistics giving an independent component. Of course, there are important issues in the design of the algorithm.

In particular what type of nonlinearities are applicable to the data at hand and the effect of noise. In fact, we can see that, in the derivation of Eq. (6.68), the choice of nonlinearity is directly related to the pdf which governs $\mathbf{y}$.

6.7.4.2 ICA, BD and Noise

Eqs. (6.58) and (6.62) must be modified in the presence of noise by the addition of a noise vector, $\mathbf{n}$, to the right hand side of both equations. Under certain approximations to negentropy, the addition of Gaussian and zero mean random noise to the data can be brought into consideration during the PCA stage of ICA such that $\mathbf{W}$ is unaffected by the noise and

$$\hat{\mathbf{y}} = \mathbf{W}\left(\mathbf{z} + \mathbf{n}\right).$$

Further, by using an alternative criterion for ICA which maximizes the joint likelihood of $\mathbf{A}$ and $\mathbf{y}$, independent components can be sought which take the noise into consideration (Moulines et al., 1997). Details concerning algorithm design, including the issue of noise, may be found in Kaplan (2002).

6.7.4.3 A Synthetic Example

The point of this example is to investigate the application of the ICA technique to the estimation of a seismic wavelet from short, noisy data sets. The reason why we consider short records, we use merely 100 points in this example, is because the seismic wavelet is nonstationary. Since we use a wavelet of 30 points, the nonstationarity would not affect the result in this example.

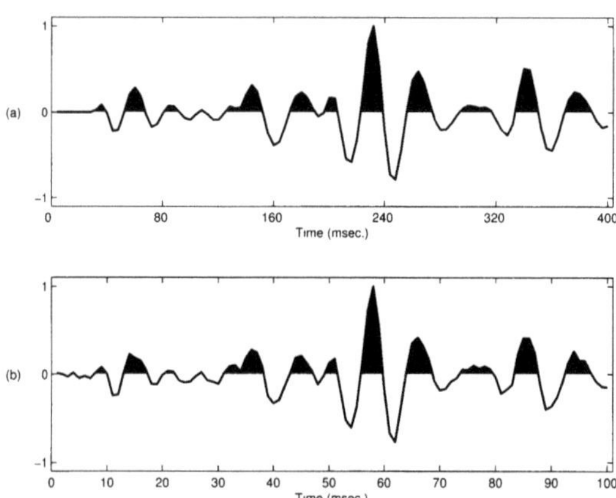

Figure 6.34: Inputs to BD example. (a) Noiseless input. (b) Input with SNR=10.

Fig. 6.34 shows the input data. No noise data are shown in panel (a), whereas the noisy data, with a SNR=10, where the SNR is computed as a ratio of maximum amplitudes, are shown in panel (b). The results of the BD using the data shown in Fig. 6.34 are illustrated in Figs. 6.35 and 6.36. The actual reflectivity and non minimum phase wavelet are shown for comparison. The noiseless results show the resolution

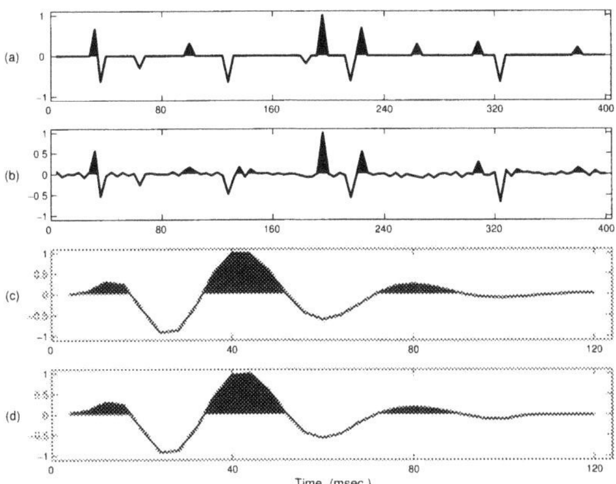

Figure 6.35: Noiseless example. (a) Actual reflectivity. (b) Recovered reflectivity. Actual wavelet. (d) Recovered wavelet.

that may be achieved in the optimum environment. Of course, it is noise that has spelt the doom of many proposed BD techniques and for this reason we consider an example with a healthy (or unhealthy) amount of noise. The results shown in Fig. 6.36 have been achieved without any preprocessing or any explicit consideration of the noise in the algorithm. As such, the results are encouraging. Extensions of the algorithm which incorporate both the presence of noise and prior information that takes into account the specific nature of the deconvolution problem are being actively pursued.

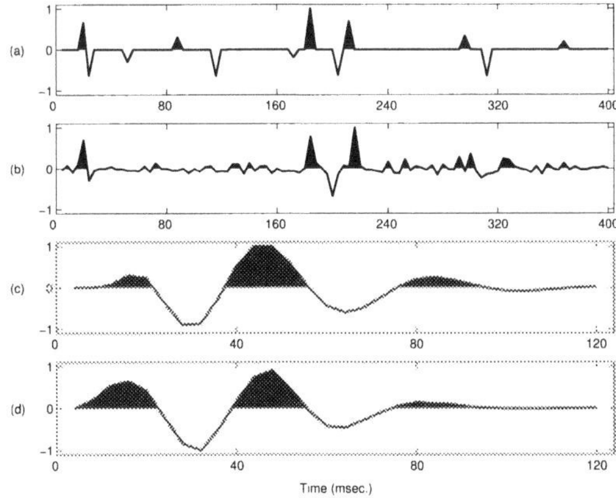

Figure 6.36: Noisy example, SNR=10. (a) Actual reflectivity. (b) Recovered reflectivity. Actual wavelet. (d) Recovered wavelet.

6.7.4.4 Remarks

Deconvolution via ICA is very rich in promise. It is a technique whose surface we have only scratched, thus only synthetic examples are included. Much work remains to be done in investigating the appropriate measure of nonGaussianity, the effect of noise and the possibility of the relaxation of the independence criterion. Of particular significance, in our opinion, is the fact that reasonable results, using noisy data, have been obtained on short data sets with a complicated source wavelet. The IEEE literature, which is replete with blind deconvolution algorithms, appears to concentrate on very long data sets and very short wavelets.

6.8 Discussion

The objective of this chapter was to present various issues involved with the ubiquitous task of deconvolution. We dealt with deconvolution when the impulse response was assumed known, a situation often incurred in practice, for example when one wishes to remove the effect of a known filter. In seismology, the problem is much more complex, since neither the impulse response of the Earth, nor the source function are known. Estimating the seismic wavelet is of great importance in seismic processing and interpretation. Thus, the most comprehensive methods of multiple attenuation are data-driven processing techniques that require such an estimate, the accuracy of which is vital for successful implementations. Of course, multiple attenuation is central to all methods that image primaries including velocity-independent imaging and, therefore, so is the estimate of the wavelet. Impedance inversion requires not only accurate wavelet estimation, but also the extrapolation of band-limited spectra, and various, different methods have been presented that attempt to fulfill this objective. The techniques that have been explored rely heavily on both linear and nonlinear inversion as well as time series modelling. The approach outlined in Section 6.7.3 is very versatile and deserves to be explored in much more depth than we have had occasion to peruse.

Chapter 7

A Potpourri of Some Favorite Techniques

7.1 Introduction

A book, by its very nature of limited length, cannot contain, in organized fashion at least, all that is in the minds of the authors. Some topics are things of beauty, however, and we have decided to present a few of our favorites, in isolated splendor. We have attempted to give just enough of an introduction to each topic to make it somewhat self contained and if, hopefully, of interest to the reader, the contained references will provide additional material. The topics that we present are quite varied in nature and are meant to introduce interesting ideas for further exploration.

7.2 Physical Wavelet Frame Denoising

We have discussed, in the preceding chapters, various approaches for the attenuation of both coherent and random noise. The development of the wavelet transform (WT)[1] theory and practice, has found a very wide range of applications in the fields of engineering and image-processing (Meyer, 1993). The WT approach has also been widely used in seismic data processing, for data compression (Luo and Schuster, 1992; Bosman and Reiter, 1993), ground roll attenuation (Deighan and Watts, 1997), frequency-time decomposition (Chakraborty and Okaya, 1995), characterization of seismic traces (Grubb and Walden, 1997), seismic inversion (Li et al., 1996), well log processing (Saito and Coifman, 1997), noise attenuation (Ioup and Ioup, 1998; Miao and Cheadle, 1998), and filtering (Du and Lines, 1998; Nguyen and Mars, 1999).

In practice, the WT can be implemented by a discrete wavelet transform (DWT). The DWT is a special case of wavelet frames (Daubechies, 1992) which, in turn, are special cases of frames. Frame theory is a general theory about the analysis and reconstruction of discrete signals, and was originated in the work of Duffin and Schaeffer in the early 1950's (Duffin and Schaffer, 1952). While the basic element in

[1]Although we do not deal, explicitly, with the WT in this book, we will direct the reader to some of the literature.

WT is the mother wavelet, the basic element in a frame is called the atom. The term atom is used by mathematicians in "atomic decomposition" (Meyer, 1993, Chapter 1). It is important to emphasize that the basic element in frames is not necessarily a wavelet. Wavelet frames represents a powerful approach to the analysis of data. In as much as Fourier analysis regards data to be composed of harmonic basis functions the purpose of Fourier analysis, frame analysis uses atoms or wavelets as basis functions. Clearly, the more similar are the frames (or analysis) atoms and data to each other, the more parsimonious and therefore efficient in terms of information retrieval is the method of signal representation.

Since seismic data exhibit high trace-to-trace coherence, it is important that denoising methods explore this property fully. Approaches, including wavelet applications, that are based on individual traces alone, will not attenuate the noise in an optimal manner. Conventional 2D wavelet denoising methods use a 2D wavelet (Miao and Cheadle, 1998; Nguyen and Mars, 1999) and attempt to take into account the coherence among traces. Since these 2D implementations typically use a tensor multiplication of 1D wavelets, the emphasis is on the vertical, horizontal and diagonal dependence in the data. If an angle parameter is used, directions corresponding to this specific angle can also be taken into account. In this case, the two independent wavelet variables become separable ($\psi(t,x) = u(t)v(x)$). Although separability leads to theoretical simplicity and computational efficiency, it may be undesirable in the processing of seismic data where the signal to be retrieved is represented by events following some curve in the $x - t$ domain which, for prestack data, is generally a hyperbola (Yilmaz, 1988)[2].

In order to take into account the hyperbolic nature of the signal, Zhang and Ulrych (2002), whose treatment of the topic we follow closely here[3], constructed a 2D wavelet where the independent variables (time and space) are not separable, as they are in conventional design. In this case, orthogonality properties no longer apply and Zhang and Ulrych (2002) use frame theory that does not require either the separability or orthogonality assumption.

7.2.1 Frames and Wavelet Frames

Mallat (1998) defines frames as

> *"Frames is a theory for the analysis of the completeness, stability and redundancy of linear discrete signal representations."*

This complex subject is well explored in the literature. Here, we present only a brief introduction to frames and wavelet frames. A detail treatment may be found in Mallat (1998, Chapter 5).

Let $\{\psi_k, k \in J\}$ be a given sequence in a Hilbert space, $\mathcal{H}$, where $J = \mathcal{Z}$ or $J = \mathcal{N}^*$, and $\mathcal{Z}$ is the set of integers and $\mathcal{N}^*$ the set of non-negative integers. The projection of a signal $f \in \mathcal{H}$ on ψ_k is $C_k =< f, \psi_k >$ which is the inner product of f

[2]The trajectories are usually more complicated due to inhomogeneities, anisotropy etc., but the hyperbolic assumption is often adequate.

[3]Since one of us is an author of this paper.

and ψ_k. The question is whether f can be uniquely determined from C_k. To answer this question in the frame context, we introduce the frame definition
Definition: For a sequence $\{\psi_k, k \in J\}$, in $\mathcal{H}$, if there exist constants $A > 0, B < \infty$ such that for all $f \in \mathcal{H}$ equation (7.1) holds, then $\{\psi_k, k \in J\}$ is called a frame in $\mathcal{H}$. Equation (7.1), the frame condition, is given by

$$A\|f\|^2 \leq \sum_{k \in J} |< f, \psi_k >|^2 \leq B\|f\|^2 \tag{7.1}$$

If $\{\psi_k, k \in J\}$ is a frame, f can be uniquely determined from C_k, and if $A \approx B$, the reconstruction formula is

$$f = \frac{2}{A+B} \sum_{k \in J} < f, \psi_k > \psi_k \tag{7.2}$$

If $A \approx B$ does not hold then there exist iterative algorithms to reconstruct f. The convergence of the iteration is assured by the frame property (Daubechies, 1992, Chapter 3).

For a frame, the form of the atom ψ_k can be arbitrary as long as it satisfies the frame condition. For a wavelet frame, the atom ψ_k has the form

$$\psi_{s,b}(t) = |s|^{-\frac{1}{2}} \psi(\frac{t-b}{s})$$

or

$$\psi_{m,n}(t) = a_0^{-\frac{m}{2}} \psi(a_0^{-m} t - n b_0)$$

where s and b are the scale and the translation parameters, respectively, m and n are integers, a_0 and b_0 are real numbers called the increments of scale and translation, and $\psi(t)$ is the mother wavelet. The reconstruction formula (7.2) becomes

$$f = \frac{2}{A+B} \sum_s \sum_b < f, \psi_{s,b} > \psi_{s,b}$$

For 2D wavelets,

$$\psi_{s_x b_x, s_t b_t}(x,t) = |s_x s_t|^{-\frac{1}{2}} \psi(\frac{x-b_x}{s_x}, \frac{t-b_t}{s_t})$$

and the corresponding reconstruction formula is

$$f = \frac{2}{A+B} \sum_{s_x} \sum_{b_x} \sum_{s_t} \sum_{b_t} < f, \psi_{s_x b_x, s_t b_t} > \psi_{s_x b_x, s_t b_t}. \tag{7.3}$$

In the special case that the $\psi_{s,b}$ are orthogonal, then,

$$A = B = 1$$

7.2.2 Prestack Seismic Frames

Although an atom of a frame can be chosen somewhat arbitrarily, the closer that the atom resembles the assumed signal (perhaps in a least squares sense), the more compact the frame. The characteristics of prestack seismic data are well-known. In time, the basic element is a seismic wavelet, in space or offset-time, the basic shape is a hyperbola. Although the seismic wavelet is not generally known, a fair approximation often used in seismic modelling is that of a Ricker wavelet. In wavelet analysis the equivalent wavelet is called the Mexican Hat wavelet Daubechies (1992) (actually, a normalized Ricker). The time and space functions described above are combined into a 2D seismic wavelet, that Zhang and Ulrych (2002) call a physical wavelet, given by

$$\begin{aligned}\psi_{s_x b_x, s_t, b_t}(x,t) &= |s_x s_t|^{-\frac{1}{2}} \left[1 - \left(\frac{t-b_t}{s_t} - \sqrt{(\frac{x-b_x}{s_x})^2 + H^2}\right)^2\right] \frac{2(2a)^{\frac{1}{4}}}{\sqrt{3\pi}} \\ &\exp\left[-\frac{1}{2}\left(\frac{t-b_t}{s_t} - \sqrt{(\frac{x-b_x}{s_x})^2 + H^2}\right)^2 - a(x-b_x)^2\right] \quad (7.4)\end{aligned}$$

where H is a constant and a is a positive real number (usually, H is set to the maximum time of the trace), s_x and s_t are scale parameters in space and time and b_x and b_t are the corresponding translation parameters. We note that the two independent variables x and t are not separable ($\psi(t,x) \neq u(t)v(x)$) as they are for a conventional 2D wavelet.

Fig. 7.1 illustrates the two kinds of wavelets. The pivotal difference between the two wavelets is that when the scale parameters are changed, the new wavelet alters both in shape with time and curvature. The conventional wavelet, however, alters in amplitude and symmetrically in shape. Fig. 7.2 illustrates Rongfeng Zhang's vision of the new wavelet, a vision that has inspired us to call this wavelet the Falcon wavelet. Fig. 7.3 shows the 2D Fourier transform of the Falcon wavelet. Both the real and imaginary parts are compact, to some degree, and symmetric about the two coordinate axes. The fact that Eq. (7.4) is a frame is proved by Zhang and Ulrych (2002) and will not be repeated here. The proof is important however, since, if the proposed wavelet does not constitute a frame, a unique and stable reconstruction will not be obtained. In this regard, Daubechies (1992) points out that in order to obtain wavelet frames, really strong conditions on the wavelet and the scaling and translation factors need not be imposed. In other words, if $\psi(t)$ is at all a "reasonable" function (it has some decay in time and frequency and $\int \psi(t)dt = 0$) then there exists a large number of a_0's and b_0's such that $\{\psi_{m,n}\}$ constitute a frame.

7.2.3 Noise Suppression

Denoting the data by

$$y = s + n$$

2D Physical Wavelet

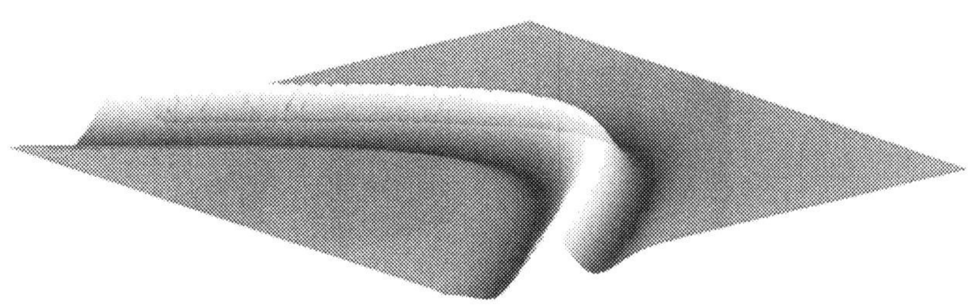

2D Conventional Wavelet

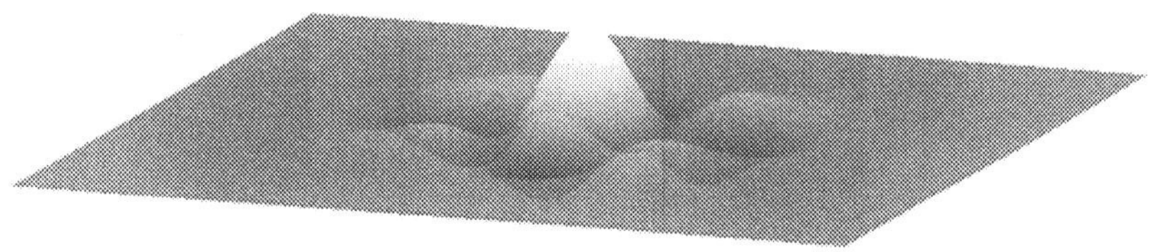

Figure 7.1: Comparing 2D wavelets.

where s represents the signal and n the noise. y is projected onto a space spanned by the bases $\{\psi_k,\ k \in Z\}$ where, hopefully, s and n are separable, i.e.,

$$< y, \psi_k >=< s, \psi_k > + < n, \psi_k >$$

If the space is spanned by a wavelet basis, wavelet thresholding can be applied. The point behind the thresholding is that the projection of s is large, the projection of n is (hopefully) small, and zeroing values below the threshold will suppress the noise. In most cases, of course, the signal coefficients are both large and small

$$< y, \psi_k >=< s, \psi_k >_{large} + < s, \psi_k >_{small} + < n, \psi_k >_{small}$$

and when noise is removed, the signal will also be affected. The new physical wavelet frame, however, has the advantage that, due to redundancy, a hyperbolic signal will always map to large values in some particular scale space. In other words, any hyperbolic event with whatever curvature (velocity) and apex position in a gather, will shrink to, or be concentrated in, one scale space of the frame (see real data example). Denoising becomes especially simple and effective. Zhang and Ulrych (2002) emphasize the important point that velocity information is not required. We illustrate the approach with synthetic and real data examples.

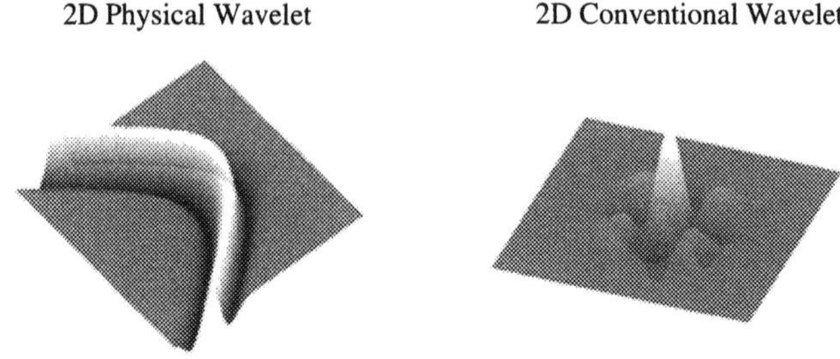

Figure 7.2: Rongfeng's metaphor.

7.2.4 Synthetic and Real Data Examples

Fig. 7.4, shows the noiseless data, data plus noise with signal to noise ratio equal to 0.5 and the denoised result, respectively. Certainly, simple bandpass filtering would have been totally inadequate in this case. Fig. 7.5 illustrates a striking example, where the noise masks the signal entirely. Although the denoised result contains distortions and artifacts, the five true events are clearly recognizable.

Zhang and Ulrych (2002) present two real data examples, one of which is reproduce here. Fig. 7.6 shows the results of noise suppression for a single-side CMP gather. As can be seen, not only is noise suppressed effectively, but in addition, the gaps in the original section have been filled to some extent. Furthermore, the ground roll, which possesses characteristics that are quite different to those of the signal, has also been attenuated.

7.2.5 Discussion

Zhang and Ulrych (2002) address the important question that relates to the filtering of events that are not strictly hyperbolic which often occur in field data. It turns out that, in as much as a particular wavelet basis used in data reconstruction is generally not identical to the signal itself, so also, the basis used by Zhang and Ulrych (2002) need not be exactly equivalent to the contained signal. Of course, the effectiveness

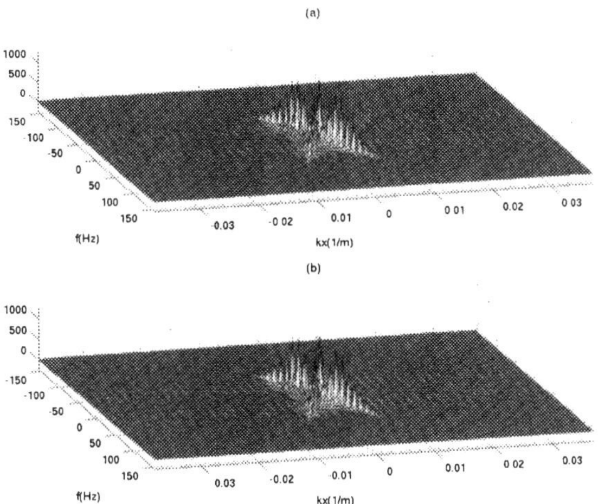

Figure 7.3: 2D Fourier transform of the new wavelet. (a) Real part of the Fourier transform. (b) Imaginary part of the Fourier transform.

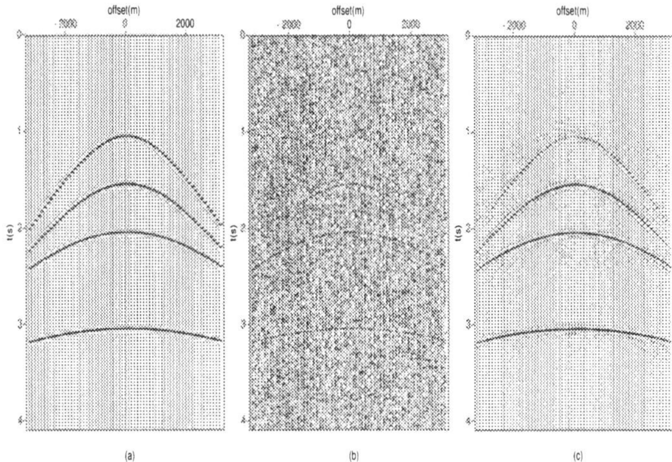

Figure 7.4: Synthetic example. (a) Data without noise (b) Data with ample random noise (c) Result after denoising by PWFD.

of the filtering technique is increased the closer the basis is to the signal. It is true, however, that departure from the hyperbolic form affects the result, particularly with regard to sharp variations.

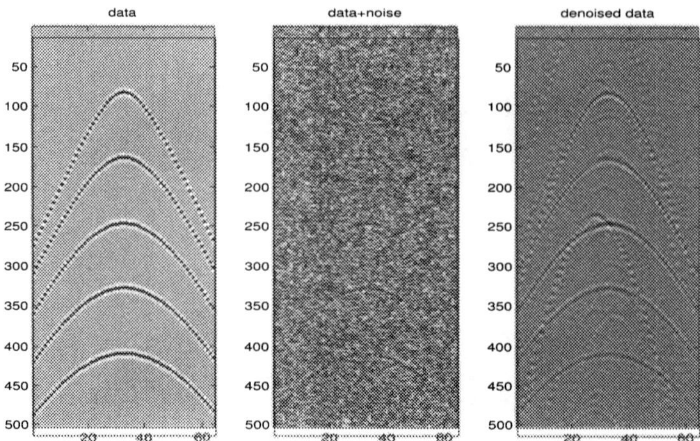

Figure 7.5: Strong noise example. (a) Data without noise (b) Data with very ample noise (c) The denoised result.

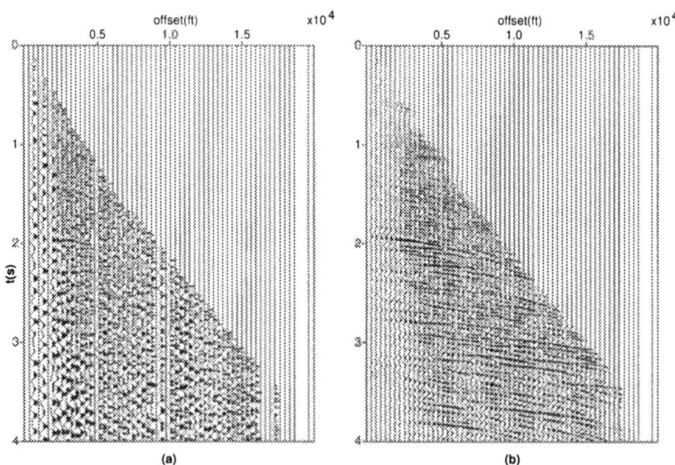

Figure 7.6: PWF processing of a real CMP gather. (a) Original CMP gather (kindly supplied by Mike Graul). (b) CMP gather processed by PWFD.

7.3 Stein Processing

As we have accentuated throughout, the separation of signal and noise is a central issue in seismic data processing. The noise is both random and coherent in nature, the coherent part often masquerading as signal. Without doubt, the process that has moved our quest for extracting information from noisy observations, by a quantum leap, is stacking. The problem with stacking, as we know, is that it may destroy important signal attributes. Nevertheless, let us investigate this process in some detail.

7.3.1 Principles of stacking

We begin by repeating Eq. (5.1), that defines the problem that we are dealing with

$$\mathbf{d} = \mathbf{s} + \mathbf{n}_c + \mathbf{n}_r \tag{7.5}$$

where $\mathbf{d}$ are the data, $\mathbf{s}$ is the signal (however defined) and $\mathbf{n}_c$ and $\mathbf{n}_r$ represent the coherent and incoherent noise components, respectively, with $\mathbf{n}_c + \mathbf{n}_r = \mathbf{n}$. Let us suppose that $\mathbf{n}_c = 0$, $\mathbf{s}$ is perfectly aligned and, most fortuitously, $\mathbf{n}$ is such that at every time sample its average value, taken spatially, is zero. Then, even with noise of Himalayan proportions, $\mathbf{s}$ may be recovered exactly by the simple process of stacking. In a less perfect world, signal stacks constructively and random noise destructively such that, for the special case of Gaussian noise, the noise amplitude is attenuated by the square root of the number of traces. Stacking involves a very special estimation of the first moment of a probability distribution. It is the maximum likelihood estimator, that we write as

$$\boldsymbol{\delta}^0(\mathbf{x}) = \frac{1}{M}\sum_{i=1}^{M} x_i \tag{7.6}$$

where the x_i are M samples of the random variable $\mathbf{x}$. $\boldsymbol{\delta}^0$ is a very special estimator indeed. Statisticians have shown that, given $\mathbf{x}|\mu \overset{\text{ind}}{\sim} N(\mu, 1)$, (symbolically representing that $\mathbf{x}$ is independently distributed and comes from a Gaussian probability distribution with mean μ and variance 1), $\boldsymbol{\delta}^0$ has lowest risk of any linear or nonlinear unbiased estimator, a characteristic that should surely appeal to all. We will return to this fascinating topic again later.

7.3.1.1 Trimmed Means

An important issue in stacking is the concept of robustness. It occasionally happens that our data are infected with a few large errors and, consequently, the tails of the underlying distribution are heavier than those of the Gaussian distribution. In such circumstances, $\boldsymbol{\delta}^0$ is much influenced by such errors and the estimate is said to be not resistant or non robust. Overcoming such effects requires the concept of order statistics. The most widely known of such statistics is the median, computed as the middle value of $M + 1$ ordered numbers. Clearly, the median is considerably less sensitive to outliers and for this reason is often used in robust data processing. A point worth noting is that, for symmetric distributions, the median is equal to the mean.

Consider first the concept of a trimmed mean that is identified by the proportion that is trimmed from each end of the ordered sample. Thus, the 10% trimmed mean of a sample of 20 points is the mean of the remaining 16 points. Note that the median is approximately a 50% trimmed mean. Sometimes, a specified trimming might entail a fraction of an observation. In this case, a weight is assigned to the remaining partially trimmed data and the resulting estimator is called an $\alpha-$trimmed mean that has seen application in seismic data processing.

Let $y_1 \leq y_2 \ldots \leq y_M$ be the ordered data points. Define k to be the integer that is less or equal to αM where $0 \leq \alpha \leq 0.5$ and assign $r = \alpha M - k$. The α-trimmed mean, $T(\alpha)$, is defined by

$$T(\alpha) = \frac{1}{M(1-2\alpha)}\left[(1-r)(y_{k+1} + y_{M-k}) + \sum_{i=k+2}^{M-k-1} y_i\right]$$

For the sake of completeness, we point out that the α-trim mean, just like the median, is a L-estimator which is a weighted estimator of order statistics. L-estimators, in particular L-moments that are robust estimates of the moments of pdf's, are seeing much press in recent times. An excellent review of some robust data analysis techniques may be found in Kleiner (1980). For those seeking further L-moment detail, there is Ulrych et al. (2000).

7.3.1.2 Weighted stack

At this point we wish to mention the weighted stack and, in particular, the version introduced by Schoenberger (1996). The idea is very appealing. Let us represent a seismic section consisting of M traces with N points/trace by the N row by M column matrix, $\mathbf{D}$. The application of conventional stacking may be written as $\mathbf{Dw} = \hat{\mathbf{s}}$, where $\mathbf{w}$ is a vector of M weights, each equal to $1/M$, and $\hat{\mathbf{s}}$ is the stack, a N element vector that is an estimate of the signal. Schoenberger suggested determining $\mathbf{w}$ so that, firstly, $\hat{\mathbf{s}}$ is an unbiased estimate and, secondly, $\mathbf{n}_c$ is attenuated due to the fact that the weights are designed to reject those frequencies at which the coherent energy is present. The former leads to $\sum w_i = 1$, the latter to a matrix equation similar in form to the least squares normal equations.

7.3.2 The Stein Estimator

We have dealt, very briefly to be sure, with some basics of stacking data. This section is aimed, primarily, for cerebral stimulation and for a somewhat more in depth discussion concerning risk. Applications to seismic processing are not quite obvious to us at present, but we believe in serendipity.

There is no doubt that, given a realization of M samples from $N(\mu, 1)$, the only admissible estimator of μ is $\boldsymbol{\delta}^0$ defined in Eq. (7.6). The question is, is this also true if more than one realization is available? An illuminating example, from the scientific discipline of baseball, that will serve to illustrate the discussion is given by Efron and Morris (1977). It goes like this. After the first 45 at bats in the 1970 season, Roberto Clemente obtained 18 hits and his average at that point in the season was, therefore, .400. Another great, Thurmon Munson, was in an early slump and managed only an average of .178. Faced with the question, what will the respective averages be at the end of the season, the only admissible answer is .400 and .178. Clemente and Munson, however, were batting in the majors with many other batters. James and Stein in the early 60's proved a very controversial theorem. An average estimated by $\boldsymbol{\delta}^0$ after forty at bats, given that there are more than two other batters in the league, is not an admissible estimator. In other words, Clemente's and Munson's average is better estimated taking into account what other batters are doing, providing that there are three or more batters. Put in a different but equivalent manner, for this case, $\boldsymbol{\delta}^0$ is not the estimator with lowest risk. Let us pay some attention to this notion of risk.

Life, as some great sage once said, is an inverse problem. In particular it is an ill-posed inverse problem. Our task, it seems, is to try and find solutions which, from the infinity of possibilities, are ones that offer lowest risk when pursued. Statisticians use two terms when dealing with this subject. Risk and efficiency. We have considered the MLE estimator for estimating baseball averages at the end of the season. The properties of $\boldsymbol{\delta}^0$ are well known. It is an unbiased estimator of the population mean with variance σ^2/M, where σ^2 is the actual variance associated with the pdf. The question we pose is whether there is not some other estimator that can obtain an estimate with lower variance (or with lower risk) than $\boldsymbol{\delta}^0$. Perhaps the median? It turns out that the variance of the sample median for a Gaussian distribution is $\pi\sigma^2/2M$. Since the efficiency of the estimator is defined in terms of the size of sample required to achieve a certain accuracy, the efficiency of the median compared to the mean is $2/\pi = 64\%$. In terms of risk, the risk of the median compared to that of the mean is the ratio of the respective variances. i.e. 1.57.

Let us now look at the J-S (for James-Stein) estimator. Consider, for a given μ_i

$$\mathbf{x}_i|\mu_i \overset{\text{ind}}{\sim} N(\mu_i, 1) \quad i = 1, 2, \ldots, \geq 3 \tag{7.7}$$

The variance is taken to be unity for convenience through an appropriate scale transformation. We wish to determine the unknown vector of means $\boldsymbol{\mu}^T = [\mu_1, \mu_2, \ldots, \mu_k]$ so as to minimize the loss, L, defined in the usual way (Section 4.5.10) as

$$L(\boldsymbol{\mu}, \hat{\boldsymbol{\mu}}) = \sum_{i=1}^{k} (\hat{\mu}_i - \mu_i)^2 \tag{7.8}$$

where $\hat{\boldsymbol{\mu}}$ is the estimate of $\boldsymbol{\mu}$. We have formally defined the risk, $R(\cdot\,, \cdot)$, for the MLE estimator in Section 4.5.10 and discussed its properties. The definition is

$$R(\boldsymbol{\mu}, \boldsymbol{\delta}^0) = E_\mu \left[\sum_{i=1}^{k} (\boldsymbol{\delta}^0(x_i) - \mu_i)^2 \right]$$

where $E_\mu[\,\cdot\,]$ is the expectation over the distribution in Eq. (7.7). It turns out that $R(\boldsymbol{\mu}, \boldsymbol{\delta}^0) = k$ for every value of $\boldsymbol{\mu}$. It is constant. This is the lowest risk that was considered possible prior to James and Stein. In 1961, James and Stein gave us a new estimator, $\boldsymbol{\delta}^1$, for $k \geq 3$, the J-S estimator, that is now much in use in finite population sampling. It is, for each score x_i

$$\boldsymbol{\delta}^1(x_i) = \boldsymbol{\delta}^0(\mathbf{x}) + \gamma(x_i - \boldsymbol{\delta}^0(\mathbf{x}))$$

γ is called the shrinkage factor and is computed as a function of the deviation of each score from the grand average $\boldsymbol{\delta}^0(\mathbf{x})$. For the normalized distribution of Eq. (7.7)

$$\gamma = 1 - \frac{(k-2)}{\sum_k (x_i - \boldsymbol{\delta}^0)^2}$$

It turns out that in this case, the risk $R(\boldsymbol{\mu}, \boldsymbol{\delta}^1) < k$ for all values of $\mathbf{x}$.

What is this magical shrinkage factor γ? Efron and Morris (1977) describe it as follows. Designating $\sum_k (x_i - \boldsymbol{\delta}^0)^2 = S$ for convenience, we see that, as S decreases, in other words, as the deviation of each score from the grand average decreases, γ tends to 0 and the individual averages are shrunk towards the grand average. As the data deviations form the grand average increase on the other hand, S increases, γ tends to 1 and not much shrinkage ensues. The shrinkage is, therefore, controlled by the data themselves and by the initial J-S hypothesis that the individual averages are not far from the grand average.

Returning now to the baseball example, in order to compare the risks associated with $\boldsymbol{\delta}^0$ and $\boldsymbol{\delta}^1$, we need to know the actual batting potentials. Of course, these are forever unknown, but we can obtain a pretty good estimate by waiting to the end of the season. By this means, Efron and Morris (1977) show that the error in the MLE estimate is 3.5 times higher than that of the J-S estimate. Clemente's average is decreased to .294 to be compared with his season's end average of .348. Munson's average, on the other hand, is increased to .247 and compares with .320 at the end of the season. For 16 of the 18 players considered by Efron and Morris, the J-S estimated average after 45 at bats is closer to the 'true' average than the conventionally estimated figure. Fascinating. Can we apply the Stein concept to seismic processing? Perhaps we can replace the batter with the seismic trace and compute a stack that is of lower risk. In fact, in shrinking x_i, the J-S estimator takes into account the effect of the variance associated with each batter and, in that sense, allows the fold to be incorporated into the calculation.

This method of stacking may, perhaps, be applied to seismic data, specifically to improve the SNR in $\mathbf{d}$ without stacking for the purpose of, say, AVO analysis. The Stein estimator could be used as a preprocessing step that allows the shrinking of the noise variance in the section. Each batter becomes a time point associated with a trace. The procedure is to consider M batters for each of the N points in the NMO corrected gather, compute the appropriate shrinkage locally and thus, the new sample value.

Fig. 7.7 illustrates results for AVO preprocessing using both NMO corrected data (Fig. 7.7 Left panel) and uncorrected data (Fig. 7.7- Right panel). The captions in this figure are descriptive of the results. We have processed the noisy data by means of light zero phase MA smoothing merely for illustration purposes. In both cases, Stein shrinking has improved the SNR by, approximately, a factor of 2 (shown in Fig. 7.7(d) and (i) as compared to Fig. 7.7(c) and (h)). Fig. 7.7(e) and (j) show eigenimage results. As explained in Section 5.3, the NMO corrected data requires only the 1st eigenimage for an excellent result, superior to the Stein output. In the uncorrected case, the best eigenimage output was obtained using the first 5 eigenimages and the result is inferior to the Stein output.

It is encouraging that, in both these simple examples, the J-S estimator has given rise to a result for which the SNR is higher than that for the input section. Applying the J-S estimator locally as we have done, is a flexible procedure, in that, if $\gamma = 1$, $\boldsymbol{\delta}^1(x_i) = \boldsymbol{\delta}^0(\mathbf{x})$ and the J-S estimator is precisely the maximum likelihood average. Our aim in introducing the Stein estimator in this chapter, is to suggest a possible new avenue for research. We do not mean to imply that such a procedure will be

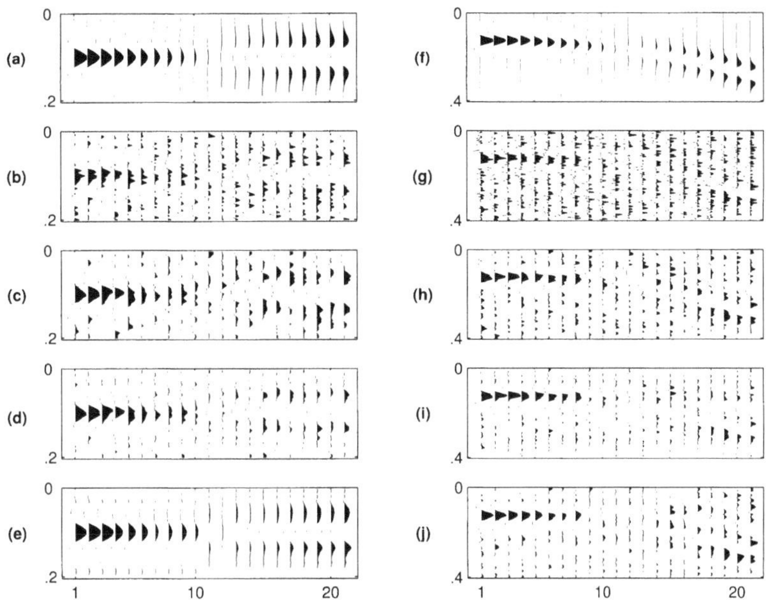

Figure 7.7: Illustrating the Stein estimator - Left panel, AVO after NMO - Right panel AVO without NMO (a) The noiseless signal. (b) The input. (c) Input following light MA smoothing. (d) Stein processed output. (e) The 1st eigenimage. (f) The noiseless signal. (g) The input. (h) Input following light MA smoothing. (i) Stein processed output. (j) Eigenimage processed output.

superior to, say, eigenimage decomposition (with NMO corrected data) or wavelet threshold filtering. We think, however, that the mathematical justification of the J-S estimator, and the fact that it may be easily and inexpensively implemented, makes it worthy of attention.

7.4 The Bootstrap and the EIC

We are, almost always, in parameter estimation, faced with the task of estimating confidence intervals for these parameters. Statistical parameters, variance for example, are often impossible to define analytically. In such instances, we turn to methods like the bootstrap, described in Section 7.4.1 below. Another problem that we often encounter in time series analysis, is the estimation of the order of the model that we choose to represent the data. When a likelihood function can be associated with the model, we can use criteria such as the AIC (Section 3.5), or the ABIC (Section 4.5.6). When this is not the case, however, as will be illustrated in our discussion in Section 7.4.2.1, a technique that is very useful, and one that uses the bootstrap approach, is the extended information criterion, EIC.

7.4.1 The Bootstrap Method

The bootstrap procedure, originally developed by Efron (1979) to compute standard errors, is a computer intensive technique for assigning measures of accuracy to statistical estimates. The technique can be also used to assign confidence intervals (Efron and Tibshirani, 1993). In general, bootstrap methods are well suited for problems where the statistics of interest cannot be estimated by analytical means. The basic idea in bootstrapping is that the actual data are resampled to produce a large number of data sets. The individual estimates of a given parameter may be then used to compute its standard errors and/or confidence intervals.

Suppose that we have observed N samples of a generic variable $\mathbf{x} = [x_1, x_2, x_3, \ldots, x_N]$, from which we compute a statistic of interest, $S(\mathbf{x})$. A bootstrap sample, $\mathbf{x}^* = [x_1^*, x_2^*, x_3^*, \ldots, x_N^*]$, is obtained by randomly sampling the original data with replacement,

$$x_i^* = x_k \quad i = 1, 2, \ldots, N$$

where k is a random uniform variable that can take values $1, 2, \ldots, N$. The data may be numbers, vectors, matrices or any other structure depending on the problem (Efron and Tibshirani, 1993). The bootstrap procedure to compute the standard error of the statistic, $S(\mathbf{x})$, is summarized as follows

1. Compute B bootstrap samples, $\mathbf{x}_1^*, \mathbf{x}_2^*, \mathbf{x}_3^*, \ldots, \mathbf{x}_B^*$, each consisting of N values drawn with replacement from $\mathbf{x}$.

2. Evaluate the statistic of interest associated with each bootstrap sample

$$\hat{\theta}_i^* = S(\mathbf{x}_i^*), \quad i = 1, 2, \ldots, B$$

3. Estimate the standard deviation using

$$\sigma^* = [\frac{1}{B-1} \sum_{i=1}^{B} (\theta_i^* - < \theta^* >)^2]^{1/2}$$

where $< \theta^* > = \sum_{i=1}^{B} \theta_i^* / B$

The statistic, $S(\mathbf{x})$, may not have an analytical form. Moreover, it may be obtained as a cascade of different numerical procedures. In figure (7.8), a flowchart illustrates the bootstrap procedure for determining standard errors. It is important to point out that the individual estimators, $\theta_i^*, \quad i = 1, 2, \ldots, B$, can be used to compute other statistics, i.e., confidence intervals, a histogram of θ, or a Gaussian kernel density estimator which is a smooth representation of the histogram (Silverman, 1986).

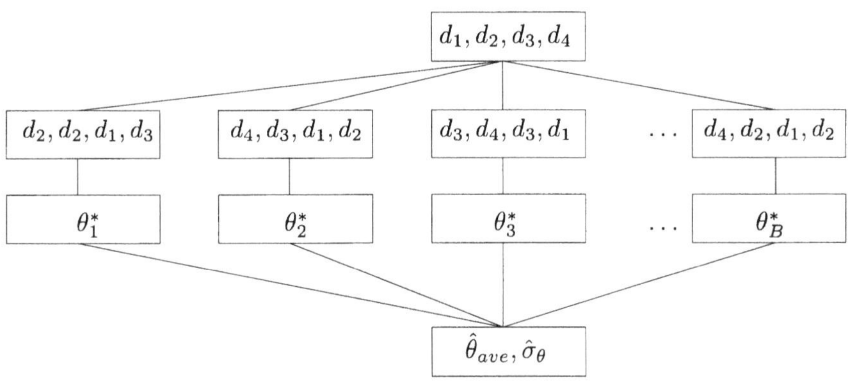

Figure 7.8: Schematic representation of the bootstrap method. The original data $[x_1, x_2, x_3, x_4]$ are resampled by randomly selecting 4 elements with replacement. This is repeated B times to form the bootstrap estimate, $\theta_i^*, \quad i = 1, 2, \ldots, B$. The B estimates are used to compute the mean and the variance.

7.4.2 The Extended Information Criterion

We now extend our discussion concerning the bootstrap method by applying this technique to the important problem of the estimation of the model order in time series analysis. In previous discussions of this problem, we have dealt with the application of both the AIC (Section 3.5) and the ABIC (Section 4.5.6). In both these instances, the assumption that an appropriate likelihood function is available, is implicit. In some situations, however, as illustrated below, the likelihood is either unavailable, or does not apply, and we seek an alternate approach. Such an approach has been proposed by Ishiguro and Sakamoto (1991) and applied to tomographic reconstruction by Nishizawa and Noro (1995). Like the AIC and the ABIC, this new criterion, which is called the extended information criterion or EIC, is also based on sound information theoretic reasoning. It depends on the correction of the bias between the expected log likelihood function associated with the residual distribution, which is unknown, and the computed log likelihood which describes the data. The bias is estimated using bootstrap statistics as suggested by Wong (1983).

7.4.2.1 The Expected Log Likelihood and the EIC

As is well known, the log likelihood, which is determined from observed data, contains the uncertainties of the estimated parameters and is in general a biased estimate of the expected log likelihood which would be observed as the number of data tend to infinity. It is for this reason that residual variance by itself is not a sufficient criterion for parameter estimation. The AIC has been such a successful criterion precisely because it addresses this question. When the estimated parameters, frequencies, amplitudes etc., are not directly determined from the likelihood function of the data, the AIC is not applicable. It turns out however that, by assuming that residuals, which are a function of both uncertainties in the data themselves as well as in the parameters, are Gaussian distributed, and using a bootstrap approach, we can determine the bias mentioned above and consequently use the expected likelihood of the residuals for model selection.

We consider $[x_1, x_2, \ldots, x_n]$ to be a discrete data set, X, belonging to a "true", continuous pdf, $g(\mathrm{Z})$. Now consider a m parameter vector $\hat{\boldsymbol{\theta}}$ which is estimated from X. On the basis of these estimates we obtain a predictive pdf, $f(\mathrm{X})$, which is an approximation to $g(\mathrm{Z})$. We evaluate the goodness of fit by means of the well known Kullback-Leibler measure (Section 3.3) $I(g, f)$, where

$$I(g,f) = \int g(\mathrm{Z}) \log \left[\frac{g(\mathrm{Z})}{f(\mathrm{Z})}\right] d\mathrm{Z} \tag{7.9}$$

$$= \int (g(\mathrm{Z}) \log g(\mathrm{Z}) d\mathrm{Z} - \int g(\mathrm{Z}) \log f(\mathrm{Z}) d\mathrm{Z} \tag{7.10}$$

$I(g, f)$ is a measure of the "distance" of $f(\mathrm{Z})$ from the true pdf $g(\mathrm{Z})$. The first term on the right hand side of Eq. (7.10) contains terms in $g(\mathrm{Z})$ only and it is the second term, which is called the expected log likelihood, which measures the closeness of $f(\mathrm{Z})$ to the true distribution. Since it is generally correct to replace the integral in Eq. (7.10) with the discrete form of the expected value on $g(\mathrm{Z})$ (Nishizawa and Noro, 1995), we can write the expected likelihood as

$$l_E(\mathrm{X}|\hat{\boldsymbol{\theta}}) = E_g[\log f(x_i|\hat{\boldsymbol{\theta}})]$$

where $E_g[\,\cdot\,]$ denotes expectation with respect to $g(\mathrm{Z})$ and $l_E(\mathrm{X}|\hat{\boldsymbol{\theta}})$ is conditional on the estimated model parameters. Since $p(\mathrm{Z})$ is unknown, we can only compute an estimate of $l_E(\,\cdot\,)$ by means of $1/N$ times the log likelihood

$$\hat{l}_E(\mathrm{X}|\hat{\boldsymbol{\theta}}) = \frac{1}{N}\sum_{i=1}^{N} \log f(x_i|\hat{\boldsymbol{\theta}}) \tag{7.11}$$

Eq. (7.11) is a biased estimate and the bias, B_g, is expressed by

$$B_g = E_g\left[\frac{1}{N}\sum_{i=1}^{N} \log f(x_i|\hat{\boldsymbol{\theta}}) - E_G \log f(q_i|\hat{\boldsymbol{\theta}})\right] \tag{7.12}$$

where the last term in Eq. (7.12) indicates that the expected value is computed with respect to data sets, $[q_1, q_2, \ldots, q_N]$, which are from the same distribution as X, but independent of it. However, since only one set, X, is available from $g(\mathrm{Z})$, it is at this stage that the bootstrap approach is required to simulate the term $f(q_i)$ to enable the computation of B_g to proceed. Specifically, we produce a bootstrap sample from X and compute $g^*(x_i)$ to use in place of $g(z)$. Labelling all bootstrap quantities with *, we proceed as follows (Nishizawa and Noro, 1995).

Let the observed data be $\mathrm{Y} = [y_1, y_2, \ldots, y_N]$, and data computed using our time series model with order m be designated as $\mathrm{Y}^m = [y_1^m, y_2^m, \ldots, y_N^m]$. A resampled error, e_k, is randomly selected from the model residual, $\mathrm{U} = [u_1, u_2, \ldots, u_N]$, where $u_i = y_i - y_i^m$, and is added to y_k^m to form the new data set. This set is called the bootstrap sample, $\mathrm{Y}^* = [y_1^*, y_2^*, \ldots, y_N^*]$, where $y_k^* = y_k^m + e_k$. The bootstrap residual, U^*, and the model parameter vector, $\hat{\boldsymbol{\theta}}^*$, are determined using the bootstrap sample, Y^*, in the particular algorithm that is used to model the time series. We now compute the bias term in Eq. (7.12) as

$$B_{\hat{g}} = E_{\hat{g}}\left[\frac{1}{N}\sum_{i=1}^{N}\log f(x_i^*|\hat{\boldsymbol{\theta}}^*) - \frac{1}{N}\sum_{i=1}^{N}\log f(x_i|\hat{\boldsymbol{\theta}}^*)\right]$$

where $E_{\hat{g}}$ denotes the expectation on $\hat{g}(x_i)$, and the second term on the right hand side is an estimate of $E_{\hat{g}}[\log f(y_i^*|\hat{\boldsymbol{\theta}}^*)]$ (Nishizawa and Noro, 1995). It is clarifying to point out that, in this procedure, we have the following correspondence between observed data and bootstrap samples

$$\begin{array}{ccc} g(\mathrm{Z}) & \rightarrow & \hat{g}(x_i) \\ x_i & \rightarrow & x_i^* \end{array}$$

and

$$\hat{\boldsymbol{\theta}} \rightarrow \hat{\boldsymbol{\theta}}^*$$

7.4.2.2 Extended Information Criterion, EIC

Following Wong's suggestion (Wong, 1983) that the bias, B_g, may be determined using bootstrap statistics as outlined above, Ishiguro and Sakamoto (1991) have defined the EIC, a new information theoretic criterion, in a manner very similar to that of Akaike's AIC. Specifically

$$\begin{aligned} EIC &= -2l(X|\hat{\boldsymbol{\theta}}) + 2 \times Bias\ Correction \\ &= -2l(X|\hat{\boldsymbol{\theta}}) + 2NB_{\hat{g}} \end{aligned}$$

Assuming Gaussian distributions for the residuals, we generate M bootstrap samples and compute the EIC as

$$EIC = N[\log(2\pi\sigma^2) + 1] + \frac{1}{M}\sum\left[-N + \frac{1}{\sigma^{2*}}\sum_{i=1}^{N}(y_i - y_i^{m^*})^2\right] \qquad (7.13)$$

where σ^2 and σ^{2^*} are the variances of the residuals X and X^* respectively. The minimum EIC computed using Eq. (7.13) is taken as an indication of the correct order.

7.4.2.3 Application of the EIC to Harmonic Retrieval

We apply the EIC to the important problem of the estimation of the number of harmonics actually present in observed data. This issue is of importance in various scenarios, as we have indicated in previous chapters. Thus, in Section 2.5.1, we described the special ARMA(p,p) model which corresponds to harmonics with additive noise and is used in high resolution spectral analysis (Pisarenko, 1973). More recently, as described in Section 5.2.1, Sacchi and Kuehl (2001) used this special ARMA model to effect SNR improvement in seismic sections.

The procedure used in the Pisarenko harmonic spectral estimator, is to determine the frequencies of the spectral lines from the roots of the eigenvector of the Toeplitz autocovariance matrix corresponding to the minimum eigenvalue. Once these frequencies have been determined, the problem becomes linear and the amplitudes are determined by means of a least squares fit. Clearly, as pointed out by Pisarenko (1973), if we overestimate the number of harmonics by assuming order m where $m > p$, the actual order of the ARMA(p, p) model, $m - p$ harmonics will appear with fictitious amplitudes (and frequencies), the amplitudes depending on the signal to noise ratio in the data. Since the roots are constrained to lie on the unit circle in the complex z plane, this estimator is applicable when the harmonics present in the signal do not exhibit either growth or decay. When this does occur, the appropriate estimator results from a slightly modified version of the Pisarenko approach which is known as the Sompi method (Kumazawa et al., 1990), a method that has been discussed by Ulrych and Sacchi (1995) in relationship to the EIC.

In order to estimate the number of harmonics in their model, Kumazawa et al. (1990) applied the AIC criterion, but the results were not encouraging. One of the reasons for this, as pointed out by Osamu Nishizawa (personal communications), is that the Sompi method is not, in reality, a maximum likelihood method in the sense that the model parameters are not determined using the likelihood function (indeed, the frequencies are determined from the roots of a polynomial). We tackle this problem here by applying the EIC. We chose the Sompi model because it is more complex than the similarly posed Pisarenko model. Variance estimates, which are in general difficult to obtain, can be obtained rather simply as a byproduct of the computation of the EIC, since we have M estimates of the model parameter vector $\hat{\boldsymbol{\theta}}^*$ at the minimum EIC value. In the next section we present some results of these computations.

We illustrate the application of the EIC to the Sompi spectral estimator by considering as an example a similar problem to the one considered by Kumazawa et al. (1990). Specifically, the input is composed of five equi-amplitude harmonics with frequencies equal to 0.1Hz, 0.15Hz, 0.2Hz, 0.3Hz and 0.4Hz. The growing rates are alternately -1.00 and $1.00/10^{-3}\text{sec}^{-1}$ the phases are random and the added noise level is 5% of the maximum amplitude of the signal. We generate bootstrap samples 250 times and produce 50 realizations. The EIC computed from these realizations, as a function of the assumed number of harmonics, is plotted in Fig. 7.9a. Fig. 7.9b shows the behavior of the average EIC computed from these realizations. Unlike the AIC itself, the EIC, at its minimum, correctly identifies the number of harmonics present. The variances associated with all the parameters are evaluated from the realizations of the bootstrap of the bootstrap procedure. Thus in this case, for the estimates of

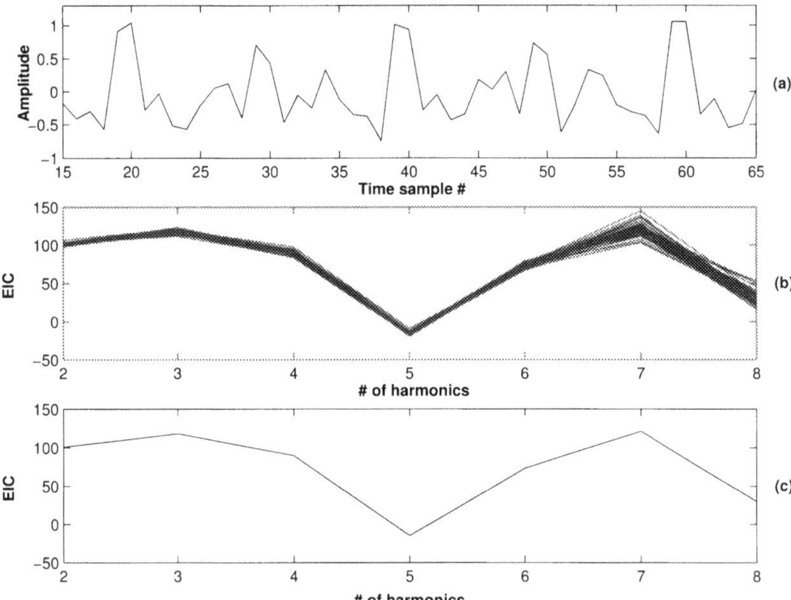

Figure 7.9: Illustrating the EIC. (a) The input signal. (b) EIC computed from bootstrap realizations. (c) Average EIC.

the frequencies, we obtained harmonics with frequencies equal to 0.0996 ± 0.0014Hz, 0.1488 ± 0.0032Hz, 0.1999 ± 0.0013Hz, 0.3002 ± 0.0004Hz and 0.3997 ± 0.0005Hz.

7.4.2.4 Discussion

The EIC, based on the use of bootstrap statistics, is particularly useful when the likelihood function that is required in the computation of its famous cousin, the AIC, is not available or does not apply. Such is the case for Sompi or Pisarenko spectral estimation where the parameters are not estimated by means of the maximum likelihood method. The bootstrap approach allows one to simulate fluctuations in the data and thereby compute the bias between the log likelihood and the expected log likelihood of the residuals (Wong, 1983). The EIC is then computed by adding the computed bias to the log likelihood in the spirit of the AIC. An important aspect of the bootstrap approach which we have presented is that variance estimates of the model parameters are immediately available. In the example that we have illustrated, the task was to estimate the number of harmonics present in the signal. Another, particularly clear example of the usefulness of the EIC, is the work of Nishizawa and Noro (1995) in velocity tomography.

Important issues arise that we have not discussed here. Examples are the effect of the number of bootstrap samples per realization and the number of realizations on the computation of the EIC and variance estimates. The interested reader is directed to the excellent paper by Nishizawa and Noro (1995) for some of these considerations.

7.5 Summary

This chapter has presented three topics that, we hope, have been of interest to the reader. We have chosen these topics so that they reflect, in some manner, the material that we have presented in the previous chapters. Thus, the 2D wavelet noise attenuation filter extends our treatment of transformations and the choice of suitable basis functions. The J-S estimator makes use of many of the ideas that we have presented in relationship to Bayesian estimation. Finally, bootstrap statistics are invaluable when it comes to the ubiquitous need for estimating such quantities as variance, for example. We could have chose many other fascinating topics. We will save these, however, for a future edition.

Bibliography

Ahmed, N. and K. R. Rao (1975). *Orthogonal Transforms for Digital Signal Processing*. Springer-Verlag.

Akaike, H. (1973). Information theory and an extension of the maximum likelihood principle. In B. N. Petrov and F. Csaki (Eds.), *2nd Inter. Symp. on Information Theory and Limit Theorems of Probability Theory*, pp. 267–281. Akademiai Kiado, Budapest.

Akaike, H. (1974). A new look at the statistical model identification. *IEEE Trans. Autom. Contr. AC-19*, 716–723.

Akaike, H. (1980). Likelihood and the Bayes procedure. In J. M. Bernardo, M. H. DeGroot, D. U. Lindley, and A. F. Smith (Eds.), *Bayesian Statistics*, pp. 142–166. University Press, Valencia, Spain.

Al-Yahya, K. M. (1991). Application of the partial Karhunen-Loève transform to suppress random noise in seismic sections. *Geophys. Prosp. 39*, 77–93.

Anderson, N. O. (1978). Comments on the performance of the maximum entropy algorithm. *Proc. IEEE 66*, 1581–1582.

Anderson, T. W. (1971). *The Statistical Analysis of Time Series*. John Wiley and Sons.

Andrews, H. C. and B. R. Hunt (1977). *Digital Image Restoration*. Prentice-Hall.

Baranov, V. and G. Kunetz (1960). Film synthétique avec réfléxtions multiples: Théorie et calcul practique. *Geophys. Prosp. 8*, 315–325.

Bassrei, A. (1990). Inversion of seismic data by a relative entropy algorithm. In *Expanded Abstracts, SEG 60th Annual Int. Meeting*, Volume Paper SI1.1, pp. 158–169.

Bienvenu, G. (1883). Optimality of high resolution array processing using the eigensystem approach. *IEEE Trans. Acoust. Speech, Signal Processing ASSP-31*, 1235–1248.

Biondi, B. L. and C. Kostov (1989). High-resolution velocity spectra using eigenstructure methods. *Geophysics 54*, 832–842.

Bosman, C. L. and E. Reiter (1993). Seismic data compression using wavelet transforms. In *63rd Annual Internat. Mtg., Soc. Expl. Geophys., Expanded Abstracts*, pp. 1261–1264. Soc. Expl. Geophys.

Box, G. E. and G. M. Jenkins (1976). *Time Series Analysis: Forecasting and Control.* Holden-Day.

Bracewell, R. N. (1986). *The Fourier Transform and Its Applications.* McGraw-Hill.

Burg, J. P. (1975). *Maximum Entropy Spectral Analysis.* Ph. D. thesis, Stanford University.

Canales, L. (1984). Random noise reduction. In *64th Int. Annual Mtg. Soc, Expl. Geophys., Expanded Abstracts*, pp. 525–527. Soc. Expl. Geophys.

Cantoni, A. and P. Butler (1976). Properties of the eigenvectors of persymmetric matrices with application to communication theory. *IEEE Trans. Commun. COM-24*, 804–809.

Cary, P. (1998). The simplest discrete Radon transform. In *68th Int. Annual Mtg. Soc, Expl. Geophys., Expanded Abstracts*, pp. 1999–2002. Soc. Expl. Geophys.

Chakraborty, A. and D. Okaya (1995). Frequency-time decomposition of seismic data using wavelet-based methods. *Geophysics 60*(06), 1906–1916.

Claerbout, J. F. (1976). *Fundamentals of Geophysical Data Processing with Applications to Petroleum Prospecting.* McGraw-Hill.

Claerbout, J. F. (1992). *Earth Soundings Analysis: Processing versus Inversion.* Blackwell Scientific.

Cooley, W. W. and P. R. Lohnes (1971). *Multivariate Data Analysis.* John Wiley & Sons, Inc.

Daubechies, I. (1992). *Ten lectures on wavelets.* SIAM.

Deighan, A. J. and D. R. Watts (1997). Ground-roll suppression using the wavelet transform. *Geophysics 62*(06), 1896–1903.

Du, B. and L. R. Lines (1998). Wavelet filtering of tube waves in the Glenn Pool cross-well seismic survey. In *68th Annual Internat. Mtg., Soc. Expl. Geophys., Expanded Abstracts*, pp. 350–352. Soc. Expl. Geophys.

Duffin, R. and A. Schaffer (1952). A class of nonharmonic fourier series. *Trans. Amer. Math. Soc. 72*, 341–366.

Efron, B. (1979). Bootstrap methods: Another look at the jackknife. *Ann. Statist. 7*(1), 1–26.

Efron, B. and C. Morris (1973). Stein's estimation rule and its competitors - an empirical Bayes approach. *J. Amer. Stat. Assoc. 68*, 117–130.

Efron, B. and C. Morris (1977, May). Stein's paradox in statistics. *Scientific American* , 119–128.

Efron, B. and R. Tibshirani (1993). *An Introduction to the Bootstrap Method.* Chapman & Hall.

Fahlman, G. and T. J. Ulrych (1982). A new method for estimating the power spectrum of gapped data. *Mon. Not. Roy. Astr. Soc. 199*, 53–65.

Farquharson, C. G. and D. W. Oldenburg (1998). Non-linear inversion using general measures of data misfit and model structure. *Geophys. J. Int. 134*, 213–227.

Fougere, P. F. (1995). Ionospheric radio tomography using maximum entropy 1. theory and simulation results. *Radio Science 30*, 429–444.

Freire, S. L. and T. J. Ulrych (1988). Applications of singular value decomposition to vertical seismic profiling. *Geophysics 53*, 778–785.

Frieden, B. R. (2001). *Probability, Statistical Optics, and Data Testing* (3 ed.). Springer.

Gerbrands, J. J. (1981). On the relationship between SVD, KLT and PCA. *Pattern Recognition 14*, 375–381.

Gershenfeld, N. (1999). *The Nature of Mathematical Modelling.* Cambridge University Press.

Goldman, S. (1968). *Information Theory.* Dover Publications.

Golub, G. H. and C. F. Van Loan (1996). *Matrix Computations* (3rd ed.). The Johns Hopkins University Press.

Golub, G. H. and C. F. Van Loan (1980). An analysis of the total least squares problem. *SIAM J. Numer. Anal. 17*, 883–893.

Goupillaud, P. (1961). An approach to inverse filtering of near surface layer effects from seismic records. *Geophysics 26*, 754–760.

Grubb, H. and A. Walden (1997). Characterizing seismic time series using the discrete wavelet transform. *Geophys. Prosp. 45*(2), 183–205.

Gull, S. and J. Skilling (1985). The entropy of an image. In C. R. Smith and W. T. G. . Jr. (Eds.), *Maximum-Entropy and Bayesian Methods in Inverse Problems*, Fundamental Theories of Physics, pp. 287–301. D. Reidel Publishing Co.

Gull, S. F. and T. J. Newton (1986). Maximum entropy tomography. *Appl. Opt. 25*, 156–168.

Hamilton, W. C. (1964). *Statistics in Physical Science: Estimation, Hypothesis Testing and Least Squares.* The Ronald Press Co.

Hamming, R. W. (1983). *Digital Filters* (2nd ed.). Prentice-Hall, Inc.

Hampson, D. (1986). Inverse velocity stacking for multiple elimination. *J. Can. Soc. Expl. Geophys. 22*, 44–55.

Harris, P. E. and R. E. White (1997). Improving the performance of f-x prediction filtering at low signal-to-noise ratios. *Geophys. Prosp. 42*, 269–302.

Hemon, C. H. and D. Mace (1978). Essai d'une application de la transformation de Karhunen-Loève au traitement sismique. *Geophys. Prosp. 26*, 600–626.

Henon, M. (1976). A two-dimensional mapping with a strange attractor. *Commun. Math. Phys. 50*, 69–77.

Herrmann, P., T. Mojesky, M. Magesan, and P. Hugonnet (2000). De-aliased high resolution Radon transform. In *70th Annual Inter. Mtg Soc. Explor. Geophys.*, pp. SP 2.3. Soc. Expl. Geophys.

Huber, J. P. (1981). *Robust Statistics* (3rd ed.). John Wiley and Sons, Inc.

Hubral, P. (1978). On getting reflection coefficients from waves. *Geophys. Prosp. 26*, 627–630.

Hyvärinen, A. (1998). New approximation of differential entropy for independent component analysis and projection pursuit. In M. J. K. M. I. Jordan and S. A. Solla (Eds.), *Advances in Neural Information Processing Systems 10*, pp. 273–279. MIT Press.

Hyvärinen, A., J. Karhunen, and E. Oja (2001). *Independent Component Analysis.* John Wiley & Sons, Inc.

Ioup, J. W. and G. E. Ioup (1998). Noise removal and compression using a wavelet transform. In *68th Annual Internat. Mtg., Soc. Expl. Geophys., Expanded Abstracts*, pp. 1076–1079. Soc. Expl. Geophys.

Ishiguro, M. and Y. Sakamoto (1991). WIC: An estimator-free information criterion. *Research Memorandum, Inst. Statist. Math., Tokyo* (410), 1–22.

Jacobs, F. J. and P. A. G. van der Geest (1991). Spiking band-limited traces with a relative-entropy algorithm. *Geophysics 56*(7), 1003–1014.

Jaynes, E. T. (1982). On the rationale of maximum entropy methods. *Proc. IEEE 70*, 939–952.

Jaynes, E. T. (1986). In J. H. Justice (Ed.), *Maximum entropy and Bayesian methods in applied statistics*, pp. 1–25. Cambridge University Press.

Jeffreys, H. (1939). *Theory of Probability.* Clarendon Press, Oxford.

Jenkins, G. M. and D. G. Watts (1969). *Spectral Analysis and its applications.* Holden-Day.

Jerri, A. J. (1977). The Shannon sampling theorem-its various extensions and applications: A tutorial review. *Proc. IEEE 65*, 1565–1596.

Johnson, R. W. (1983). Algorithms for single signal and multisignal minimum-cross-entropy spectrum analysis. *Naval Research Lab. Report Report 8667.*

Johnson, R. W. and J. E. Shore (1984). Power spectrum estimation by means of relative-entropy minimization with uncertain constraints. In *IEEE Int. Conf. Acoust., Speech, Signal Processing, San Diego, Calif.*, pp. 158–169.

Jones, I. F. and S. Levy (1987). Signal-to-noise ratio enhancement in multichannel seismic data via the Karhunen-Loève transform. *Geophys. Prosp. 35*, 12–32.

Jowitt, P. W. (1979). Bayesian estimates of material properties from limited data. *Eng. Struct. 1*, 170–178.

Kanasewich, E. R. (1981). *Time Sequence Analysis in Geophysics.* The University of Alberta Press.

Kaplan, S. T. (2002). Principal and independent component analysis for seismic data. Master's thesis, University of British Columbia.

Kaplan, S. T. and T. J. Ulrych (2001-2002). From eigenfaces to eigensections. *Journal of Seismic Exploration 10*, 353–366.

Kapur, J. N. (1989). *Maximum Entropy Models in Science and Engineering.* John Wiley & Sons, NY.

Kapur, J. N. and H. K. Kesavan (1992). *Entropy Optimization Principles with Applications.* Academic Press.

Kay, S. M. (1988). *Modern Spectral Estimation: Theory and Application.* Prentice-Hall.

Key, S. C. and S. B. Smithson (1990). New approach to seismic-reflection event detection and velocity determination. *Geophysics 55*, 1057–1069.

Kirby, M. and L. Sirovich (1987, March). Low-dimensional procedure for the characterization of human faces. *J. Opt. Soc. Am. A 4*(3), 519–524.

Kirlin, R. L. (1992). The relationship between semblance and the eigenstructure velocity estimators. *Geophysics 57*, 1027–1033.

Kleiner, B. (1980). Exploratory data analysis in the geophysical sciences. *Rev. Geophys. Space Phys., 18*, 699–717.

Kostov, C. (1990). Toeplitz structure in slant stack inversion. In *60th Int. Annual Mtg. Soc, Expl. Geophys., Expanded Abstracts*, pp. 618–1621. Soc. Expl. Geophys.

Kumazawa, M., Y. Imanishi, Y. Fukao, M. Furumoto, and A. Yamamoto (1990). A theory of spectra; analysis based on the characteristic property of a linear dynamic system. *Geophys. J. Int. 101*, 613–630.

Kunetz, G. (1964). Generalization des opérateurs d'antireésonance a un nombre quelconque de réflecteurs. *Geophys. Prosp. 12*, 283–289.

Kunetz, G. and I. D'Ecerville (1962). Sur certaines propriétés d'une onde acoustique plane de compression dans un milieu stratifié. *Ann. Géophys. 18*, 351–359.

Lanczos, C. (1956). *Applied Analysis.* Dover, NY.

Lathi, B. P. (1968). *An Introduction to Random Signals and Communication Theory.* International Textbook Co.

Levinson, N. (1947). The Wiener rms (root mean square) error criterion in filter design and prediction. In *Extrapolation, Interpolation and Smoothing of Stationary Time Series by N. Wiener (Appendix B.* Technology Press, MIT, Cambridge, Mass.

Levy, S. and D. W. Oldenburg (1987). The deconvolution of phase shifted wavelets. *Geophysics 47*, 1285–1294.

Li, X.-G., M. D. Sacchi, and T. J. Ulrych (1996). Wavelet transform inversion with prior scale information. *Geophysics 61*(05), 1379–1385.

Lo, T. W., G. L. Duckworth, and M. N. Toksöz (1990). Minimum cross entropy seismic diffraction tomography. *JASA 87*, 748–756.

Loéve, M. (1977). *Probability Theory* (4 ed.). Springer-Verlag, Berlin.

Luo, Y. and G. T. Schuster (1992). Wave packet transform and data compression. In *62nd Annual Internat. Mtg., Soc. Expl. Geophys., Expanded Abstracts*, pp. 1187–1190. Soc. Expl. Geophys.

Lupton, R. (1993). *Statistics in Theory and Practice.* Princeton University Press.

Makhoul, J. (1981). On the eigenvectors of symmetric Toeplitz matrices. *IEEE Trans. Acoust., Speech, Signal Processing ASSP-29*, 868–872.

Mallat, S. G. (1998). *A wavelet tour of signal processing.* Academic Press.

Marden, M. (1966). *Geometry of Polynomials.* American Mathematical Society.

Marple, S. L. (1987). *Digital Spectral Analysis With Applications.* Prentice-Hall.

Mastronardi, N., P. Lemmerling, and S. Van Huffel (2000). Fast structured total least squares algorithm for solving the basic deconvolution problem. *SIAM J. Matrix Anal. Appl. 22*, 533–553.

Matsuoka, T. and T. J. Ulrych (1986). Information theory measures with application to model identification. *IEEE Trans. Acoust., Speech, Signal Processing ASSP-34*, 511–517.

May, R. M. (1976). Some simple mathematical models with very complicated dynamics. *Nature 261*, 459–467.

Menke, W. (1989). *Geophysical Data Analysis: Discrete Inverse Theory.* Academic Press.

Meyer, Y. (1993). *Wavelets : Algorithms & Applications.* Society for Industrial and Applied Mathematics.

Miao, X. and S. P. Cheadle (1998). Noise attenuation with wavelet transforms. In *68th Annual Internat. Mtg., Soc. Expl. Geophys., Expanded Abstracts*, pp. 1072–1075. Soc. Expl. Geophys.

Moon, F. C. (1987). *Chaotic Vibrations.* John Wiley & Sons.

Moulines, E., J. Cardoso, and E. Gassiat (1997). Maximum likelihood for blind separation and deconvolution of noisy signals using mixture models. *IEEE Trans. Acoust., Speech and Signal Processing 5*, 3617–3620.

Nguyen, M.-Q. and J. Mars (1999). Filtering surface waves using 2-D discrete wavelet transform. In *69th Annual Internat. Mtg., Soc. Expl. Geophys., Expanded Abstracts*, pp. 1228–1230. Soc. Expl. Geophys.

Nikias, C. L. and M. R. Raghuveer (1987). Bispectrum estimation: A digital signal processing framework. *IEEE Proc. 75*, 869–891.

Nishizawa, O. and H. Noro (1995). Bootstrap statistics for velocity tomography: Application of a new information criterion to velocity tomography. *Geophys. Prospect. 43*, 157–176.

Oldenburg, D. W., T. Scheuer, and S. Levy (1983). Recovery of the acoustic impedance from reflection seismograms. *Geophysics 48*, 1318–133.

Pakula, L. and S. Kay (1983). Simple proofs of the minimum phase property of the prediction error operator. *IEEE Trans. Acoust., Speech, Signal Processing ASSP-31*, 501–502.

Papoulis, A. (1962). *The Fourier Integral and its Applications.* McGraw-Hill.

Papoulis, A. (1977). *Signal Analysis.* McGraw-Hill.

Papoulis, A. (1984). *Probability, Random Variables and Stochastic Processes.* McGraw-Hill.

Peacock, K. L. and S. Treitel (1969). Predictive deconvolution - theory and practice. *Geophysics 34*, 155–169.

Pearson, K. (1901). On lines and planes of closest fit to points in space. *Philos. Mag. 2*, 559–572.

Pentland, A. and M. Turk (1991). Eigenfaces for recognition. *Journal of Cognitive Neuroscience 3*(1), 71–86.

Pisarenko, V. P. (1973). The retrieval of harmonics from a covariance function. *Geophys. J. R. Astr. Soc. 33*, 347–366.

Porsani, M. J. and B. Ursin (2000). Mixed-phase deconvolution and wavelet estimation. *The Leading Edge 19*, 76–79.

Press, S. J. (1989). *Bayesian Statistics: Principles, Models and Applications.* John Wiley.

Priestley, M. B. (1981). *Spectral Analysis and Time Series.* Academic Press.

Ready, P. J. and P. A. Wintz (1973). Information extraction, SNR improvement, and data compression in multispectral imagery. *IEEE Trans. Comm. COM-21*(10), 1123–1131.

Richards, J. A. (1993). *Remote Sensing Digital Image Analysis.* Springer-Verlag.

Robinson, E. A. (1957). Predictive decomposition of seismic traces. *Geophysics 22*, 767–778.

Robinson, E. A. (1967). *Statistical Communication and Detection.* Hafner Publishing Co., N.Y.

Robinson, E. A. and S. Treitel (1978). The fine structure of the normal incidence synthetic seismogram. *Geophys. J. R. astr. Soc. 53*, 289–310.

Robinson, E. A. and S. Treitel (2002). *Geophysical Signal Analysis.* Society of Exploration Geophysicists.

Robinson, E. R. (1964). Recursive decomposition of stochastic processes. In H. O. A. Wold (Ed.), *Econometric Model Building*, pp. 111–168. North-Holland Publishing Co.

Rosen, J. B., H. Park, and J. Glick (1996). Total least norm formulation and solution for structured problems. *SIAM J. Matrix Anal. 17*, 110–126.

Rutty, M. J. and G. M. Jackson (1992). Wavefield decomposition using spectral matrix techniques. *Explor. Geophys. 23*, 293–298.

Sacchi, M. D. (1998). A bootstrap procedure for high-resolution velocity analysis. *Geophysics 65*, 1716–1725.

Sacchi, M. D. and H. Kuehl (2001). Arma formulation of fx prediction error filters and projection filters. *Journal of Seismic Explor. 9*, 185–197.

Sacchi, M. D. and M. J. Porsani (1999). Fast high resolution Radon transform. In *69th Int. Annual Mtg. Soc, Expl. Geophys., Expanded Abstracts.* Soc. Expl. Geophys.

Sacchi, M. D. and T. J. Ulrych (1995a). High resolution velocity gathers and offset space reconstruction. *Geophysics 60*(11), 1169–1177.

Sacchi, M. D. and T. J. Ulrych (1995b). Improving resolution of radon transform operators using a model re-weighted least squares procedure. *J. Seismic Explor. 4*, 315–328.

Sacchi, M. D., T. J. Ulrych, and C. J. Walker (1998). Fourier transform estimation and data extrapolation using prior information. *IEEE Trans. Signal Process. 46*, 31–38.

Sacchi, M. D., D. R. Veliz, and A. H. Cominguez (1994). Minimum entropy deconvolution with frequency domain constraints. *Geophysics 59*, 938–946.

Saito, N. and R. R. Coifman (1997). Extraction of geological information from acoustic well-logging waveforms using time-frequency wavelets. *Geophysics 62*(06), 1921–1930.

Sakamoto, Y., M. Ishigure, and G. Kitagawa (1986). *Akaike Information Criterion Statistics.* D. Reidel Publishing Co.

Scales, J. A., A. Gersztenkorn, and S. Treitel (1988). Fast lp solution of large, sparse, linear systems: Application to seismic travel time tomography. *Journal of Comp. Phys. 75*, 314–333.

Scales, J. A. and R. Sneider (1997). To Bayes or not to Bayes. *Geophysics 62*, 1045–1046.

Scales, J. A. and L. Tenorio (2001). Prior information and uncertainty in inverse problems. *Geophysics 66*, 389–397.

Scargle, J. (1977). Absolute value optimization to estimate phase properties of stochastic time series. *IEEE Trans. Inform. Theory 23*, 140–143.

Schoenberger, M. (1996). Optimum weighted stack for multiple suppression. *Geophysics 61*, 891–901.

Shannon, C. E. (1948). A mathematical theory of communications. *Bell System Technical Journal 27*, 379–423,623–656.

Shore, J. E. (1981). Minimum cross-entropy spectral analysis. *IEEE Trans. Acoust., Speech, Signal Processing 29*, 230–237.

Shore, J. E. and R. W. Johnson (1980). Axiomatic derivation of the principle of maximum entropy and the principle of minimum cross-entropy. *IEEE Trans. Information Theory IT-26*, 26–37.

Silverman, B. W. (1986). *Density Estimation for Statistics and Data Analysis.* Chapman & Hall.

Sivia, D. S. (1996). *Data Analysis: A Bayesian Tutorial.* Clarendon Press - Oxford.

Skaggs, T. H. and A. J. Kabala (1994). Recovering the release history of a groundwater contaminant. *Water Resour. Res. 30*, 71–79.

Soubaras, R. (1994). Signal preserving noise attenuation by the f-x projection filter. In *64th Annual Inter. Mtg. Soc. Explor. Geophys.*, pp. 1576–1579. Soc. Expl. Geophys.

Späth, H. (1986). Orthogonal least squares fitting with linear manifolds. *Numer. Math. 48*, 441–445.

Spitz, S. (1983). Seismic trace interpolation in the f-x domain. *Geophysics 56*, 785–749.

Stoica, P. and A. Nehorai (1987). On stability and root location of linear prediction models. *IEEE Trans. Acoust., Speech, Signal Processing ASSP-35*, 582–584.

Strang, G. (1988). *Linear Algebra and its Applications.* Harcourt Brace & Company.

Takens, F. (1981). Detecting strange attractors in turbulence. In D. A. Rand and L. S. Young (Eds.), *Dynamical Systems and Turbulence*, pp. 366–381. Springer-Verlag.

Taner, T. and R. Sheriff (1997). Application of amplitude, frequency and other attributes to stratigraphic and hydrocarbon determination. In C. E. Petro (Ed.), *Seismic stratigraphy: Applications to hydrocarbon exploration*, Number 26, pp. 389–416. Am. Assoc. Petr. Geol.

Tarantola, A. (1987). *Inverse Problem Theory: Methods for Data Fitting and Model Parameter Estimation.* Elsevier.

Tarits, P., V. Jouanne, M. Menvielle, and M. Roussignol (1994). Bayesian statistics of non-linear problems: Example of the magnetotelluric 1-d inverse problem. *Geophys. J. Int. 119*, 353–368.

Thomas, G. B. and R. L. Finney (1979). *Calculus and Analytic Geometry.* Addison-Wesley Publishing Co.

Thorson, J. R. and J. F. Claerbout (1985). Velocity stack and slant stack stochastic inversion. *Geophysics 50*, 2727–2741.

Tikhonov, A. N. (1963). Resolution of ill-posed problems and the regularization method. *Dokl. Akad. Nauk SSSR 151*, 501–504.

Tipping, M. E. and C. M. Bishop (1999). Probabilistic principal component analysis. *J. Royal Stat. Soc., Series B 21*, 611–622.

Trad, R. D., T. Ulrych, and M. D. Sacchi (2002). Accurate interpolation with high-resolution Radon transforms. *Geophysics 67*, 644–656.

Trad, R. D., T. Ulrych, and M. D. Sacchi (2003). Latest views of the sparse Radon transform. *Geophysics 68*, 386–399.

Treitel, S., J. L. Shanks, and C. M. Frazier (1967). Some aspects of fan filtering. *Geophysics 32*, 789–800.

Turner, R. E. and D. S. Betts (1974). *Introductory Statistical Mechanics.* Sussex University Press.

Ulrych, T. J., A. Bassrei, and M. Lane (1990). Minimum relative entropy inversion of 1d data with applications. *Geophys. Prosp. 38*, 465–487.

Ulrych, T. J. and T. N. Bishop (1975). Maximum entropy spectral analysis and autoregressive decomposition. *Revs. Geop. and Space Phys. 13*, 183–200.

Ulrych, T. J. and R. W. Clayton (1976a). Time series modelling and maximum entropy. *Phys. Earth Planetary Interiors 12*, 188–200.

Ulrych, T. J. and R. W. Clayton (1976b). Time series modelling and maximum entropy. *Phys. Earth Planetary Interiors 12*, 188–200.

Ulrych, T. J. and J. F. Fokkema (1995). Infinite series, limits, relationships and aliasing. *Revista Brasileira de Geofísica 13*(2), 103–109.

Ulrych, T. J. and T. Matsuoka (1991). The output of predictive deconvolution. *Geophysics 56*, 371–377.

Ulrych, T. J. and M. D. Sacchi (1995). Sompi, pisarenko and the eic. *Geophys. J. Int. 122*(1), 719–724.

Ulrych, T. J., M. D. Sacchi, and S. L. M. Freire (1999). Eigenimage processing of seismic sections. In R. L. Kirlin and W. J. Done (Eds.), *Covariance Analysis for Seismic Signal Processing*, pp. 241–274. Society Exploration Geophysicists.

Ulrych, T. J., M. D. Sacchi, and A. D. Woodbury (2001). A bayes tour of inversion: A tutorial. *Geophysics 66*, 55–69.

Ulrych, T. J., D. E. Smylie, O. G. Jensen, and G. K. C. Clarke (1973). Predictive filtering and smoothing of short records by using maximum entropy. *J. Geophysical Res. 78*, 4959–4961.

Ulrych, T. J. and S. Treitel (1991). A new proof of the minimum phase property of the unit prediction error operator - revisited. *IEEE Trans. Acoust., Speech, Signal Processing ASSP-39*, 252–254.

Ulrych, T. J., D. R. Velis, A. D. Woodbury, and M. D. Sacchi (2000). L-moments and c-moments. *Stoch. Envir. Res, and Risk Assessment 414*, 50–68.

Ulrych, T. J. and C. Walker (1984). A modified algorithm for the autoregressive recovery of the acoustic impedance (short note). *Geophysics 49*, 2190–2197.

Ursin, B. and M. J. Porsani (2000). Estimation of an optimal mixed-phase inverse filter. *Geophys. Prosp. 48*, 663–672.

Van Huffel, S. and J. Vandewalle (1991). *The Total Least Squares Problem: Computational Aspects and Analysis.* Society for Industrial and Applied Mathematics (SIAM).

Van Huffel, S., H. Park, and J. B. Rosen (1996). Formulation and solution of structured total least norm problems for parameter estimation. *IEEE Trans. Signal. Process. 10*, 2464–2474.

Velis, D. R. and T. J. Ulrych (1996). Simulated annealing wavelet estimation via fourth-order cumulant matching. *Geophysics 61*, 1939–1948.

Wagner, B. J. (1992). Simultaneous parameter estimation and continuous source characterization for coupled ground water flow and contaminant transport modelling. *J. Hydrol. 135*, 275–300.

Walker, C. and T. J. Ulrych (1983). Autoregressive recovery of the acoustic impedance. *Geophysics 48*, 1338–1350.

Wiggins, R. A. (1978). Minimum entropy deconvolution. *Geoexploration 16*, 12–35.

Wiggins, R. A. and S. P. Miller (1972). New noise-reduction techniques applied to long-period oscillations from the Alaskan earthquake. *BSSA 62*, 417–479.

Wong, H. W. (1983). A note on the modified likelihood for density estimation. *J. Am. Stat. Assoc. 78*, 461–463.

Woodbury, A. D. and T. J. Ulrych (1993). Minimum relative entropy: Forward probabilistic modelling. *Water Resources Research 29*(8), 2847–2860.

Woodbury, A. D. and T. J. Ulrych (1996). Minimum relative entropy inversion: Theory and application to recovering the release history of a groundwater contaminant. *Water Resources Research 32*(9), 2671–2681.

Woodbury, A. D. and T. J. Ulrych (1998). Mre and probabilistic inversion in groundwater hydrology. *Stochastic Hydrology and Hydraulics 12*, 317–358.

Woodbury, A. D. and T. J. Ulrych (2000). A full bayesian approach to the groundwater inverse problem for steady state flow. *Water Resources Research 36*(8), 2081–2093.

Woodbury, A. D., T. J. Ulrych, E. A. Sudicky, and R. Ludwig (1998). 3d plume source reconstruction using mre. the release history of a groundwater contaminant. *J. of Contaminant Hydrology 32*, 131–158.

Yalamov, P. and J. Y. Yuan (2003). A successive least squares method for structured total least squares problems. *J. Comp. Math. 22*, 467–472.

Yilmaz, O. (1988). *Seismic data processing.* Soc. Expl. Geophys. Video course available from SEG Publications Department.

York, D. (1980). Least squares fitting of a straight line. *Canad. J. Phys. 44*, 1079–1086.

Young, T. Y. and T. W. Calvert (1974). *Classification, Estimation and Pattern Recognition.* Elsevier, New York.

Yule, G. U. (1927). On a method of investigating periodicities in disturbed series, with special reference to wolfer's sunspot numbers. *Philos. Trans. R. Soc. London, ser. A 226*, 267–298.

Zhang, R. and T. J. Ulrych (2002). Physical wavelet frame denoising. *Geophysics 68*(1), 225–231.

Zhou, B. and S. A. Greenhalgh (1994). Linear and parabolic $\tau - p$ revisited. *Geophysics 59*, 1133–1149.

Index

www.ingramcontent.com/pod-product-compliance
Ingram Content Group UK Ltd.
Pitfield, Milton Keynes, MK11 3LW, UK
UKHW021440280726
14060UKWH00001BA/167

9 780080 447216